ASTEROSEISMOLOGY ACROSS THE HR DIAGRAM

Proceedings of the Asteroseismology Workshop
Porto, Portugal
1–5 July 2002

Edited by:
MICHAEL J. THOMPSON
Imperial College, London, United Kingdom

MARGARIDA S. CUNHA
Centro de Astrofísica da Universidade do Porto, Porto, Portugal

MÁRIO J.P.F.G. MONTEIRO
Centro de Astrofísica da Universidade do Porto, Porto, Portugal

Reprinted from *Astrophysics and Space Science*
Volume 284, No. 1, 2003

KLUWER ACADEMIC PUBLISHERS
DORDRECHT / BOSTON / LONDON

A.C.I.P. Catalogue record for the book is available from the Library of Congress.

ISBN 1-4020-1173-3

Published by Kluwer Academic Publishers,
P.O. Box 17, 3300 AA Dordrecht, The Netherlands.

Sold and distributed in the U.S.A. and Canada
by Kluwer Academic Publishers,
101 Philip Drive, Norwell, MA 02061, U.S.A.

In all other countries, sold and distributed
by Kluwer Academic Publishers,
P.O. Box 322, 3300 AH Dordrecht, The Netherlands.

Printed on acid-free paper

All Rights Reserved
© 2003 Kluwer Academic Publishers
No part of the material protected by this copyright notice may be reproduced or
utilized in any form or by any means, electronic or mechanical,
including photocopying, recording or by any information storage and
retrieval system, without written permission from the copyright owner.

Printed in the Netherlands

ASTROPHYSICS AND SPACE SCIENCE / Vol. 284 No. 1 2003

Special Issue:
ASTEROSEISMOLOGY ACROSS THE HR DIAGRAM

Proceedings of the Asteroseismology Workshop
Porto, Portugal
1–5 July 2002

Guest Editors:
MICHAEL J. THOMPSON, MARGARIDA S. CUNHA
and MÁRIO J.P.F.G. MONTEIRO

Foreword	xi
Committees and Sponsors	xiii
List of Participants	xv

1. PRESENT OBSERVATIONAL STATUS

1.1 Space and Ground Based Data

H. KJELDSEN / Space and Ground Based Data for Asteroseismology	1
R. ALONSO, J.A. BELMONTE and T.M. BROWN / STARE Results on a Single Field: Tens of New Pulsating Stars	13
I.W. ROXBURGH and F. FAVATA / The Eddington Mission	17

1.2 Present Status for Different Pulsators

F. BOUCHY and F. CARRIER / Present Observational Status of Solar-Type Stars	21
D.W. KURTZ / Present Observational Status of the Intermediate Mass Stars	29
P. DE CAT / Present Observational Status of High Mass Pulsating Stars	37
M.S. O'BRIEN / Old Pulsators: White Dwarfs and their Immediate Precursors	45
C. CATALA / δ Scuti Pulsations in Pre-Main Sequence Stars	53
T.R. BEDDING / Solar-like Oscillations in Semiregular Variables	61
E.M. GREEN, K. CALLERAME, I.R. SEITENZAHL, B.A. WHITE, E.A. HYDE, M. GIOVANNI, M. REED, G. FONTAINE and R. OSTENSEN / Discovery of a New Class of Pulsating Stars: Gravity Mode Pulsators among Subdwarf B Stars	65
V. RIPEPI / Recent Observations of PMS δ Scuti Stars	69

2. ASTEROSEISMIC TECHNIQUES
2.1 Mode Identification and Frequency Determination

M. REED / Multicolour Photometry for Mode Identification	73
J. TELTING / High-Resolution Spectroscopy for Mode Identification	85
A. PAMYATNYKH / Theoretical Clues for Mode Identification – Instability Ranges and Rotational Splitting Patterns	97
T. APPOURCHAUX / Peakbagging for Solar-Like Stars	109
L.A. BALONA / Mode Identification from Line Profiles Using the Direct Fitting Technique	121
J. DE RIDDER, G. MOLENBERGHS and C. AERTS / Statistical Revision of the Moment Method	125
M.-A. DUPRET, R. SCUFLAIRE, A. NOELS, A. THOUL, R. GARRIDO, A. MOYA, J. DE RIDDER, P. DE CAT and C. AERTS / An Improved Method of Photometric Mode Identification: Applications to Slowly Pulsating B, β Cephei, δ Scuti and γ Doradus Stars	129
J. DASZYŃSKA-DASZKIEWICZ, W.A. DZIEMBOWSKI and A.A. PAMYATNYKH / Photometric Nonadiabatic Observables in Rotating β Cephei Models	133
M. BREGER / Gravity Modes in Delta Scuti Stars	137

2.2 Inferring Information from the Frequencies

T.S. METCALFE / Seismic Inference Using Genetic Algorithms	141
S. BASU / Stellar Inversions	153
D. GOUGH / On the Principal Asteroseismic Diagnostic Signatures	165
I.W. ROXBURGH and S. VORONTSOV / Diagnostics of the Internal Structure of Stars Using the Differential Response Technique	187

3. ASTEROSEISMIC CONSTRAINTS ON STELLAR STRUCTURE
3.1 Solar-like and Near Main-Sequence Stars

P. DEMARQUE and F.J. ROBINSON / Stellar Convection	193
S. VAUCLAIR / Mixing and Diffusion in Stars	205
L. BIGOT and W.A. DZIEMBOWSKI / Are Pulsation and Magnetic Axes Aligned in roAp Stars?	217
R. SAMADI, M.J. GOUPIL, Y. LEBRETON, A. NORDLUND and F. BAUDIN / Seismic Diagnostics of Stellar Convection Treatment from Oscillation Amplitudes of p-Modes	221
S. TURCOTTE and O. RICHARD / Of Variability, or its Absence, in HgMn Stars	225
M.P. DI MAURO, J. CHRISTENSEN-DALSGAARD and L. PATERNÒ / A Study of the Solar-Like Properties of β Hydri	229

T.C. TEIXEIRA, J. CHRISTENSEN-DALSGAARD, F. CARRIER, C. AERTS, S. FRANDSEN, D. STELLO, T. MAAS, M. BURNET, H. BRUNTT, J.R. DE MEDEIROS, F. BOUCHY, H. KJELDSEN and F. PIJPERS / Giant Vibrations in Dip   233

B. DINTRANS, A. BRANDENBURG, A. NORDLUND and R.F. STEIN / Stochastic Excitation of Gravity Waves by Overshooting Convection in Solar-Type Stars   237

A. THOUL, R. SCUFLAIRE, B. VATOVEZ, A. NOELS, P. MAGAIN, M. BRIQUET and M.A. DUPRET / p-Mode Oscillations of α Cen A   241

3.2 Away from the Main Sequence (Childhood and Old Age)

M. MARCONI and F. PALLA / Outstanding Issues for the Pulsation of Young Stars   245

G. FONTAINE, P. BRASSARD and S. CHARPINET / Outstanding Issues for Post-Main Sequence Evolution: Instability Strips and Excitation of Modes in Compact Pulsators   257

R. OREIRO, F. PÉREZ HERNÁNDEZ, M. MANTEIGA, A. ULLA, J.M. GONZÁLEZ PÉREZ, M.R. ZAPATERO OSORIO, R. GARCÍA LÓPEZ, J. MACDONALD, P. THEJLL, A. FERRIZ-MAS, R.A. SAFFER and V. ELKIN / Hot Subdwarfs: Magnetic, Oscillatory and other Physical Properties   269

R. NOWAKOWSKI and W. DZIEMBOWSKI / Nonlinearity of Nonradial Modes in Evolved Stars   273

4. SUMMARY

J. CHRISTENSEN-DALSGAARD / Summary: Problems, Connections and Expectations   277

5. POSTERS (CDROM)

5.1 Present Observational Status: Stars Near and In the Main Sequence

R. ALONSO, J.A. BELMONTE, T. BROWN, R. GARRIDO and E. MICHEL / STARE Observations of Potential COROT Fields   P295

L.A. BALONA, C.D. LANEY and W. ZIMA / Spectroscopy of roAp Stars   P299

T.R. BEDDING, R.P. BUTLER, C. McCARTHY, H. KJELDSEN, G.W. MARCY, S.J. O'TOOLE, C.G. TINNEY and J. WRIGHT / Oscillations in α Cen A Observed with UCLES at the AAT   P303

F. BOUCHY and F. CARRIER / Prospects in Doppler Ground-Based Asteroseismology   P307

F. CARRIER, F. BOUCHY and P. EGGENBERGER / p-Mode Observations on the K0 IV Star δ Eridani   P311

F. CARRIER, F. BOUCHY and P. EGGENBERGER / New Seismological Results on the G0 IV η Bootis   P315

V. COSTA, A. ROLLAND, P. LÓPEZ DE COCA, I. OLIVARES and J.M. GARCÍA-PELAYO / Simultaneous uvby Photometry of the δ Scuti Star ε Cepphei	P319
J.-Y. HU, J.-J. WANG, S.-N. ZHANG and J. GUO / SVOM-A Proposal for a Micro-Satellite Used as Space-Based Multi-Band Variable Objects Monitor	P323
S. JANKOV, P. MATHIAS, C. AERTS, P. DE CAT and A. DOMICIANO DE SOUZA / Time Variability of NRP Modes Along Orbital Phase in ε Per	P327
E. KAMBE, B. SATO, Y. TAKEDA, H. IZUMIURA, S. MASUDA and H. ANDO / Solar-Type Oscillations in Procyon: Test Observations with the Iodine Cell and the HIgh Dispersion Echelle Spectrograph (HIDES)	P331
S.-L. KIM, H. JIN, S.-G. KWON, C.-U. LEE, J.-H. YOUN and H. LEE / Photometric Study of ROTSE-I δ Scuti Type Stars	P335
G. KOPACKI, M. JERZYKIEWICZ, J. MOLENDA-ZAKOWICZ and Z. KOLACZKOWSKI / Variable Stars in NGC 2169 from CCD Photometry	P339
J. KRZESINSKI, S. KLEINMAN and A. NITTA / The Sloan Digital Sky Survey: A Million Redshifts Pulsators	P343
J.D. LANDSTREET, O. KOCHUKHOV, F. KUPKA, T. RYABCHIKOVA and W. WEISS / Observations of Rapid Radial Velocity Variations of Spectral Lines in Rapidly Oscillating Ap (roAp) Stars	P347
H. LEHMANN and D.E. MKRTICHIAN / RZ Cas: First Spectroscopic Detection of Short-Term Pulsations in an Algol System	P351
P. LÓPEZ DE COCA, S.-L. KIM, M.A. HOBART, A. ROLLAND, H. LEE, J.M. GARCÍA-PELAYO and S. CHO / Multiperiodicity of 21 Mon	P355
P. MATHIAS, E. CHAPELLIER, J.-M. LE CONTEL, J.-P. SAREYAN, R. GARRIDO, E. RODRÌGUEZ, E. PORETTI, A. ARELLANO FERRO, L. ALVAREZ, L. PARRAO L. EYER, C. AERTS, W. WEISS and A. ZHOU / Multisite Multitechnique Monitoring of a Large Sample of γ Doradus Candidates	P359
J. MOLENDA-ZAKOWICZ / Multiperiodicity of Be and Bn Stars	P363
B. MOSSER and J.-P. MAILLARD / Asteroseismology with a Fourier Transform Seismometer	P367
A. PICCIONI, C. BARTOLINI, S. BERNABEI, I. BRUNI, A. GUARNIERI, F. GIOVANNELLI, L. SABAU-GRAZIATI and R. SILVOTTI / BD +22° 1268: a Probable δ Scuti Star	P371
A. PIGULSKI and Z. KOLACZKOWSKI / Discovery of β Cephei Stars in the Magellanic Clouds	P375
P. REEGEN / The δ Scuti Star HR 6290 High-Precision Robotic Photometry	P379
A. RETTER, T.R. BEDDING, D. BUZASI and H. KJELDSEN / Evidence for Solar Like Oscillations in Arcturus (α Boo)	P383
F. RODLER, M. BREGER, W. ZIMA, K.M. BISCHOF, V. ANTOCI, G. HANDLER, A.A. PAMYATNYKH, P. REEGEN, D. LORENZ, B. STEININGER and R. GARRIDO / Recent Work of the Delta Scuti Network	P387

E. RODRÍGUEZ, V. COSTA and M.J. LÓPEZ-GONZÁLEZ, A.Y. ZHOU, J.M. GARCÍA, J.Y. WEI and Y. FAN / SAO 32177: a New Multiperiodic γ Dor Variable — P391

E. RODRÍGUEZ, G. HANDLER, V. COSTA and J.M. GARCÍA / HD 205: a New Multiperiodic δ Sct Variable with a Very Complex Spectrum — P393

A. ROLLAND, P.J. AMADO, M.A. HOBART, V. COSTA, P. LÓPEZ DE COCA and E. RODRÍGUEZ / Strömgren Photometry of the δ Scuti Star 21 Vul: 1995 and 1999 — P397

J. TELTING, K. UYTTERHOEVEN and I. ILYIN / A Survey for Line-Profile Variability in Early B Type Main-Sequence Stars — P401

S. TURCOTTE, C. AERTS and P. KNOGLINGER / Searching for Line Profile Variability in HgMn Stars — P405

K. UYTTERHOEVEN, J.H. TELTING and C. AERTS / Analysis of the Frequency Spectrum of the Spectroscopic Binary β Cephei Star λ Scorpii — P409

5.2 Present Observational Status: Away from the Main Sequence

M. BILLÉRES and G. FONTAINE / Search for EC 14026 Pulsators: Results of a Survey in the Northern Hemisphere — P413

Z.E. DIND, T.R. BEDDING, S.J. O'TOOLE and S.D. KAWALER / Applying Weights to Whole Earth Telescope Data: Results for Four Stars — P417

J. HUBER, T.R. BEDDING and S.J. O'TOOLE / Colour Ratios of Pulsation Amplitudes in Red Giants — P421

M.J. IRELAND, P.G. TUTHILL, T.R. BEDDING, J.G. ROBERTSON and A.P. JACOB / Multi-Wavelength Angular Diameters of Pulsating Red Giants with MAPPIT — P425

L.L. KISS, B. CSÁK, A. DEREKAS, E.J. ALFARO, I.B. BÍRÓ and G.A. BAKOS / A Surface-Brightness Analysis of Selected Northern Cepheids — P429

L.L. KISS and K. SZATMÁRY / R Cygni: a Mira Star Pulsating Chaotically — P433

A. NOELS, R. SCUFLAIRE, P. MAGAIN, G. PARMENTIER and A. THOUL / Variable Blue Stragglers and the EASE Scenario — P437

S.J. O'TOOLE, T.R. BEDDING, H. KJELDSEN and M.A.S.G. JØRGENSEN / Time-Resolved Spectroscopy and Spectrophotometry of the Pulsating sdB Star PG 1605+072 — P441

S.K. RANDALL, I.P. LOPES and A.E. LYNAS-GRAY / Pulsation Period Variation in the Subdwarf B Star EC 20117-4014 — P445

F.-X. SCHMIDER, J. GAY, C. JACOB, E. FOSSAT, J.-C. VALTIER, B. MOSSER, D. MEKARNIA, T. GUILLOT and J. PROVOST / SYMPA: A Dedicated Instrument and a Network for Seismology of Giant Planets — P449

A. STANKOV, M. FRIDLUND and I. ILYIN / Spectroscopic Analysis of BW Vulpeculae — P453

K. ZWINTZ and W.W. WEISS / Pulsating Pre-Main Sequence Stars in NGC 6383? — P457

5.3 Asteroseismic Techniques

C. AERTS, J. CUYPERS, M.A. DUPRET, J. DE RIDDER, R. SCUFLAIRE and L. EYER / Photometric Mode Identification in the Two γ Doradus Stars HD12901 and HD 48501 — P461

G. BERTHOMIEU and T. APPOURCHAUX, on behalf of the COROT Seismology Working Group / Hare & Hound Exercise with Simulated COROT Data — P465

M. BRIQUET and C. AERTS / A New Version of the Moment Method, Optimized for Mode Identification in Multiperiodic Stars — P469

T. KALLINGER, A. KAISER, CH. STUETZ, W.W. WEISS, K. ZWINTZ and L. BIGOT / MOST and COROT High Precision Photometry Simulations of the roAp Star 10 AQUILAE — P473

A. MAZUMDAR and I. ROXBURGH / The Asteroseismic Diagram for $\ell = 0, 1$ p-Modes — P477

M. PAPARÓ and L.G. BALÁZS / Multivariante Analyses of Multicolour Photometry for Mode Identification — P481

F.P. PIJPERS / Selecting Asteroseismic Targets — P485

W. ZIMA, U. HEITER, P.L. COTTRELL, H. LEHMANN, P. MATHIAS, E. PORETTI and M. BREGER / The 2002 DSN Campaign of FG Vir: Mode Identification by High Resolution Spectroscopy - Preliminary Results — P489

5.4 Asteroseismic Constraints on Stellar Structure and Evolution

C. AERTS, H. LEHMANN, R. SCUFLAIRE, M.A. DUPRET, M. BRIQUET, J. DE RIDDER and A. THOUL / Mode Identification and Seismic Modelling of the β Cep Star EN(16)Lac — P493

J. BALLOT, S. TURCK-CHIÉZE, R.A. GARCÍA and P.A.P. NGHIEM / What Can We Learn from Global Acoustic Modes About the Structure of the Sun and α Centauri A? — P497

P. BRASSARD and G. FONTAINE / Constraints on the $^{12}(\alpha, \gamma)O^{16}$ Reaction Rate from White Dwarf Asteroseismology? — P501

M.S. CUNHA / roAp Stars Versus noAp Stars: a Theoretical Approach — P505

P. DEMARQUE, L.H. LI, F.J. ROBINSON, S. SOFIA and D.B. GUENTHER / Three-Dimensional Radiative Hydrodynamical Simulations of the Outer Layers of Sun-Like Stars — P509

J. FERNANDES and M.J.P.F.G. MONTEIRO / Observational Constraints on Models of β Hydri — P513

G. FONTAINE, P. BRASSARD, S. CHARPINET, E.M. GREEN and B. WILLEMS / On the Potential of Tidal Excitation of Gravity Modes in Hot B Subdwarfs — P517

L. FOX-MACHADO, F. PÉREZ HERNÁNDEZ, J.-C. SUÁREZ and E. MICHEL / Searching for Mode Identification in the δ Scuti Stars of the Pleiades Cluster — P521

J. KUBÁT and J. KRTICKA / Instability Caused by a Multicomponent Nature of Radiatively-Driven Stellar Winds	P525
W. LÖFFLER / g-Mode Pulsations in Doradus Stars: The Frozen-Flux Approximation and the Conservation of Energy	P529
T.S. METCALFE and G. HANDLER / The $^{12}C(\alpha, \gamma)^{16}O$ Nuclear Reaction Rate from Asteroseismology of the DBV White Dwarf CBS 114	P533
A. MIGLIO, J. CHRISTENSEN-DALSGAARD, M.P. DI MAURO, M.J.P.F.G. MONTEIRO and M.J. THOMPSON / Seismological Analysis of the Helium Ionization Zones in Low- and Moderate-Mass Stars	P537
P. MITTERMAYER and W.W. WEISS / Atmospheric Parameters and Abundances of the δ Scuti Star FG Virginis	P541
M.H. MONTGOMERY and D.O. GOUGH / On the Effect of a Starspot on the Modes of Oscillation of a Toy Ap Star Model	P545
A. MOYA, R. GARRIDO and M.A. DUPRET / Non-Adiabatic Pulsational Observables in δ Scuti and γ Doradus Stars: a Comparison of Two Different Numerical Codes	P549
F.J.G. PINHEIRO, M. MARCONI, V. RIPEPI, D.F.M. FOLHA, F. PALLA and M.J.P.F.G. MONTEIRO / Re-Visiting the Seismic Properties of the PMS δ Scuti Star V346 Ori	P553
J. PROVOST, G. BERTHOMIEU, F. THÉVENIN, P. MOREL, F. BOUCHY and F. CARRIER / Revisited Calibration of α Cen Binary System	P557
F. RINCON and M. RIEUTORD / Shear-Alfvén Waves in Magnetic Stars: the Spherical Shell Model	P561
M.R. TEMPLETON and S. BASU / Nonadiabatic Pulsations in Pre-Main Sequence δ Scuti Stars	P565
S. THÉADO and S. VAUCLAIR / Rotation Induced Mixing and Diffusion in Stellar Radiative Zones: 2D Simulation	P569
S.E. THOMPSON and J.C. CLEMENS / The Inclination of G185-32	P573

FOREWORD

We stand at the threshold of an exciting era of Asteroseismology. In a few months' time, the Canadian small-satellite asteroseismology mission MOST will be launched. Danish and French missions MONS and COROT should follow, with the ESA mission Eddington following in 2007/8. Helioseismology has proved spectacularly successful in imaging the internal structure and dynamics of the Sun and probing the physics of the solar interior. Ground-based observations have detected solar-like oscillations on alpha Centauri A and other Sun-like stars, and diagnostics similar to those used in helioseismology are now being used to test and constrain the physics and evolutionary state of these stars. Multi-mode oscillations are being observed in an abundance of other stars, including slowly pulsating B stars (SPB stars), delta Scuti stars, Ap stars and the pulsating white dwarfs. New classes of pulsators continue to be discovered across the Hertzsprung-Russell diagram. For good reason it was decided to entitle our conference 'Asteroseismology Across the HR Diagram'.

Yet the challenges still to be faced to make asteroseismology across the HR diagram a reality are formidable. Observation, data analysis and theory all pose hard problems to be overcome. In conceiving this meeting, the aim of the organisers was to facilitate a cross-fertilization of ideas and approaches between researchers working on different pulsators and with different areas of expertise. We venture to suggest that in this the conference was a great success. Not only did those working in many different types of stellar pulsators participate, so too did researchers working in helioseismology and in the theory of stellar evolution. This led to a highly productive interchange of ideas which we believe is conveyed by the papers in these proceedings.

Many people and organisations contributed to making the meeting a great success. We gratefully acknowledge the invaluable direction provided by the Scientific Organising Committee and the tireless work of the Local Organising Committee. The beautiful surroundings of the Jardins do Palácio de Cristal and the superbly apportioned conference venue in the Almeida Garrett Public Library inspired and facilitated our discussions, and we thank the Porto Town Council for making them available for our use. The meeting was organized by the Centre for Astrophysics of the University of Porto, with the invaluable contributions from several sponsors (see list below).

Of course the whole success of the meeting was only possible thanks to the participants themselves and in particular the review speakers and those who presented papers. In all, for 118 participants, there were 41 oral and 72 poster presentations, the vast majority of which are represented in these proceedings. (The poster papers are reproduced on CDRom.)

We did not solve all the problems of Asteroseismology in one week, and much observational and theoretical work remains to be done. Most excitingly, we look forward to the abundance of data that the forthcoming asteroseismic space missions will surely bring, and the new insights and understanding of stellar evolution that will be gained. These proceedings may provide a roadmap for that journey of discovery.

Michael J. Thompson, Margarida S. Cunha, and Mário J.P.F.G. Monteiro
18 October 2002

Centro de Astrofísica da Universidade do Porto
Rua das Estrelas
4150-762 Porto, Portugal
http://www.astro.up.pt

COMMITTEES AND SPONSORS

Scientific Organising Committee

Conny Aerts	*Leuven, Belgium*
Michel Breger	*Wien, Austria*
Jørgen Christensen-Dalsgaard	*Aarhus, Denmark*
Margarida S. Cunha	*Porto, Portugal*
Wojciech A. Dziembowski	*Warsaw, Poland*
Douglas O. Gough	*Cambridge, U.K.*
Steve D. Kawaler	*Ames, U.S.A.*
Donald W. Kurtz	*Preston, U.K.*
Eric Michel	*Paris, France*
Mário J.P.F.G. Monteiro	*Porto, Portugal*
Michael J. Thompson (Chairman)	*London, U.K.*

Local Organising Committee

Margarida S. Cunha
Daniel F.M. Folha
Mário J.P.F.G. Monteiro
Fernando J.G. Pinheiro

Sponsors

Biblioteca Municipal Almeida Garrett
    Câmara Municipal do Porto (*www.cm-porto.pt*)
Fundação para a Ciência e a Tecnologia (*www.fct.mct.pt*)
    Ministério da Ciência e do Ensino Superior
Fundação Luso-Americana para o Desenvolvimento (*www.flad.pt*)
Universidade do Porto (*www.up.pt*)
and
Acidados2 – HP Partner (*www.acidados.pt*)
Faculdade de Arquitectura da Universidade do Porto (*www.arq.up.pt*)

LIST OF PARTICIPANTS

Aerts, Conny	*Belgium*	125, 129, 233, P323, P327, P359, P405, P409, **P461**, P469, **P493**
Alonso, Roi	*Spain*	**13**, **P295**
Appourchaux, Thierry	*The Netherlands*	**109**, P465
Ballot, Jérôme	*France*	**P497**
Balona, Luis A.	*South Africa*	**121**, **P299**
Basu, Sarbani	*U.S.A.*	**153**, P565
Baudin, Frederic	*France*	221
Bedding, Tim R.	*Australia*	**61**, **P303**, P383, P417, P421, P425, P441
Belmonte, Juan Antonio	*Spain*	13, P295
Berthomieu, Gabrielle	*France*	**P465**, P557
Bigot, Lionel	*Denmark*	**217**, P473
Billéres, Malvina	*Germany*	**P413**
Bossi, Michele	*Italy*	
Bouchy, François	*Switzerland*	**21**, 233, **P307**, P311, P315, P557
Brassard, Pierre	*Canada*	P257, **P501**, P517
Breger, Michel	*Austria*	**137**, P387, P489
Briquet, Maryline	*Belgium*	241, **P469**, P493
Cardini, Daniela	*Italy*	
Carrier, Fabien	*Switzerland*	21, 233, P307, **P311**, **P315**, P557
Catala, Claude	*France*	**53**
Christensen-Dalsgaard, Jørgen	*Denmark*	229, 233, **275**, P537
Costa, Victor	*Spain*	**P319**, P391, P393, P397
Cunha, Margarida S.	*Portugal*	**P505**
Daszyńska-Daszkiewicz, Jadwiga	*Poland*	**133**
De Cat, Peter	*Belgium*	**37**, 129, P327, P359
De Ridder, Joris	*Belgium*	**125**, 129, P461, P493
Demarque, Pierre	*U.S.A.*	**193**, **P509**
Di Mauro, Maria Pia	*Italy*	**229**, P537
Dintrans, Boris	*France*	**237**
Dupret, Marc-Antoine	*Belgium*	**129**, 241, P461, P493, P549
Dziembowski, Wojciech A.	*Poland*	133, 217, 273
Fernandes, João	*Portugal*	**P513**

Folha, Daniel F.M.	*Portugal*	P553
Fontaine, Gilles	*Canada*	65, **257**, P413, P501, **P517**
Fox-Machado, Lester	*Spain*	**P521**
Garrido, Rafael	*Spain*	129, P295, P387, P549
Gough, Douglas O.	*U.K.*	**165**, P545
Goupil, Marie Jo	*France*	221
Green, Elizabeth M.	*U.S.A.*	**65**, P517
Hobart, Marco Antonio	*Mexico*	P355, P397
Hu, Jing-Yao	*China*	**P323**
Kambe, Eiji	*Japan*	**P331**
Kawaler, Steve D.	*U.S.A.*	P417
Kilkenny, David	*South Africa*	
Kim, Seung-Lee	*Korea (South)*	**P335**, P355
Kiss, László L.	*Hungary*	**P429, P433**
Kjeldsen, Hans	*Denmark*	**1**, 233, P303, P383, P441
Krzesinski, Jerzy	*U.S.A.*	**P343**
Kubát, Jirí	*Czech Republic*	**P525**
Kurtz, Donald W.	*U.K.*	**29**
Landstreet, John D.	*Canada*	**P347**
Le Contel, Jean-Michel	*France*	P359
Lehmann, Holger	*Germany*	**P351**, P489, P493
Lochard, Jérémie	*France*	
Löffler, Wolfgang	*Switzerland*	**P529**
Lopes, Ilídio P.	*U.K.*	P445
López de Coca, Pilar	*Spain*	P319, **P355**, P397
López-González, Maria José	*Spain*	P391
Marconi, Marcella	*Italy*	**245**, P553
Marques, João P.C.	*Portugal*	
Martinez, Andrea P.	*Italy*	
Martins, Adriana	*Portugal*	
Martins, Jorge H.	*Portugal*	
Matthews, Jaymie M.	*Canada*	
Mazumdar, Anwesh	*France*	**P477**
Metcalfe, Travis S.	*Denmark*	**141, P533**
Michel, Eric	*France*	P295, P521
Miglio, Andrea	*Denmark*	**P537**
Mittermayer, Peter	*Austria*	**P541**
Molenda-Zakowicz, Joanna	*Poland*	P339, **P363**

Monteiro, Mário J.P.F.G.	*Portugal*	P513, P537, P553
Montgomery, Michael H.	*U.K.*	**P545**
Moreira, Olga	*The Netherlands*	
Moskalik, Pawel A.	*Poland*	
Mosser, Benoît	*France*	**P367**, P449
Moya, Andrés	*Spain*	129, **P549**
Novais, Liliana	*Portugal*	
Nowakowski, Rafal M.	*Poland*	273
O'Brien, M. Sean	*U.S.A.*	45
Oreiro, Raquel	*Spain*	269
Palla, Francesco	*Italy*	245, P553
Pallé, Pere L.	*Spain*	
Pamyatnykh, Alexei A.	*Poland*	**97**, 133, P387
Paparó, Margit	*Hungary*	**P481**
Paternò, Lucio	*Italy*	229
Piccioni, Adalberto	*Italy*	**P371**
Pigulski, Andrzej	*Poland*	**P375**
Pijpers, Frank P.	*Denmark*	233, **P485**
Pinheiro, Fernando J.G.	*Portugal*	**P553**
Pinsonneault, Marc H.	*U.S.A.*	
Provost, Janine	*France*	P449, **P557**
Randall, Suzanna K.	*U.K.*	**P445**
Reed, Mike D.	*U.S.A.*	65, **73**
Reegen, Peter	*Austria*	**P379**, P387
Rincon, François	*France*	**P561**
Ripepi, Vicenzo	*Italy*	**69**, P553
Rodler, Florian	*Austria*	**P387**
Rodríguez, Eloy	*Spain*	P359, **P391, P393**, P397
Rolland, Angel	*Spain*	P319, P355, **P397**
Roxburgh, Ian W.	*U.K.*	**17, 187**, P477
Schmider, François-Xavier	*France*	**P449**
Stankov, Anamarija	*The Netherlands*	**P453**
Teixeira, Teresa C.	*Denmark & Portugal*	233
Telting, John H.	*Spain*	85, **P401**, P409
Théado, Sylvie	*France*	**P569**
Thompson, Michael J.	*U.K.*	P537
Thompson, Susan E.	*U.S.A.*	**P573**

Thoul, Anne	*Belgium*	129, **241**, P437, P493
Turck-Chiéze, Sylvaine	*France*	P497
Turcotte, Sylvain	*U.S.A.*	**233, P405**
Uytterhoeven, Katrien	*Belgium*	P401, **P409**
Vauclair, Gérard P.	*France*	
Vauclair, Sylvie	*France*	**203**, P569
Vorontsov, Sergei V.	*U.K.*	187
Wang, Jun-Jie	*China*	P323
Zima, Wolfgang	*Austria*	P299, P387, **P489**
Zwintz, Konstanze	*Austria*	**P457**, P473

SPACE AND GROUND BASED DATA FOR ASTEROSEISMOLOGY

H. KJELDSEN

Institut for Fysik og Astronomi, Aarhus Universitet, Denmark
Teoretisk Astrofysik Center, Danmarks Grundforskningsfond, Denmark

Abstract. Measuring eigenfrequencies and identifying eigenmodes provide the observational basis for a significant improvement in our understanding of stellar evolution and structure. Development throughout the last few years show that we may be at the dawn of a 'Golden Age' for Asteroseismology. For this to become a reality we only need two things: Better data and better models. In this paper I describe some aspects of how one detects stellar oscillations.

Keywords: Techniques: asteroseismology; Techniques: High-precision; Stars: evolution; Stars: structure

1. Observational Asteroseismology

The study of the solar core through Helioseismology is founded on a huge amount of high-quality data supported by detailed state-of-the-art theoretical models (see Di Mauro et al., 2002). It is however difficult to imagine that asteroseismology will reach a level similar to that in which we find helioseismology today, simply due to the fact that a lot of the observed solar oscillation frequencies will be impossible to observe in other stars, where one can only marginally resolve the surface.

However we are still able, for some stars, to extract information by doing seismology on stars (see e.g. Vauclair, 1997; Kawaler, 1998; Breger and Montgomery, 2000; Kjeldsen and Bedding, 2001; and Brown and Gilliland, 1994). Observational asteroseismology focus on the following two activities:
 – Measuring Eigenfrequencies,
 – Identifying Eigenmodes.
Those activities aim at providing the observational basis for a significant improvement in our understanding of stellar evolution and structure. Oscillation frequencies are the most accurate properties one can measure for a star, and this has motivated a huge observational effort with the aim of doing asteroseismology in a similar way as helioseismology is used to study details of the solar interior. For most stars, as stated above, one is not yet at a stage where asteroseismology can be used to gain information of the stellar interiors, simply because the lack of adequate quality of data and models. However, development throughout the last few years show that we may be at the dawn of a 'Golden Age' for Asteroseismology. For this to become a reality we only need two things:
 – Better data,

- Better models.

2. Data for Asteroseismology

The basic question concerning observing oscillations is; how can we detect the oscillations? And which observing technique provide the most accurate measure?

In order to answer this question, let us evaluate the observational properties of the low amplitude solar p-mode oscillations and use this to illustrate how we can observe the oscillations.

A single 5 min oscillation mode at the solar surface, oscillates with a surface velocity of 20–25 cm/s. With a period of 5 min this corresponds to a change in radius of 10–13 m ($dR/R = 0.017$ ppm). Since density in the solar atmosphere will oscillate with an amplitude proportional to the ratio between oscillations velocity and the sound speed ($d\rho/\rho = v/v_{sound}$), and since density oscillations will cause temperature oscillations, we find that the temperature in the solar photosphere oscillates at a level of 1 ppm (dT/T) corresponding to a temperature amplitude (dT) of 5–6 mK. Due to the link between temperature and surface flux and since the change in radius is insignificant compared to the change in temperature (this is due to the high order of the solar oscillations), we find the amplitude of the solar luminosity oscillations to be at the level of 4 ppm (dL/L). Assuming a solar temperature blackbody spectrum we can easily transform the solar oscillation amplitude to photometric amplitudes in the R, V and B spectral filters:
- $dR = 3.8$ ppm
- $dV = 4.5$ ppm
- $dB = 5.5$ ppm
- $d(B - R) = 1.7$ ppm

One may also detect the oscillations through the effect on temperature sensitive spectral absorption lines such as the Balmer lines and metal lines from Na I and Fe I. The oscillation in the equivalent width of absorption lines is in the order of: $dEQW/EQW = 6$ ppm (Fe I, H I, Na I).

3. How is This Observed?

In the following I will concentrate on the observational effects caused by radial oscillations. The observational properties will be illustrated through a number of figures illustrating the effect of oscillations.

3.1. VELOCITY

The most fundamental way of observing oscillations is through velocity (see Figures 4 and 5). Velocity is sensitive to:

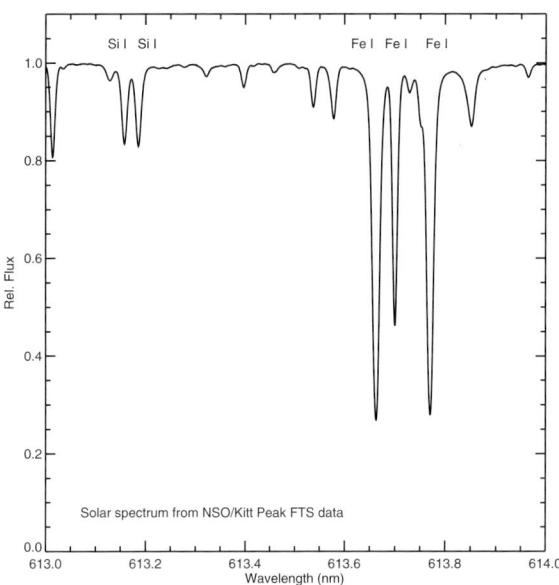

Figure 1. A small part of the solar spectrum (613–614 nm). This spectrum will be used in this paper to demonstrate how one may observe stellar oscillations.

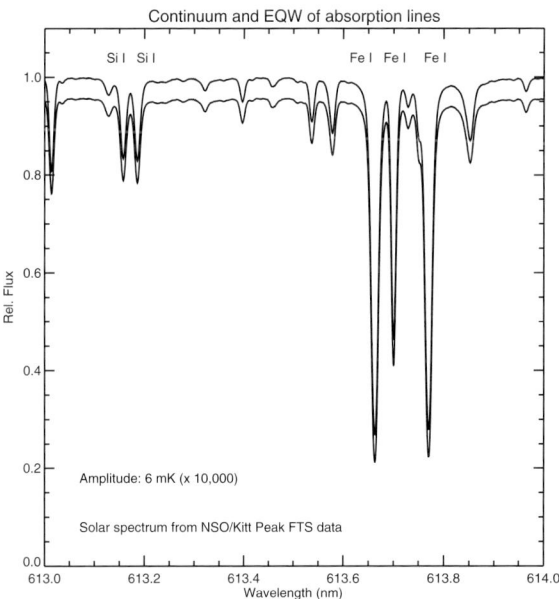

Figure 2. The effect of an oscillation on the stellar spectrum (due to temperature). The amplitude is 10,000 times solar.

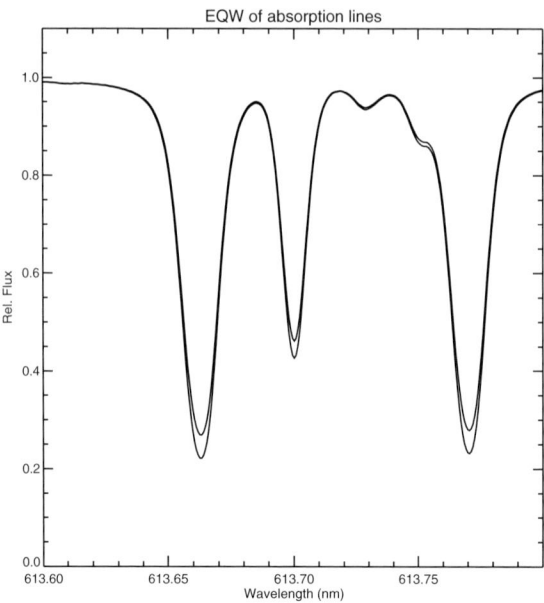

Figure 3. Same as Figure 2, but after normalizing the continuum. The selected spectral region is between 613.6 nm and 613.8 nm.

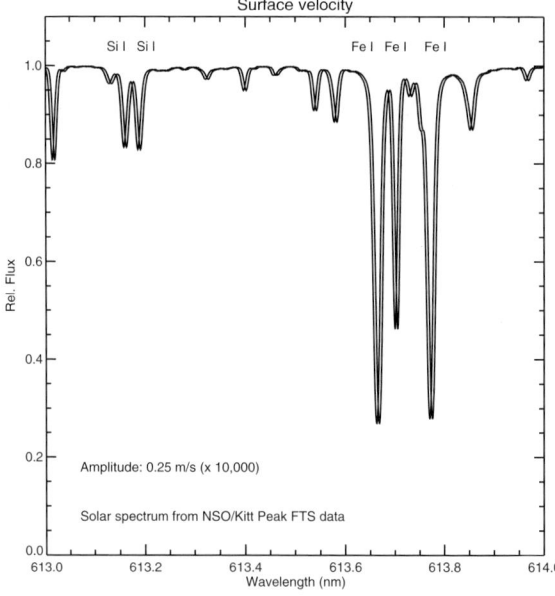

Figure 4. Velocity shift due to an oscillation 10,000 times larger than a single mode in the Sun.

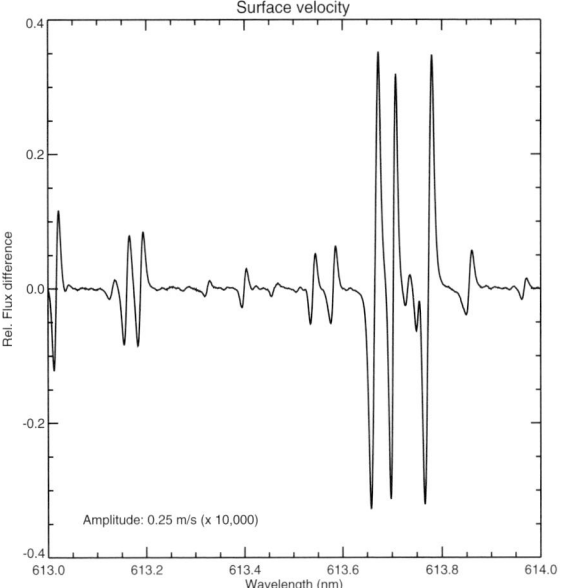

Figure 5. Same as Figure 4, but showing the relative flux difference. This flux difference indicates the sensitivity to velocity for different parts of the spectrum.

- Number of absorption lines,
- Line widths,
- Stable reference (short/long term).

This technique can reach the highest sensitivity for F, G and K stars. The precision one is able to reach depend, however, on the rotation velocity for the star. The highest demonstrated precision for bright solar-type stars are at the following levels:

- For non-rotating stars: 2 m/s (per min),
- For $v \sin(i)$ = 10-15 km/s: 10 m/s (per min),
- For $v \sin(i)$ = 50 km/s: 40 m/s (per min).

A more detailed discussion of this method can be found in Bouchy (these proceedings), Bouchy and Carrier (2002) and Bedding et al. (2001).

3.2. INTENSITY AND COLOUR

The oscillations may also be detected through their effect on the stellar brightness (intensity) and temperature (colour). This is illustrated in Figures 2, 3, 6 and 7.

Intensity is sensitive to:
- Atmospheric Scintillation,
- Instrumental effects (CCD flat field, gain, ...).

Colour is sensitive to:
- Atmospheric Scintillation,

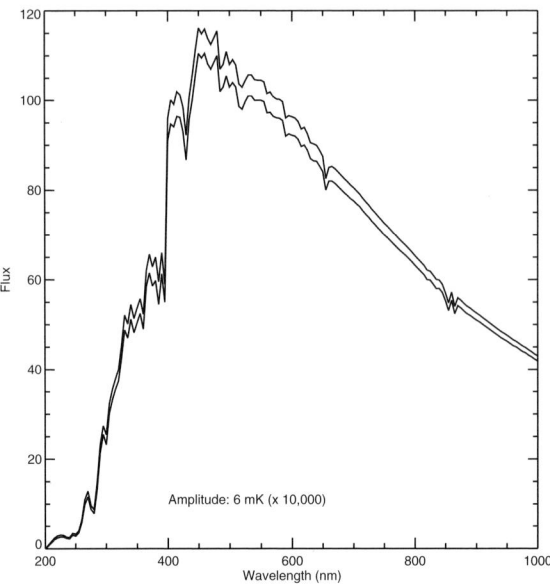

Figure 6. A solar flux spectrum and the change in the spectrum due to an oscillation 10,000 times solar.

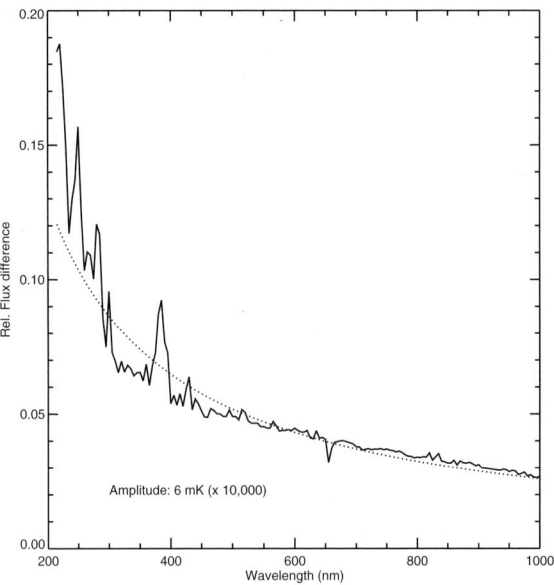

Figure 7. Same as Figure 6, but showing the relative flux difference. The dotted line corresponds to the blackbody approximation. The real spectrum show higher temperature sensitivity than the black-body approximation in the UV and lower in the blue (300–550 nm).

- Filter separation,
- Instrumental effects (Colour stability).

Due to atmospheric scintillation one will gain significantly by observing intensity and colour observations from space. The highest demonstrated precision for bright solar-type stars on ground is at the level of 250 ppm/min in differential photometry (Gilliland et al., 1993).

3.3. EQUIVALENT WIDTH

Equivalent width measurements of temperature sensitive spectral lines are not sensitive to atmospheric scintillation. However instrumental effects (e.g. flexure) will contribute to the noise budget (see e.g. Kjeldsen et al., 1995, 1999).

3.4. INSTRUMENTAL NOISE SOURCES

For ground-based time series data one is limited by the following noise sources:

Photons:	Poisson statistics.
Stellar:	Starspots, activity, Granulation, Variability.
Instrument:	Shutter, Stray light, Tracking, Temperature variability, Focus variability.
Detector:	Sub-pix structure, Global Flat Field (incl. Colour), Sensitivity Stability, Cosmic Rays, Radiation damage (CTE), Read-out-noise, Detector Linearity.
Gain:	Electronic drift, Temperature drift.
Software:	Non-linearities.
Observed Field:	Stellar crowding.
Atmospheric:	Scintillation, Extinction.

4. Measuring Eigenfrequencies

The main task for time series analysis is the frequency determination in the power spectrum. This task involves considering the following two aspects:
- Extracting frequencies,
- Detection sensitivity at low signal-to-noise.

4.1. NOISE IN AMPLITUDE SPECTRUM

In the following I show a number of examples on how the noise in the amplitude spectrum will affect the detection of solar-like oscillations (observed at low S/N). I show two types of examples. One set of examples deals with coherent modes and one set deals with non-coherent modes (mode lifetime: 2 days). The parameters

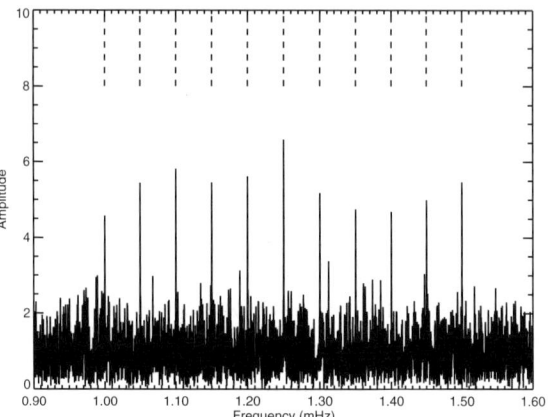

Figure 8. A simulated p-mode structure at S/N = 5. All modes are coherent. Dashed lines indicate the position of the input frequencies.

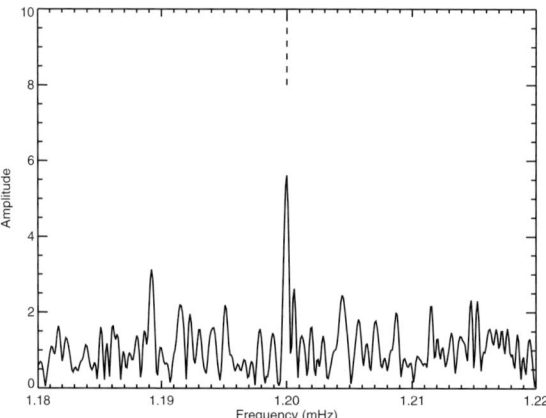

Figure 9. A part of Figure 8.

used in the examples are (for coherent modes): amplitude: 5, p-mode separation: 50 μH; and for non-coherent modes: mode lifetime: 2 days, amplitude: 5, p-mode separation: 50 μHz.

Precision for measured frequencies, phases and amplitudes can be estimated using the calculations by Montgomery and O'Donoghue (1999). Amplitude [$\sigma(a)$], phase [$\sigma(\phi)$] and frequency [$\sigma(f)$] is:

$$\sigma(a) = \sqrt{2/\pi} \cdot <a> \approx 0.80 <a> \qquad (1)$$

where $<a>$ is the 1σ noise level in the amplitude spectrum.

$$\sigma(\phi) = \sigma(a)/a = \sqrt{2/\pi} \cdot \frac{<a>}{a} \approx 0.80 \frac{<a>}{a} \qquad (2)$$

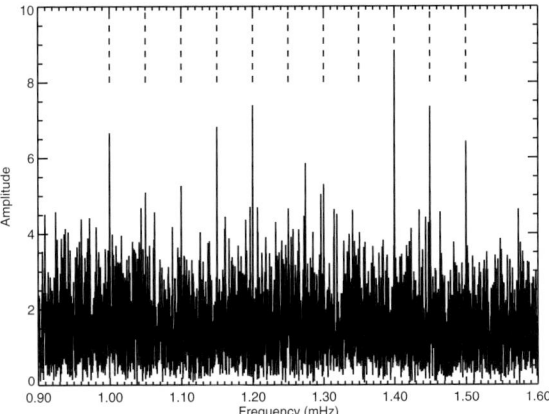

Figure 10. A simulated p-mode structure at S/N = 3. All modes are coherent. About half of the modes can be detected.

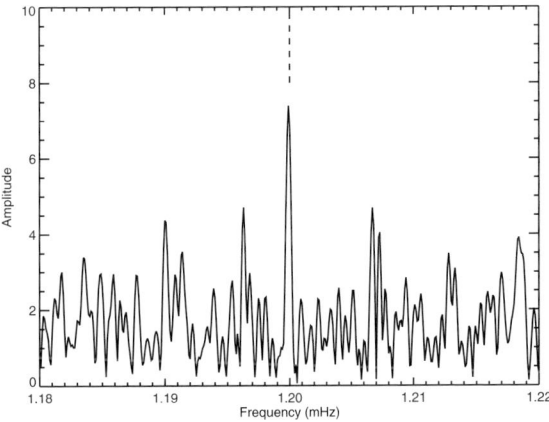

Figure 11. A part of Figure 10.

$$\sigma(f) = \sigma(\phi) \cdot \frac{\sqrt{3}}{\pi T} = \sqrt{6/\pi^3} \cdot \frac{<a>}{a} \cdot \frac{1}{T} \approx 0.44 \frac{<a>}{a} \cdot \frac{1}{T} \qquad (3)$$

T is the length of the observing run. As illustrated in figure 8-14 one can detect oscillations at S/N=3 (some modes). In case of non-coherent modes (as for the solar-like oscillations), the required S/N will be higher to ensure a detection (depending on the mode life-time).

4.2. WINDOW SIDEBANDS

If we concentrate on low Signal-to-Noise we will find that for a single-site campaign one may detect wrong frequencies (due to beating between sidebands and noise). If we, as an example, take a main peak at S/N = 4 and sidebands at S/N

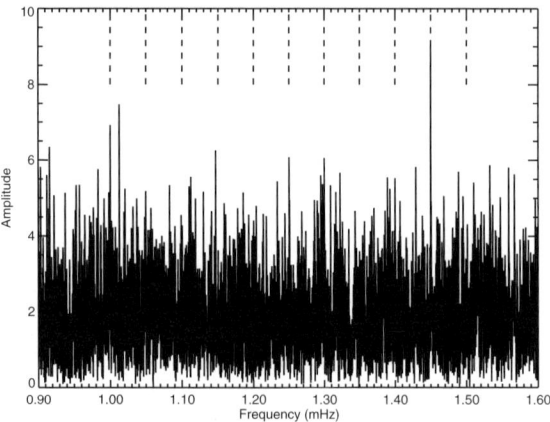

Figure 12. A simulated p-mode structure at S/N = 2.5 All modes are coherent. Only a few modes can be detected.

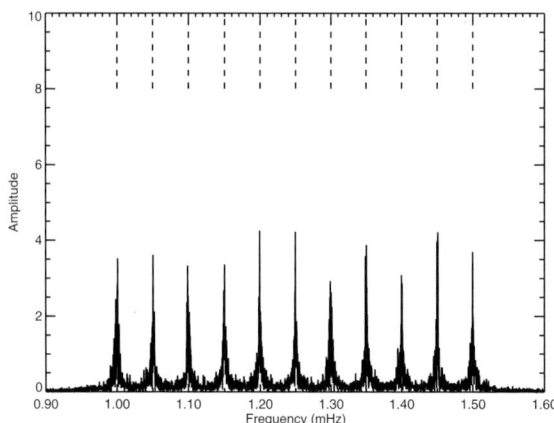

Figure 13. A simulated p-mode structure using non-coherent oscillations (mode life-time is 2 days). The amplitude is below input (=5) due to the damping and re-excitation of the modes. This will make detection more difficult.

= 3 (single site observations), one may risk detecting the sidebands instead of the central peaks in a significant number of cases. We find that for a S/N = 4 (central peak) we will find higher amplitude in one of the sidebands in 30% of the cases, while for S/N = 3.5 one will find higher amplitude in one of the sidebands in 50% of the cases.

The only solution to this is to use multi-side campaigns or uninterrupted space missions.

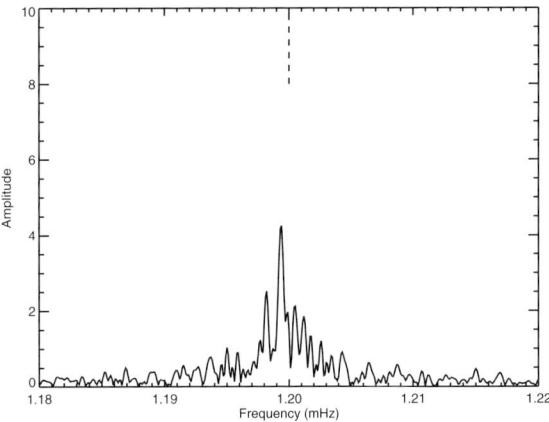

Figure 14. A part of figure 13 (mode life-time is 2 days).

4.3. IDENTIFYING EIGENMODES

Identifying eigenmodes is essential in order to use the frequencies for seismology. In the case of solar-like stars a number of different methods aiming at identifying the mode structure of the high-order p-modes have been developed. The reader can e.g. consult Bouchy and Carrier (2002) for an example on how some of those methods are being used. In the case of classical pulsating stars it is in general more difficult to identify oscillation modes. A large number of papers at this conference deals with questions on mode identification (see e.g. Reed; Telting; Balona; De Ridder; and Dupret).

5. The Future

Ground-based data are and will be limited due to the non-uniform data quality (Single site, airmass, differences in instrumentation between sites). Velocity is optimized for cool and non-rotating stars, and the ultimate solution will be space missions such as: MOST, COROT, MONS and EDDINGTON. In those cases we will see no scintillation and other atmospheric effects and combined with the high duty cycle those data will form a uniform set of data. But not all noise sources will disappear in space! Some will be worse, e.g. field crowding and blind reduction procedures.

The Space missions are expected to form the basis for 'the Golden Age' concerning the data. Frequencies will be determined to 0.05 μHz precision allowing separations in solar-like stars to be measured to 0.01 μHz precision. We are at present at the dawn of a 'the Golden Age' for Asteroseismology, but we will still require better data and better models. This conference may not have solved those two requirements. However, it guided us in the right direction and can as such be seen as a roadmap, allowing us to begin the voyage towards 'the Golden Age'.

Acknowledgements

The author wishes to thank the financial support of the Danish National Science Foundation, through the establishment of the Theoretical Astrophysics Center.

References

Bedding, T.R., Butler, R.P., Kjeldsen, H. et al.: 2001, *ApJ* **549**, L105.
Bouchy, F. and Carrier, F.: 2002, *A&A* **390**, 205.
Breger, M. and Montgomery, M. (eds.): 2000, *Sixth Vienna Workshop in Astrophysics: Delta Scuti and Related Stars*, A.S.P. Conf. Ser. **210**.
Brown, T.M. and Gilliland, R.L.: 1994, *ARA&A* **33**, 37.
Di Mauro, M.P., Christensen-Dalsgaard, J., Rabello-Soares, M.C. and Basu, S.: 2002, *A&A* **384**, 666.
Gilliland, R.L., Brown, T.M., Kjeldsen, H. et al.: 1993, *AJ* **106**, 2441.
Kawaler, S.: 1998, in: F.L. Deubner, J. Christensen-Dalsgaard and D.W. Kurtz (eds.), *Proc. IAU Symp. 185, New Eyes to See Inside the Sun and Stars*, Kluwer Academic Publishers, Dordrecht, p. 261.
Kjeldsen, H. and Bedding, T.R.: 2001, in: *Proc. SOHO 10/GONG 2000 Workshop: Helio- and Asteroseismology at the Dawn of the Millennium ESA SP*-**464**, 361.
Kjeldsen, H., Bedding, T.R., Viskum, M. and Frandsen, S.: 1995, *AJ* **109**, 1313.
Kjeldsen, H., Bedding, T.R., Frandsen, S. and Dall, T.H.: 1999, *MNRAS* **303**, 579.
Montgomery, M. and O'Donoghue, D.: 1999, *Delta Scuti Star Newsletter*, No. 13.
Vauclair, C.: 1997, in: J. Provost and F.X. Schmider (eds.), *Proc. IAU Symp. 181, Sounding Solar and Stellar Interiors,* Kluwer Academic Publishers, Dordrecht, p. 367.

STARE RESULTS ON A SINGLE FIELD: TENS OF NEW PULSATING STARS

ROI ALONSO and JUAN ANTONIO BELMONTE
Instituto de Astrofísica de Canarias, E-38200 La Laguna, Spain; E-mail: ras@ll.iac.es

TIM BROWN
High Altitude Observatory/National Center for Atmospheric Research, 3450 Mitchell Lane, Boulder, CO 80307, USA

Abstract. We present preliminary results on variable stars of a STARE's three month observational run centered at the Cygnus constellation. A total amount of aprox. 14000 stars with $9 < R < 12.5$ magnitude, in STARE's $6.1 \times 6.1^{\circ 2}$ FOV, have been analyzed to obtain lightcurves for each of these stars. The data spans for ~ 90 nights. In this single field, we detect more that 40 stars with pulsation modes between 5 and 40 c/d, the vast mayority previously unknown to be variables.

Keywords: variable stars, asteroseismology, δ Scuti, β Cepheid, δ Cepheid, γ Dor

1. Introduction

Photometric exoplanet transit searches are a new field o observational astronomy, possible since the development of wide field CCDs and telescopes. A huge amount of data comes out from projects like VULCAN (Borucki et al., 2001), OGLE (Udalski et al., submitted to *AcA*), EXPLORE (Mallén-Ornelas et al., submitted to *ApJ*), STARE (Brown and Kolinski, 1999), and many others. Apart from the valuable information on transiting planets, many lightcurves of previously unknown variable stars appear in these data. To show the capabilities of these projects to detect and characterize variable stars, we present results of the observations of one single field centered at the Cygnus constellation, with the STARE (STellar Astrophysics and Research on Exoplanets) telescope.

2. Observations and Data Reduction

The telescope is a semi-automated Schmidt camera with an aperture of 9.9 cm, 286 mm of focal length and a 2040x2048 pixels Pixel Vision CCD attached. This combination provides a $6.1 \times 6.1^{\circ 2}$ FOV, with 10.8 arcsec/pixel. It was designed and built at the High Altitude Observatory (Boulder, CO, USA), and moved to the Observatorio del Teide (Spain) in July 2001. The observations cover ~ 40 nights from July 27 to October 26, 2001. With 107s of exposure time, we obtain lightcurves of ~ 14000 stars in the field, in R filter and brightnesses $9 < R <$

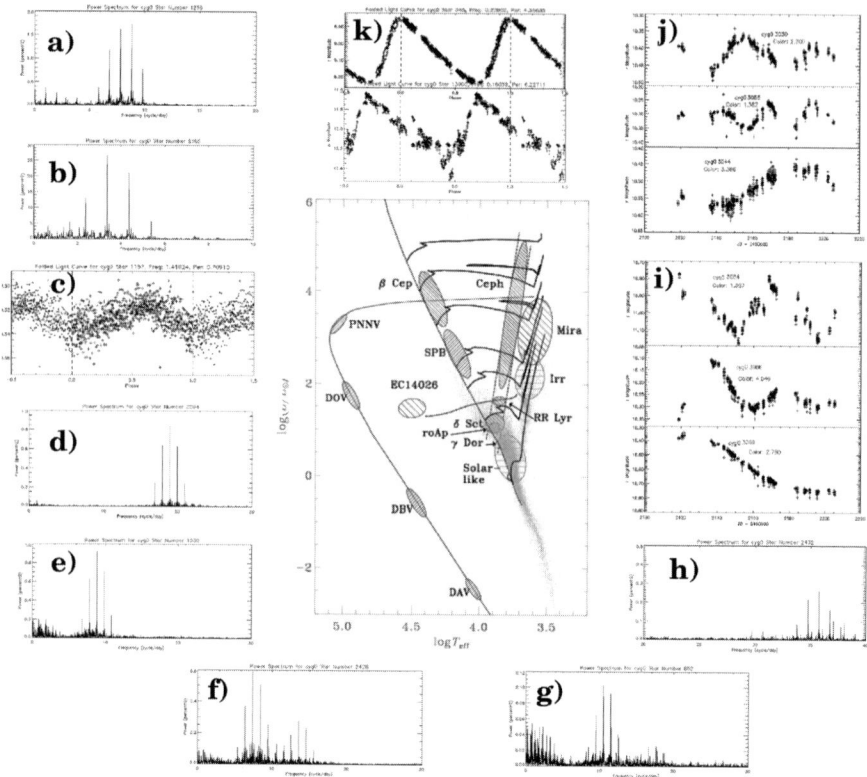

Figure 1. Different examples of variability across the H-R diagram, from STARE data. a) Power spectrum of HD227977, probably a βCepheid star. b) and c) Two stars with oscillations in the range [1–5] cycles/day, thus probably γ Doradus or SPBs. d), e), f), g), h) Power spectra of stars with oscillation periods in the range [6–40] cycles/day; d) shows a mono-periodic behaviour, f) and g) show a complex spectrum, and h) shows frequencies higher than 30 cycles/day. i) and j) are plots of the lightcurves of long period or irregular red giant stars. k) are two folded lightcurves of classical Cepheids: top – V547Cyg; bottom – V402Cyg. The figure at the center is from Christensen-Daslgaard (1999).

12.5, sampled in a 2 min cadence. Our reduction algorithms (see Brown et al., in preparation) provide us with the lightcurves sampled in a 15 min cadence, thus reducing the noise of the data. We achieve nightly precisions (measured with the nightly rms) better than 1% for $\sim$ 7000 of the stars.

3. Variable Stars

We made a search in the Fourier space for stars with periodic pulsations with periods in the range [6–50] cycle/day. A Scargle algorithm (Scargle, 1982) was employed to make the Fourier spectra. More than 40 stars were found to be clearly pulsating in this range of periods. We plot in Figure. 1 six of them. Only one of

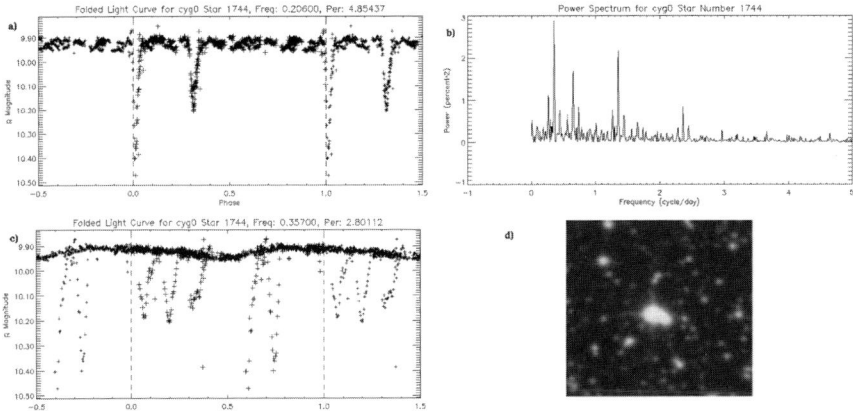

Figure 2. The case of HD 227269: a) Folded lightcurve with the eclipsing binary period. b) Power spectrum after the eclipses were removed from the lightcurve. c) Folded lightcurve with the highest peak frequency of b). d) DSS-1 R image of the star, showing an apparent close companion.

these stars appeared in SIMBAD database as a variable star: V1821Cyg (Delgado et al., 1984). In Figure. 1a) we plot the power spectrum of the star HD 227977, catalogued as B2 type in this database, thus very probably a β Cepheid. Overlaping the same region in the H-R diagram as the δ Scuti stars we find the Gamma Doradus stars. Two candidates of this kind of pulsation found in this fiel are plotted in Figure. 1b) and 1c). Also in the instability strip, with higher intrinsic luminosity, are the δ Cepheid stars. Figure 1k) plots the folded lightcurves of two of these stars. Red giants clearly dominate our sample of variable stars, with amplitudes of up to tenths of magnitudes, and periodicities of the order of days, when found to be periodic. Some examples of the lightcurves of these stars appear in Figures 1i) and 1j).

An interesting case is that of HD 227269 (Figure 2). This high eccentricity eclipsing binary seems to show pulsating behaviour in the range of few cycles/day when the eclipses are removed from the lightcurve. As shown in Figure 2c), the eclipses are not in phase with the pulsation. A look at a DSS image shows a close ($\sim 9''$) companion with similar brightness, that can't be resolved with STARE's angular resolution. Whether this pulsation is real or just an artifact of the crowding needs more observations to be determined.

4. Conclusions and Future Work

More than 40 stars with periods have been found in a single STARE FOV; in the period range of the search, there was only one of these stars catalogued as variable. A deeper analysis of these stars is currently under way. More data coming from this and similar projects will increase the list of variable stars dramatically.

Acknowledgements

The STARE project is an international cooperation supported by NASA's Origins of the Solar System program. This research has made use of the SIMBAD database operated at CDS, Strasbourg (France).

References

Borucki, W.J., Caldwell, D., Kock, D.G. and Webster, L.D.: 2001, *PASP* **113**, 439.
Brown, T.M. and Kolinski, D.: 1999, http://www.hao.ucar.edu/public/research/public/stare/stare.html
Christensen-Dalsgaard: 1999, *Ap&SS* **261**, 1.
Delgado, A.J., Alfaro, E.J., García-Pelayo, J.M., Garrido, R. and Vidal, S.: 1984, *A&AS* **58**, 447.
Scargle, J.D.: 1982, *ApJ* **263**, 835.

THE *EDDINGTON* MISSION

IAN ROXBURGH[1,2] and FABIO FAVATA[3]
[1]*Astronomy Unit, Queen Mary, University of London, London E1 4NS, UK*
[2]*Observatoire de Paris, Place Jules Janssen, 92195 Meudon, France*
[3]*Astrophysics Division, ESTEC, ESA, Noordwijk, The Netherlands*

Abstract. The *Eddington* mission was given full approval by the European Space Agency on the 23rd May 2002, as part of the new 'Cosmic Vision' Science programme, with launch scheduled for 2007/8. Its twin science objectives are asteroseismology and planet finding. In its current design it consists of 4×60 cm folded Schmidt telescopes, each with $6^o \times 6^o$ field of view and its own CCD array camera. The current observing plan is to spend 2 years primarily devoted to asteroseismology with 1–3 months on different target fields monitoring up to 50,000 stars per field, and 3 years continuously on a single target field monitoring upwards of 100,000 stars as required for planet searching. The asteroseismic goal is to be able to detect oscillations frequencies with a precision 0.1–0.3 μHz.

Keywords: space missions: *Eddington*; stars: asteroseismology; stellar evolution; planets: planet finding, earth like planets, habitable planets

1. The *Eddington* Mission

The *Eddington* mission was proposed in early 2000 to ESA in response to the 'Call for mission proposals for two flexi-missions (F2 and F3)', released in October 1999 in the context of the Horizons 2000 programme. The proposal was submitted by an international scientific team led by Roxburgh, Christensen-Dalsgaard and Favata (2000). It built on the work done over two decades on previous proposals that had been studied within ESA and CNES (*EVRIS, PRISMA, PRISMA2, STARS*). The mission has two complementary primary scientific aims, to produce the data on stellar oscillations necessary for understanding the interior structure and evolution of stars, and to detect and characterize habitable planets around other stars.

Both scientific goals will be achieved by performing highly accurate time-resolved photometry on a large number of stars, with a simple payload which is essentially a wide field broad-band photometer. The accurate light curves will be used to determine stellar oscillation frequencies, and thus allow the use of asteroseismic tools to probe the interior structure of stars, quantitatively determining, e.g., the size of convective regions, the structure of regions with steep changes in chemical composition and internal rotation and to therefore determine very accurate stellar ages. At the same time, planets as small as the Earth orbiting the target stars in the habitable zone will be found through the temporary drops in the stellar light caused by their transits.

Following the favourable reviews by ESA's scientific advisory bodies (the Astronomy Working Group and the Space Science Advisory Committee) *Eddington* was selected in March 2000 as one of the 4 missions for which an assessment study was carried out in the course of the spring and summer of 2000 (Favata et al., 2000). Following presentation of the studies in September 2000, *Eddington* was selected with a 'reserve status'. Study activities continued in the course of 2001 and 2002, and in May 2002 *Eddington* was approved by ESA's Science Programme Committee to be implemented in the framework of the *Herschel* and *Planck* projects in ESA's rescoped science programme **Cosmic Vision**.

Data rights and observing programme

Eddington is a facility-type mission, for which the observing plan will be the result of a combination of an open, competitive proposal cycle with a broad community consulting process, and for which the resulting data are proposed to be available to the scientific community, without proprietary rights for individual proposers.

Payload configuration and data product

The primary *Eddington* data product is a set of relatively calibrated photometric light curves for each star in the field of view down to a predefined magnitude limit. To maximize the field of view and collecting area the payload is baselined to consist of 4 parallel very similar (possibly identical) instruments, each of them independently producing a light curve, which will be merged a posteriori. Each of the 4 instruments consists of 3 basic components: (1) a wide field optical telescope, (2) a mosaic CCD camera (named 'EddiCam') and (3) a data processing unit (DPU).

Mission procurement

ESA will have overall responsibility for the mission, and will in particular be responsible for the procurement of the spacecraft and of the ground segment, for the integration of the payload and spacecraft units, for the launch and operations, the acquisition of the data, their reduction, archiving and their distribution to the holders of the data rights. The payload is planned to be procured through a partnership between ESA and a consortium of scientific institutes.

ESA is planned to operate the Scientific Operations Centre (SOC) which will process and archive the data from the instrument and deliver the scientific data products (relatively calibrated light curves) to the holders of the data rights.

2. The Current Mission Concept

The final form of the mission will be decided following detailed industrial studies and advice from the *Eddington* Science Team. The current working concept is as follows:

4 × 0.6m folded Schmidt telescopes each with $6^o \times 6^o$ field of view.
Mosaic CCD 'panoramic' cameras 8 E2V 42-C0 chips.
Orbit at L2 – 95% duty cycle – 5 year mission.
3 years on 1 field (planet finding + asteroseismology).
1 – 3 months on asteroseismology fields (total 2 years).
Launch: 2007/8.

Asteroseismology fields
Max. no. of stars per field: 2000 ($m_v < 11$), 50,000 ($m_v < 15$).
Precision on frequencies: 0.1 μHz (goal), 0.3 μHz (bottom line).
Frequency range: 1 μHz – 100 mHz (goal), 1 μHz – 10 mHz (bottom line).
Magnitude range: 3 – 15 (goal), 5 – 15 (bottom line).
Time sampling: $\leq$ 30 sec (baseline), $\leq$ 5 sec (some targets).
Chromatic information: to be determined.

Planet finding field (also available for asteroseismology)
No. of stars: > 100,000 ($m_v < 17$).
Time sampling: 30 sec (goal), 600 sec (bottom line).

Acknowledgements

We wish to record our thanks to the many scientists who have contributed over two decades to convincing ESA to fly an asteroseismology mission by working on the precursor proposals *PRISMA, PRISMA2* and *STARS*, and especially to those who worked for the selection of *Eddington*. In particular we record the contribution to *Eddington* made by the Assessment Study Team, the successor *Eddington* Science Team, and technical support from ESA.

References

Favata, F., Roxburgh, I.W. and Christensen-Dalsgaard, J. (eds.): 2000, with Aerts, C., Antonello, E., Catala, C., Deeg, H., Gimenez, A., Grenon, M., Pace, O., Penny, A., Schneider, J., Waltham, N., *Eddington, A Mission to Map Stellar Evolution through Oscillations and to Find Habitable Planets*, Report of Assessment Study, ESA-SCI(2000)8.

Roxburgh, I.W., Christensen Dalsgaard, J. and Favata, F. (eds.): 2000, with Antonello, E., Baade, D., Badiali, M., Baglin, A., Bedding, T., Brown, T., Catala, C., Collier, A., Dziembowski, W., Gilmore, G., Gimenez, A., Gough, D., Horne, K., Kjelsden, H., Leger, A., Penny, A., Preite-Matinez, A., Rivinus, Th., Schneider, J., Stefl, Z., Sterken, C., Weiss, W., *Eddington – A Stellar Physics and Planet Finder Explorer*, Proposal submitted to ESA in response to the ESA F2/F3 call for proposals.

PRESENT OBSERVATIONAL STATUS OF SOLAR-TYPE STARS

FRANÇOIS BOUCHY and FABIEN CARRIER
Observatoire de Genève, 1290 Sauverny, Switzerland

Abstract. Observing stellar oscillations provides a powerful probe for studying stellar interiors. The frequencies of these modes depend on the properties of the star and give strong constraints on stellar models and evolution theories. The five-minute oscillations in the Sun, induced by stochastic excitation of its convective zone, have provided a wealth of information about the solar interior and has led to significant revisions to solar models. Until recently, the Sun was the only star in which solar-like oscillations were clearly established and characterized. The most important difficulty lies in the extremely small amplitude of the acoustic modes. Thanks in great part to high precision ground based Doppler measurements, solar-like oscillations have been now clearly detected in a growing list of main sequence and subgiant stars (Procyon, β Hyi, ζ Her A, α Cen A, δ Eri and η Boo). In some of them, p-modes were identified and characterized. New results and prospects in this field are presented.

1. P-Mode Oscillations on Solar-Type Stars

Solar-type stars correspond to relatively cool stars with intermediate masses (up to around 1.7 $M_\odot$), near the main sequence, in the phase of central hydrogen fusion and with a significant outer convective zone (F, G and K dwarfs and sub-giants). The convective zone near the stellar surface excite stochastically pressure modes (also called p-modes) in a broad frequency range. These modes are characterized by high radial orders n and, when the stellar disk is not resolved, low angular degrees l. P-mode oscillations produce a characteristic comb-like structure in the power spectrum (see Figure 1) with frequencies fairly regularly spaced and well approximated by the simplified asymptotic relation:

$$\nu_{n,l} = \Delta\nu(n + l/2 + \epsilon) - l(l+1)\delta\nu_{02}/6 \tag{1}$$

On the Sun, such oscillations have typical amplitude of about 20 cm s^{-1} in radial velocity and about 4 or 5 ppm in photometry, with frequencies around 3 mHz, a large separation between modes of same angular degree $\Delta\nu = 135$ μHz and a small separation between modes l=0 and l=2 equal to $\delta\nu_{02} = 9$ μHz. Low-degree helioseismology has been used for several decades to study in detail the solar interior. It is not the case for asteroseismology of solar-type stars.

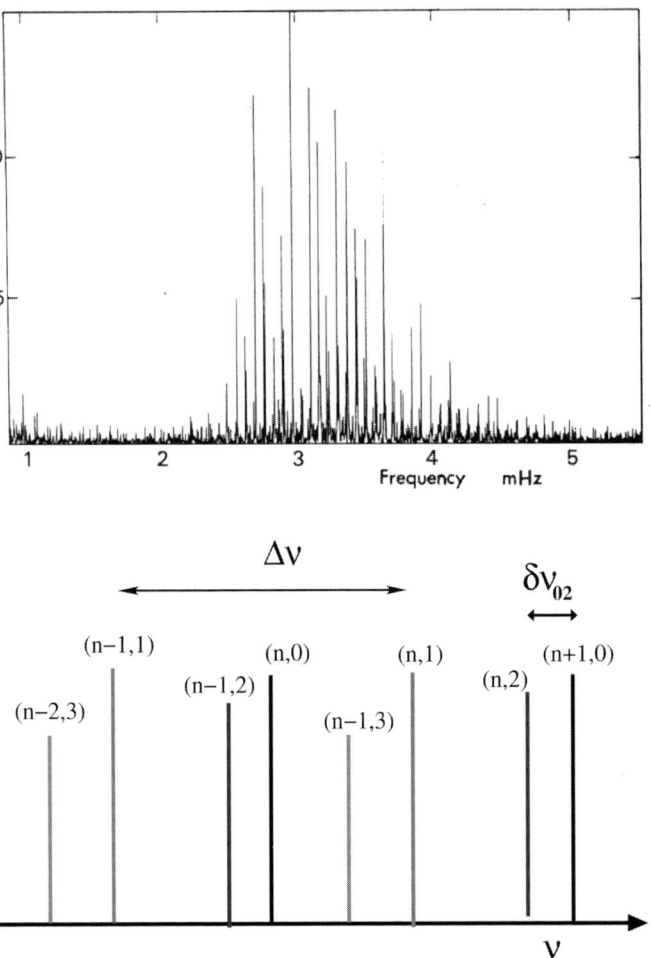

Figure 1. (Top) Acoustic spectrum of the full solar disk (Grec et al., 1983) showing the presence of several tens of p-mode frequencies. (Bottom) Characteristic pattern of solar-type oscillation frequencies.

2. How to Detect and to Characterize Solar-Like Oscillations?

The first difficulties with solar-like oscillations lies in their extremely small amplitudes. The level to reach in photometry (2–20 ppm) requires a space mission or maybe extreme atmospherical conditions (as expected in the south pole). The level to reach in radial velocity (10–100 cm s^{-1}) is now available from the ground with high-stabilized echelle spectrograph thanks to improvements in high-precision Doppler measurements driven by the extra-solar planet searches. The two main radial velocity methods are described by Bouchy and Carrier (these proceedings).

Observation and characterization of p-modes requires to cover the frequency range 0.5–5 mHz, hence to realize intensive observations with short exposures and high cyclic rate. The accuracy needed on the frequencies determination requires a resolution of at least 1 μHz, hence to observe more than 10 nights with the highest duty cycle. This means large programmes on small ground-based telescopes or dedicated space missions.

3. Observational Status

3.1. PROCYON

The F5 subgiant Procyon was observed by Brown et al. (1991) with the AFOE spectrograph and an excess of power was detected around 1 mHz. Recent observations were conducted by Martic et al. (1999) with the ELODIE spectrograph and the analysis of the power spectrum gives an excess of power between 0.5 and 1.5 mHz, a maximum amplitude of about 50 cm s^{-1} and a most probable frequency spacing equal to 55 μHz. This excess of power was observed and confirmed by several other groups (Carrier et al., 2002; Bouchy et al., 2002; Kambe et al., these proceedings). No modes were identified due to (1) the low S/N detection, (2) the single-site window function, (3) the probably short time-life of the modes.

3.2. β HYDRI

The G2 subgiant β Hydri was observed by Bedding et al. (2001) with the UCLES spectrograph. Analysis gives an excess of power between 0.7 and 1.3 mHz, with a maximum amplitude of 50 cm s^{-1} and a most probable frequency spacing equal to 56 μHz. Our group confirmed this result with the CORALIE spectrograph (Carrier et al., 2001). No modes were identified due to (1) the low S/N detection, (2) the single-site window function, (3) the probably short time-life of the modes (as suggested by Bedding et al., 2002).

3.3. ζ HERCULI A

The G0 subgiant ζ Her A was observed by Martic et al. (2001) with the ELODIE spectrograph and analysis shows an excess of power between 0.5 and 1.2 mHz with maximum amplitude of 90 cm s^{-1} and a most probable frequency spacing of 43 μHz. This result was confirmed by Bouchy et al. (2002). The very small S/N detection prevents identification of modes.

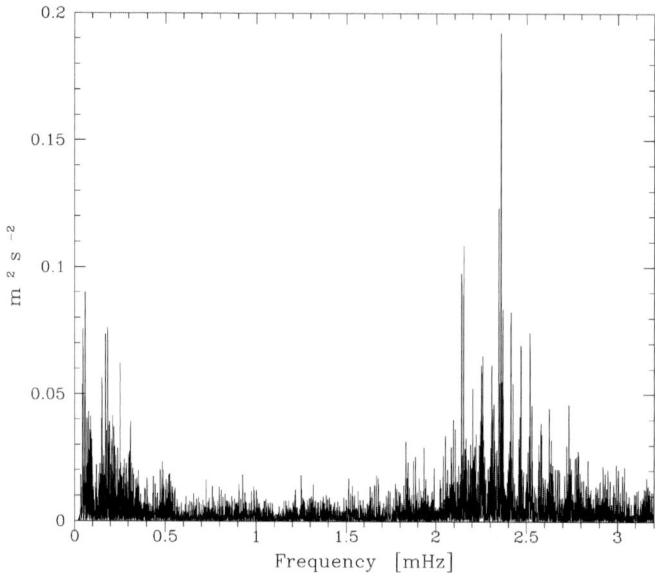

Figure 2. Power spectrum of Doppler measurements made with CORALIE on the G2V star α Cen A.

3.4. α CENTAURI A

The G2 dwarf α Cen A was observed during 13 nights with the CORALIE spectrograph (Bouchy and Carrier, 2001, 2002). 1850 high precision Doppler measurements allow to reach a noise level of only 4.3 cm s^{-1} in the power spectrum (see Figure 2). A clear comb-like structure appears on the power spectrum and yield a clear identification of modes $l=0$, 1 and 2. In total, 28 modes were identified in the range 1.8-2.9 mHz with amplitude between 12 and 44 cm s^{-1} and a large and small spacing of respectively 105.5 μHz and 5.6 μHz. Although with a S/N up to 10 in amplitude, the rotational splitting and the damping time cannot be determined. Based on the identified modes, several groups have developed new theoretical models of α Cen A (Thevenin et al., 2002; Provost et al., Thoul et al., Ballot et al., these proceedings).

α Cen A was observed simultaneously with the UCLES spectrograph (Butler et al., these proceedings) and the combined CORALIE-UCLES data are expected to improve the modes characterization.

3.5. δ ERIDANI

The K0 subgiant δ Eri was observed recently with CORALIE spectrograph (Carrier et al., these proceedings). The excess of power appears very sharped near 0.7 mHz and with an unexpected and very impressive S/N (up to 25 in amplitude). P-modes

TABLE I
Stellar fundamental parameters

Star	Spec. Type	mv	T_{eff} [K]	L [$L_\odot$]	M [$M_\odot$]	[Fe/H] [dex]	$v \sin i$ [km s^{-1}]
Procyon	F5 IV-V	0.4	6530	7.24	1.46	–0.07	6.0
β Hyi	G2 IV	2.8	5800	3.5	1.10	–0.20	4.1
ζ Her A	G0 IV	2.8	5825	6.52	1.4	0.04	4.9
α Cen A	G2 V	0.0	5790	1.52	1.10	0.20	4.4
δ Eri	K0 IV	3.5	5050	3.3	1.20	0.12	1.0
η Boo	G0 IV	2.7	6050	9.0	1.55	0.19	12.8

appear between 0.5 and 0.9 mHz with amplitude in the range 40-170 cm s^{-1} and a large spacing of 43.8 μHz. In spite of a very high S/N, identification of modes is complicated by the mono-site window function and more precisely by the second side-lobe at 23 μHz. Moreover, on this kind of evolved star, non-radial modes $l=1$ are not expected to necessarily follow the asymptotic relation.

3.6. η BOOTIS

The G0 subgiant η Boo was observed recently in multi-site with the two spectrographs CORALIE at La Silla (Chile) and ELODIE at Observatoire de Haute Provence (France) (Carrier et al., these proceedings). The power spectrum clearly presents p-modes between 0.5 and 1.0 mHz with amplitude in the range 40-80 cm s^{-1} and a large spacing of 39.6 μHz. Apart from the large spacing, this recent observation is not in full agreement with the previous claim of modes detection made by Kjeldsen et al. (1995) from measurements of Balmer-line equivalent widths. The small amplitude of modes explain the no-detection in radial velocity by Brown et al. (1997).

3.7. SUMMARY

Detection is now made without doubts on 6 bright solar-like stars and for 3 of them p-modes identification is proposed. Summary of these seismic observations are presented in the following Tables I and II.

4. Mode Identification Difficulties

Several observing campaigns, based on radial velocity measurement, have now detect solar-type oscillations in stars other than the sun with significant evidence.

TABLE II
Seismic parameters

Star	Freq. range [mHz]	A_{osc} [cm s^{-1}]	noise [cm s^{-1}]	$\Delta\nu$ [μHz]	$\delta\nu_{02}$ [μHz]
Procyon	0.5–1.5	50	12	55	–
β Hyi	0.7–1.3	50	13	56	–
ζ Her A	0.5–1.2	90	27	43	–
α Cen A	1.8–2.9	31	4.3	105.5	5.6
δ Eri	0.5–0.9	160	6.5	43.8	–
η Boo	0.5–1.0	70	12.1	39.6	–

However detection of an excess power and of a most probable large spacing of the modes not clearly identified can not conduct to sufficient informations and constraints for theoretical models. Detection of a large number of individual modes with identification and precise characterization (frequency, amplitude, width) is highly needed. We identify at least 3 difficulties for p-mode identifications and propose some solutions.

4.1. THE LOW SIGNAL-TO-NOISE OF DETECTION

On some cases of detection (see Table II), the signal-to-noise of the highest peaks in the amplitude spectrum do not exceed 3 or 4. As shown on simulations (see Figure 7 of Kjeldsen and Bedding, 1995) this level do not allow to identify properly individual modes. The S/N in the amplitude spectrum can be expressed by the following relation:

$$S/N = \frac{A_{osc} \cdot \sqrt{N_{obs}}}{\sqrt{\pi} \cdot \sigma_{obs}} \quad (2)$$

where, A_{osc} is the mode amplitude, N_{obs} the number of measurements and σ_{obs} the individual measurement error. The solutions, in order to increase S/N, consist to:
 – increase the number of observations,
 – reduce the individual Doppler measurement error,
 – select targets with the highest expected mode amplitude.

4.2. THE DAY ALIAS

With single-site ground-based campaign, the identification of oscillation modes could be complicated by the observational window limited by the diurnal cycle and meteorological conditions, especially if the mode spacing is close to the day alias (11.57 μHz) or one of its harmonics. This is typically the case for δ Eri where

the expected spacing between $l=0$ and $l=1$ modes is close to 23 μHz. The solutions consist to:
- use gap-filling or CLEAN methods (but it may be used with a very great caution and can conduct to false identification),
- select targets with expected modes spacing $\neq$ n × 11.57 μHz,
- realize multi-site campaign, develop a ground-based network or observe from the south pole,
- go to space.

4.3. THE 'COMPLICATE' MODES

In some stars, the pattern of p-modes is not as well defined and as simple as the pattern of Figure 1. Mixed modes which correspond to combination of p- and g-modes are expected on some evolved solar-type stars. These modes do not necessarily follow the asymptotic relation. The stellar rotation induce split modes with visibilities depending on the stellar axis inclination. The short-time life of the oscillations induce a broadening of the modes. The solutions are probably to:
- improve stellar models in order to expect these 'complicated' modes,
- explore the HR diagram step by step from the Sun.

5. Conclusions and Prospects

Oscillations on solar-type stars can be detected from the ground even with small telescopes (CORALIE spectrograph is based on a 1.2-m telescope). Detection is now made on 6 solar-like stars and for 3 of them p-modes identification is proposed (see Figure 3). The goal of the next campaigns is now to clearly characterize individual p-modes at a level at least comparable or better to this obtained on α Cen A. Small telescopes have a key role to play in solar-type oscillation observations. They offer the possibility of long multi-site campaigns.

Next year will see the birth of at least 3 new instruments which will conduct to important breakthroughs in solar-type oscillation characterizations. The future spectrograph HARPS (Bouchy and Carrier, these proceedings) will offer the possibility to observe stars 100 times dimmer (5 magnitude) than CORALIE and with an individual measurement uncertainty of at least 1 m s^{-1}. HARPS will be able to conduct an asteroseismological program on a sample of more than 100 solar-like stars (see Figure 3). The spectrograph UVES with the fiber link FLAMES will offer the possibility to observe simultaneously up to 7 objects up to the 9th-magnitude. The satellite MOST, which will be launched on December 2002, will yield the first results on high-precision photometric measurements from the space.

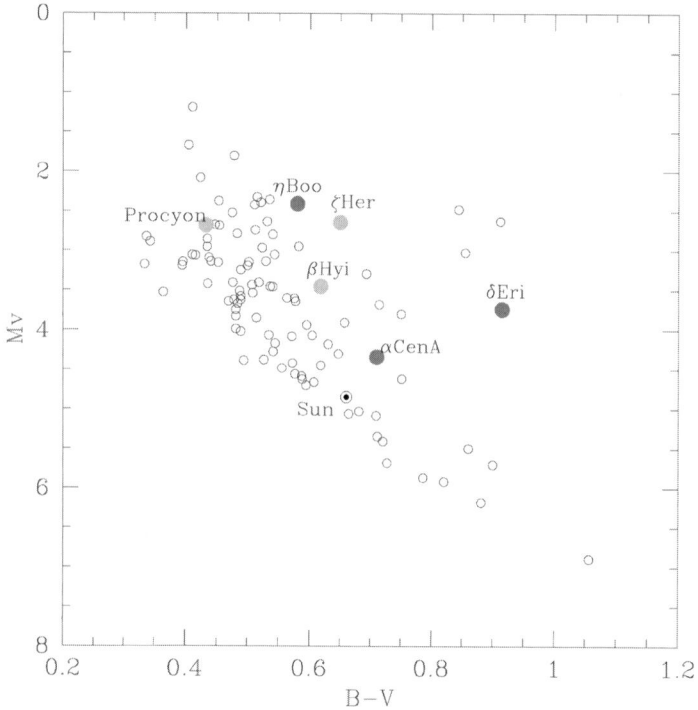

Figure 3. Location in the HR diagram of solar-type stars where p-mode oscillations were detected (in filled circle). Solar-type targets of the future ground-based spectrograph HARPS (in open circle).

References

Bedding, T.R., Butler, R.P., Kjeldsen, H., et al.: 2001, *ApJ* **549**, L105.
Bedding, T.R., Kjeldsen, H., Baldry, I.K., et al.: 2002, *A.S.P. Conf. Ser.* **259**, 464.
Bouchy, F. and Carrier, F.: 2001, *A&A* **374**, L5.
Bouchy, F. and Carrier, F.: 2002, *A&A* **390**, 205.
Bouchy, F., Schmitt, J. and Bertaux, J.-L.: 2002, *A.S.P. Conf. Ser.* **259**, 472.
Brown, T., Gilligand, R.L., Noyes, R.W., et al.: 1991, *ApJ* **368**, 599.
Brown, T., Kennelly, E., Korzennik, S., et al.: 1997, *ApJ* **475**, 322.
Carrier, F., Bouchy, F., Kienzle, F., et al., 2001: *A&A* **378**, 142.
Carrier, F., Bouchy, F., Kienzle, F., et al.: 2002, *A.S.P. Conf. Ser.* **259**, 468.
Grec, G., Fossat, E., Pomerantz, M.A., 1983: *Solar Phys.* **82**, 55.
Kjeldsen, H., Bedding, T., Viskum, M., Frandsen, S., 1995: *AJ* **109**, 1313.
Kjeldsen, H. and Bedding, T.R.: 1995, *A&A* **293**, 87.
Martic, M., Schmitt, J., Lebrun, J.-C., et al.: 1999, *A&A* **351**, 993.
Martic, M., Lebrun, J.-C., Schmitt, J., et al.: 2001, in: *Proc. SOHO 10/GONG 2000 Workshop*, ESA SP-**464**, p. 431.
Thevenin, F., Provost, J., Morel, P., et al.: 2002, *A&A* **392**, L9.

PRESENT OBSERVATIONAL STATUS OF THE INTERMEDIATE MASS STARS: δ SCT STARS, γ DOR STARS AND roAp STARS

D.W. KURTZ

Centre for Astrophysics, University of Central Lancashire, Preston PR1 2HE, UK
Laboratoire d'Astrophysique de l'OMP, CNRS UMR 5572, Observatoire Midi Pyrénées,
14 avenue Édouard Belin, 31400 Toulouse, France

Abstract. The present observational status of the δ Sct stars, γ Dor stars and roAp stars is discussed. The δ Sct stars are the most intensively observed of the three groups, but it has become clear that there are severe problems in extracting asteroseismic information from them. Dozens of frequencies are observed, but hundreds of frequencies are predicted from the models; unique matches of observation and theory still elude us. The δ Sct stars are observationally complex – some recent 'best case' campaigns are discussed. It is possible that substantial observational advances for δ Sct stars may need to await upcoming satellite missions. New γ Dor stars are being discovered frequently, and new behaviour is being found for them. They constitute an observationally young field. Their pulsational frequency range is being expanded, their position in the HR diagram is becoming better known (but is yet to be fully constrained), and the possibility exists of hybrid γ Dor – δ Sct stars that have great asteroseismic promise, although it is clear such stars are rare, if they do exist. It has been observationally challenging to extract more than a few frequencies for any γ Dor star so far. Exciting spectroscopic discoveries of new behaviour in roAp stars promise unprecedented information about the structure of the peculiar atmospheres of those stars – pulsation amplitude and phase in 3D, magnetic field structure in 3D, abundance stratification in 3D, realistic T-τ for the most peculiar stars – as well as entirely new information about the interaction of pulsation, rotation and magnetic fields. Recent theoretical work has led to new understanding of the previously inexplicable frequency spacing of HR 1217 with new Whole Earth Telescope observations supporting this theory. An 'improved oblique pulsator model' has been developed in which the pulsation axis is *not* the magnetic axis; this model has passed several observational tests and new ones are being devised to examine it further.

Keywords: stars: variable: δ Sct, γ Dor, roAp

1. Introduction

When I was an undergraduate student 35 years ago in San Diego, I went to a colloquium by an astronomer from JPL who had spent five years intensively studying the orbits of the Galilean satellites of Jupiter to determine the mass of Jupiter to the highest precision possible. This was critically needed to calculate the trajectory past Jupiter for the then-upcoming Pioneer 10 mission. (Pioneer 10 was launched 2 March 1972 and is now > 80 au away, headed towards a flyby of Aldebaran in 2 Myr; it is still being tracked – SETI at Aricebo uses it as a reference signal). I was greatly impressed with the dedication of this research to improve the precision of an important observational number. This was hard work, but satisfying for the pleasure of doing the observations and extracting the maximum information from them. Many years later when Pioneer 10 passed by Jupiter the telemetry of its actual orbital path immediately improved our knowledge of the mass of Jupiter to another couple of decimal places and the five years of hard work that I had heard about at that long-ago colloquium were history. The astronomer doing the work knew throughout his five-year project this was going to happen: Pioneer 11's orbit was guided by the mass of Jupiter determined by Pioneer 10.

With the imminent launch of MOST (Matthews, these proceedings), and the upcoming launches of the MONS, COROT and Eddington missions over the next few years, we are nearing the end of an era of ground-based photometric studies of δ Sct, γ Dor and roAp stars. These upcoming missions with their space telescopes will improve the observational precision of the photometry by one to two orders of magnitude with very high duty cycles – better than even the best multi-site ground-based campaigns are now getting. As soon as the first frequency set is derived from the satellite observations, the ground-based photometric work on that star will be history for asteroseismology; all theoretical work will use the higher-precision, richer satellite frequency set (although the ground-based observations will still have value for studies of long-term variability of frequency and amplitude).

Are we now like the astronomer calculating the mass of Jupiter, knowing it is needed but will soon be obsolete? Is it now time to stop ground-based, multi-site, observational efforts on δ Sct, γ Dor and roAp stars? My answer to that during my talk was a qualified 'yes' for δ Sct stars, except for exploratory studies of pre-main sequence stars, 'no' for the γ Dor stars which are an observationally young field, and 'no' for selected cases for the roAp stars. During the discussion Mike Breger, speaking for the Delta Scuti Network (DSN), said 'no' for the δ Sct stars (observations should continue) and 'yes' for the γ Dor stars which he feels are too difficult to get many frequencies from. Conny Aerts disagreed and said 'no' for the γ Dor stars, pointing out that their long periods mean that campaigns need time-spans longer than the satellites would be able to give to these stars, making ground-based campaigns very important still. Eric Michel, speaking for the STEPHI network, was not impressed with my suggestion that it might be time to stop. From many discussions during the meeting it was clear that the consensus is that continuing ground-based multi-site campaigns on all three classes of stars is still valuable and should be pursued.

2. The δ Sct Stars

A year ago at IAU Colloquium 185 in Leuven I heard many pessimistic comments from theoreticians about the prospects of seismology of δ Sct stars. They have the richest frequency spectra of any non-degenerate stars other than the sun with up to two dozen modes detected, hence appear to be very attractive for asteroseismology. But they are complex objects with rapid rotation, core convection and element separation in some cases. Current models find many hundreds of modes to be excited; with rotational splitting, possible differential rotation, possible mixed modes (with g-mode characteristics in the interior and p-mode characteristics at the surface) and uncertain amounts of core overshooting all allowing substantial adjustment of the model frequencies, it is not possible presently to do more than put a few constraints on the stellar structure using the observed frequencies.

A recent example is the theoretical discussion by Templeton et al. (2001) of data obtained in a large multi-site observing campaign on FG Vir by the DSN (Breger et al., 1998). Templeton et al. have many qualifications, such as: 'A close match to the observed pulsation spectrum was not obtained ...', 'Current data are not sufficient to allow an effective inversion ...' and 'Unfortunately, the possible parameter space of models required to fit a specific star can be very large.'

Given these theoretical problems, what developments are taking place observationally for δ Sct stars? What observations should be made for these stars? My first answer is that it is time to stop small efforts of δ Sct stars – single-site campaigns, or even multi-site campaigns that can only find a few frequencies for another star. There are many of these in the literature, and there is little theoretical discussion of the results, since there is little than can be said. The theoretical work concentrates on the stars with the largest number of detected frequencies and with some mode identification.

A recent example of this is the multi-site campaign of the DSN on BI CMi. This campaign obtained 1024 hr of photometric observations on 177 nights and detected 29 frequencies of which 20 were independent pulsation modes (Breger et al., 2002). But unambiguous mode identification was not possible, and the several sets of very closely spaced frequencies were found that may be the result of mixed modes, rotational splitting, or even the small splitting (Breger and Bischof, 2002). This sort of uncertainty in mode identification causes severe problems for astero seismic interpretation of the frequencies. Another recent example is the STEPHI X campaign on the Pleiades Cluster which collected 343 hr of data, yet found only two frequencies in HD 23194 and seven frequencies in V624 Tau. This was a great improvement on previous work on these stars, but cannot be put to much use asteroseismically.

With the exception of pre-main sequence stars, and some stars of particular interest for other reasons (e.g., possible δ Sct pulsation in magnetic Ap stars), all ground-based photometric work on δ Sct stars should be of the quality and quantity of that obtained by the DSN and STEPHI. Small efforts are now out-dated. The question that I posed in the introduction was: Are even the large efforts by the networks now out-dated with the impending launch of asteroseismic satellites? I worry that they might be, but the consensus is that these large multi-site campaigns will continue to be of substantial value. Breger (2002) discusses such campaigns in detail.

Of course, δ Sct stars are not just observed photometrically, but also spectroscopically. A recent multi-site spectroscopic campaign was that of the STACC network on δ Sct stars in Praesepe (Dall et al., 2002). The advantages of spectroscopy include better mode identification in some cases, different, complimentary information to the photometry, and the possibility to observe modes of higher ℓ, which is very important for asteroseismology and eventual inversion of the data. The problems and prospects of mode identification are discussed in detail in this volume (see individual papers by Reed, Telting, Balona, de Ridder, and Dupret). Spectroscopic studies of δ Sct stars are not planned for the asteroseismic space missions and are very important to have (simultaneously, if possible) with both ground-based and space-based campaigns, so there is plenty of work to be done spectroscopically. The exception I mentioned above for small, exploratory photometric work on δ Sct stars is for the pre-main sequence stars. These have possibly substantial interest asteroseismically, especially if there are detectable differences between pre-main sequence and post-main sequence stars that can be unambiguously modelled theoretically. There are few observations of pre-main sequence δ Sct stars, so even single-site exploration has value, although, as always, larger efforts with multi-site networks are much better. These stars are discussed in these proceedings in individual papers by Catala, Palla, Ripepi and Marconi.

3. γ Dor Stars

The γ Dor stars are late-A to early-F main sequence variables that pulsate in high overtone g-modes with periods of about 8 hr to days (Zerbi and Kaye, 2002). The latest instability strip for them can be found in Figure 7 of Handler and Shobbrook (2002) which shows them to have some overlap with the coolest δ Sct stars. Of five γ Dor stars that lie in the δ Sct instability strip, none shows short period δ Sct pulsation, suggesting that there may be a complete separation of the two groups. One star was previously found to be both a γ Dor and a δ Sct star, HD 209295 (Handler et al., 2002), but the longer period was found to be tidally induced. The relationship between γ Dor stars and δ Sct stars was also discussed by Breger and Beichbuchner (1996). If hybrid stars that pulsate in both p-modes and g-modes can be found, they will be rich asteroseismic targets, but so far there is no clear evidence that the two classes may exist in a single star.

Multi-site photometric campaigns for γ Dor stars are scarce, so there is a dearth of frequency information for them. One recent campaign by Poretti et al. (2002) shows light curves for HD 224838 and HD 224945 that make the multi-periodic nature and long periods of these stars beautifully clear. But the frequencies derived in that work, other than the highest amplitude peaks, are not to be taken seriously, since there are clearly severe problems in the frequency analysis – problems of resolution of the claimed frequencies, and an offset at zero frequency after prewhitening that shows a non-zero mean, hence skews the statistical tests for the reality of the detected frequencies. This work does show the potential of photometric campaigns on γ Dor stars, and it shows what hard work they are.

During the presentation of this paper I suggested that perhaps the DSN should now call itself the Gamma Dor Network and work on γ Dor stars. In the discussion Mike Breger rejected this; he pointed out that a large effort by the DSN on a γ Dor star had only detected two frequencies, therefore they were not worthwhile candidates for asteroseismology. Conny Aerts pointed out that the γ Dor stars are of great interest theoretically, because of their g-modes, and that the space asteroseismology missions will not be working on them, since very long time spans will be needed to detect more than a couple of frequencies. She therefore urged that substantial ground-based observing efforts be put into these stars. I agree with her since much less has been done on γ Dor stars than on δ Sct stars, so they are worth some multi-site campaigns to see just how many frequencies can be detected.

4. roAp Stars

Exciting new theoretical and observational studies of roAp stars have been published in the last two years. Theoretical developments concerning the interaction of pulsation and magnetic fields were published by Cunha and Gough (2000) and Bigot et al. (2000). The excitation mechanism for roAp stars was studied by Balmforth et al. (2001). Very interesting new work on the interaction of pulsation with both rotation and the magnetic field by Bigot and Dziembowski (2002) has presented an entirely new look at the oblique pulsator model – they find that the pulsation modes are *not* axisymmetric modes with pulsation and magnetic axes aligned. Instead, the pulsation axis is inclined to both the magnetic and rotation axes, and the pulsation modes are complex combinations of spherical harmonics that result in modes that, in many cases, can be travelling waves looking similar to (but are not exactly) sectoral m-modes (see Bigot, these proceedings). This new model has already passed some important observational tests, and more are being conducted.

An outstanding problem in roAp stars was the frequency spacing in HR 1217. Kurtz et al. (1989) found a set of six pulsation modes in this star, five of which were either separated by 68 μHz and were alternating even and odd modes, or were separated by 34 μHz and were consecutive overtones of the same ℓ. The Hipparcos parallax resolved this ambiguity in favour of the first interpretation (Matthews et al., 1999): the large spacing is 68 μHz and the modes are alternating even and odd ℓ, although none of them is a purely normal mode. This left the puzzle of the sixth frequency; its separation from the next lower frequency mode was $\frac{3}{4}\Delta\nu_0$ which made no sense.

Cunha and Gough (2000) and Bigot et al. (2000) found that the pulsation modes in roAp stars are magneto-acoustic, and that the coupling of the magnetic and acoustic components changes with frequency so that the magnetic perturbation to the frequencies can have a large range – from –40 to +10 μHz. These perturbations are progressive over a range of frequencies, effectively stretching the large spacing. Then, as the acoustic and magnetic coupling passes through a full cycle in their relation with each other, Cunha and Gough (2000) found that there is a big jump in the magnetic perturbation. Cunha (2001) used this to predict that there should be an additional frequency present in HR 1217 between the fifth and sixth modes found by Kurtz et al. (1989) and separated from the fifth frequency by about half of the large spacing. Kurtz et al. (2002) found that predicted frequency using a large data set from the 20th Whole Earth Telescope extended coverage campaign. Further analysis of those data show that the frequencies have been stable between the 1986 and 2000 data sets, but that there have been significant amplitude changes. New, improved information on the rotational modulation of all the pulsation modes will provide useful tests of the improved oblique pulsator model of Bigot and Dziembowski. The newest observational results for the roAp stars are spectacular high-resolution spectra of γ Equ and HR 3831 (Kochukhov and Ryabchikova, 2001a,b) that show very clearly the extreme stratification effects of abundances and the short vertical wavelength of the pulsation modes in roAp stars. For γ Equ λ6160.24 Å of Pr III and λ6145.07 Å of Nd III show significant radial velocity variations, while most other lines in the spectrum show none. A plausible interpretation of this phenomenon is that those ions are concentrated in thin layers by the effects of radiative diffusion, and that these layers lie near a vertical anti-node of the pulsation mode.

This is consistent with previous observations of strong line-depth dependence (atmospheric height dependence) of the pulsation amplitude in the Hα line found by Baldry and Bedding (2000) for α Cir and by Baldry et al. (1999) for HR 3831, and the strong drop-off of photometric amplitude with increasing wavelength explained by Medupe and Kurtz (1998). Two poster papers in these proceedings (Balona and Zima; Kochukhov et al.) show similar results. Besides the vertical stratification and the resolution of the vertical structure of the pulsation modes in γ Equ and HR 3831, the line profiles shown by Kochukhov and Ryabchikova (2001a,b) show clearly the signature of travelling waves – probably something similar to $\ell = 1, m = \pm 1$. This is qualitatively expected in the improved oblique pulsator model of Bigot and Dziembowski (2002; and these proceedings), but this has yet to be tested quantitatively.

Cunha (2002; and these proceedings) has calculated a theoretical instability strip for the roAp stars based on ideas for the excitation mechanism of Balmforth et al. (2001). One interesting result is that she expects Ap stars that are more luminous than the known roAp stars, and some slightly hotter to pulsate (see Figures 1, 3 and 4 in Cunha, 2002), but with periods longer than those in the known roAp stars – periods of 15 to 24 min consistent with the larger radii of these more luminous stars. Surveys for roAp stars have not been systematic, and current observations of noAp stars are not adequate to rule out the possibility of more luminous, somewhat longer period stars. Many searches for roAp stars have used a peculiarity of a negative Strömgren δc_1 index to select candidates. Since this index increases with luminosity, more luminous Ap stars with strong peculiarities will not show a negative δc_1 index, so may be mistaken for less luminous, less peculiar Ap stars. Also, the surveys have often used a 1-hr test time for new roAp stars. This is adequate for finding periods $\leq$ 15 min, but may easily miss 20- or 30-min, small amplitude variations which can be masked by small sky transparency variations. A search for these predicted longer period, more luminous roAp stars using CCDs with comparison stars in the same fields is highly desirable.

References

Baldry, I.K. and Bedding, T.R.: 2000, *MNRAS* **318**, 341.
Baldry, I.K., Viskum, M., Bedding, T.R., Kjeldsen, H. and Frandsen, S.: 1999, *MNRAS* **302**, 381.
Balmforth, N.J., Cunha, M.S., Dolez, N., Gough, D.O. and Vauclair, S.: 2001, *MNRAS* **323**, 362.
Bigot, L. and Dziembowski, W.A.: 2002, *A&A* **391**, 235.
Bigot, L., Provost, J., Berthomieu, G., Dziembowski, W.A. and Goode, P.R.: 2000, *A&A* **356**, 218.
Breger, M.: 2002, *A.S.P. Conf. Ser.* **256**, 17.
Breger, M. and Beichbuchner, F.: 1996, *A&A* **313**, 851.
Breger, M. and Bischof, K.M.: 2002, *A&A* **385**, 537.
Breger, M., Garrido, R., Handler, G., Wood, M.A., Shobbrook, R.R., Bischof, K.M., Rodler, F., Gray, R.O., Stankov, A., Martinez, P., O'Donoghue, D., Szabó, R., Zima, W., Kaye, A.B., Barban, C. and Heiter, U.: 2002, *MNRAS* **329**, 531.
Breger, M., Zima, W., Handler, G., Poretti, E., Shobbrook, R.R., Nitta, A., Prouton, O.R., Garrido, R., Rodriguez, E. and Thomassen, T.: 1998, *A&A* **331**, 271.
Cunha, M.: 2001, *MNRAS* **325**, 373.
Cunha, M.: 2002, *MNRAS* **333**, 47.
Cunha, M. and Gough, D.O.: 2000, *MNRAS* **319**, 1020.
Dall, T.H., Frandsen, S., Lehmann, H., Anupama, G.C., Kambe, E., Handler, G., Kawanomoto, S., Watanabe, E., Fukata, M., Nagae, T. and Horner, S.: 2002, *A&A* **385**, 921.
Handler, G., Balona, L.A., Shobbrook, R.R., Koen, C., Bruch, A., Romero-Colmenero, E., Pamyatnykh, A.A., Willems, B., Eyer, L., James, D.J. and Maas, T.: 2002, *MNRAS* **333**, 262.
Handler, G. and Shobbrook, R.R.: 2002, *MNRAS* **333**, 251.
Kochukhov, O. and Ryabchikova, T.: 2001a, *A&A* **374**, 615.
Kochukhov, O. and Ryabchikova, T.: 2001b, *A&A* **377**, L22.
Kurtz, D.W., Kawaler, S.D., Riddle, R.L., Reed, M.D., Cunha, M.S., Wood, M., Silvestri, N., Watson, T.K., Dolez, N., Moskalik, P., Zola, S., Pallier, E., Guzik, J.A., Metcalfe, T.S., Mukadam, A.S., Nather, R.E., Winget, D.E., Sullivan, D.J., Sullivan, T., Sekiguchi, K., Jiang, X., Shobbrook, R., Ashoka, B.N., Seetha, S., Joshi, S., O'Donoghue, D., Handler, G., Mueller, M., Gonzalez Perez, J.M., Solheim, J.-E., Johannessen, F., Ulla, A., Kepler, S.O., Kanaan, A., da Costa, A., Fraga, L., Giovannini, O. and Matthews, J.M.: 2002, *MNRAS* **330**, 57.
Kurtz, D.W., Matthews, J.M., Martinez, P., Seeman, J., Cropper, M., Clemens, J.C., Kreidl, T.J., Sterken, C., Schneider, H., Weiss, W.W., Kawaler, S.D. and Kepler, S.O.: 1989, *MNRAS* **240**, 881.
Matthews, J.M., Kurtz, D.W. and Martinez, P.: 1999, *ApJ* **511**, 422.
Medupe, R. and Kurtz, D.W.: 1998, *MNRAS* **299**, 371.
Poretti, E., Koen, C., Bossi, M., Rodriguez, E., Martin, S., Krisciunas, K., Akan, M.C., Crowe, R., Wilcox, M., Ibanoglu, C. and Evren, S.: 2002, *A&A* **384**, 513.
Templeton, M., Basu, S. and Demarque, P.: 2001, *ApJ* **563**, 999.
Zerbi, F.M. and Kaye, A.B.: 2002, *A.S.P. Conf. Ser.* **259**, 494.

PRESENT OBSERVATIONAL STATUS OF HIGH MASS PULSATING STARS

PETER DE CAT*

Instituut voor Sterrenkunde, Celestijnenlaan 200 B, B-3001 Leuven, Belgium

Abstract. In this short review we present the current observational status of high mass pulsating stars. We give an overview of the results of the first asteroseismic studies performed for some of the best known β Cephei and slowly pulsating B stars and discuss the asteroseismic potential of these pulsators for future *ground-* and *space-based* data-sets.

1. Introduction

Stellar pulsation is common across the HR diagram. Thanks to the successful application of asteroseismic methods first to the Sun and later on to White Dwarfs, the study of pulsating stars has gained a lot of interest in the last decade. A successful application of asteroseismic techniques requires the ability to detect and to identify as many as possible excited pulsation modes in the considered star. Therefore, high quality data with a long time-base is needed. The detection of frequency multiplets is not only crucial from an asteroseismic point of view, but it is also very helpful in the identification of the pulsation modes.

In the upper part of the main sequence, at least two classes of pulsating stars are found. The β Cephei stars (β Ceps) are early-B type *p-mode* pulsators with pulsation periods ranging from 3 hrs to 6 hrs. The Slowly Pulsating B stars (SPBs) are mid- to late-B type *g-mode* pulsators with periods of the order of days. For an overview of the characteristics of their observed brightness and line profile variations, we refer to De Cat (2002). In this short review, we discuss their asteroseismic potential. The multiperiodic spectroscopic variations of (some) rapidly rotating Be stars can also be described by nonradial pulsation models, but Be stars do not form a homogeneous class of pulsating stars and their circumstellar matter makes it much more difficult to study them. Therefore, they are not discussed here.

This paper is organised as follows. In Sections 2 and 3, we give an overview of recent results obtained for the most promising β Ceps and SPBs. We end in Section 4 with our conclusions and future prospects.

---

* Postdoctoral Fellow of the Fund for Scientific Research of Flanders

TABLE I

The observed frequencies of β Cephei and their identifications

frequency	$(\ell,|m|)$[1]	(ℓ,m)[2]
$f_1 = 5.2497104$ c d^{-1}	(0,0)	(0,0)
$f_2 = 5.385$ c d^{-1}	(2,2)	(2/1,+1) ?
$f_3 = 4.920$ c d^{-1}		
$f_4 = 5.083$ c d^{-1}		
$f_5 = 5.417$ c d^{-1}		

[1] Aerts et al. (1994), [2] Telting et al. (1997).

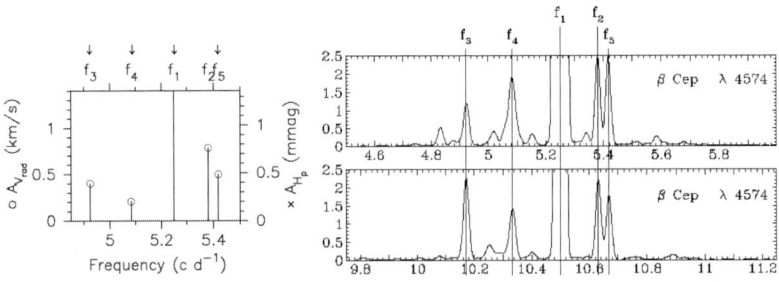

Figure 1. Left: The amplitudes for the observed frequencies in the radial velocity (o) and in the Hipparcos H$_p$ magnitudes (×) for β Cephei. The amplitudes for f_1 are much higher than shown. Right: The variational power in the CLEANed Fourier periodogram summed over the wavelength range of the Si III λ 4574 Å line for the frequency range around f_1 (top) and $2f_1$ (bottom). (Figure taken from Telting et al., 1997.)

2. β Cephei Stars

2.1. β CEPHEI

The radial velocity and brightness variations of β Cephei (HD 205021, SpT B2IIIe, m$_V$ = 3.2) were first detected respectively by Frost (1902) and Guthnick (1913). Aerts et al. (1994) were the first to unravel the multiperiodic character of this star by studying 660 high-quality spectra with a time base of 33 consecutive nights. They found 3 frequencies (f_1, f_2, f_3) while Telting et al. (1997) found another 2 frequencies (f_4, f_5) in the same dataset (Table I). The scaled amplitudes for the observed frequencies in the radial velocity (o) and in the Hipparcos H$_p$ magnitudes (×) are given in the left panel of Figure 1. f_1, f_3, f_4, and f_5 form a multiplet, pointing towards a rotionally splitted $\ell \geq 2$ mode.

This suggestion is not supported by the spectroscopic mode identification. The dominant mode within the multiplet, f_1, turns out to be a radial mode, and f_2, which is not a part of the multiplet, is attributed to an $\ell = 1$ or 2 mode (Table I). According to Telting et al. (1997), the observed multiplet can not be due to tem-

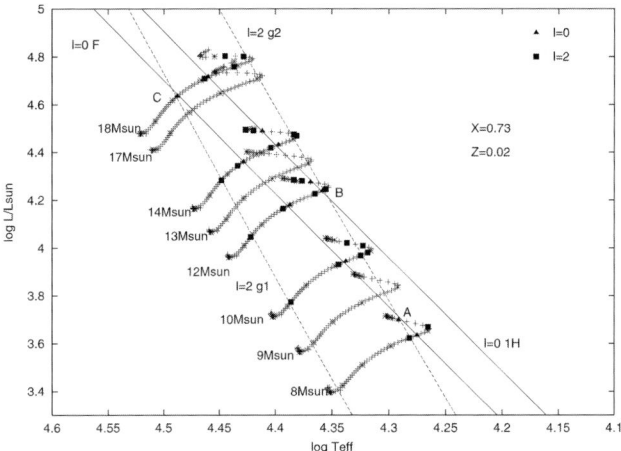

Figure 2. The HR diagram for 8–18 M$_\odot$ stars. Theoretical models for which a radial mode frequency coincides with f_1 and those for which a zonal $\ell = 2$ mode coincides with f_2 are connected by respectively full and dashed lines. β Cephei must be one of the crossing points A, B or C. (Figure taken from Shibahashi and Koshikawa, 2002.)

perature spots. They suggest that we are dealing with an oblique magnetic pulsator. Indeed, for β Cephei, a variable magnetic field with an amplitude of 90 Gauss and a period of 12 days is found (Henrichs et al., 2000). This period corresponds to twice the observed frequency splitting within the multiplet. The oblique pulsator model was further studied by Shibahashi and Aerts (2000), who based their analysis on the ratios of the CLEAN power components (Figure 1, top right panel). Note that these CLEAN ratios are not the same as those found by sinusoidal fitting of the radial velocity variations (Figure 1, left panel). The ratios of the latter are similar to those found by CLEAN around the first harmonic of the main frequency (Figure 1, bottom right panel).

Shibahashi and Koshikawa (2002) used the observed frequencies f_1 and f_2 to place β Cephei in the HR diagram. On Figure 2, theoretical models for which a radial mode frequency coincides with f_1 and those for which a zonal $\ell = 2$ mode coincides with f_2 are connected by respectively full and dashed lines. β Cephei must be one of the crossing points A, B or C. The luminosity derived from the Hipparcos parallax is not compatible with C, so β Cephei must be a $\sim$8 M$_\odot$ star in its late stage of hydrogen core-burning or $\sim$12 M$_\odot$ star near the turning point.

2.2. 12 Lacertae

12 Lacertae (HD 214993, SpT B2III, m$_V$ = 5.2) is a β Cephei star whose radial velocity and brightness variations were detected in the beginning of the last century (Adams, 1912; Stebbins, 1917). For this star, 6 frequencies are commonly present

TABLE II
The observed frequencies of 12 Lacertae and their identifications

frequency	ℓ^1	$(\ell,m)^2$	$(\ell,m)^3$
$f_1 = 5.179015(4)$ c d^{-1}	1	(2,0)	(2,-1)
$f_2 = 5.06640(1)$ c d^{-1}	1/2 ?	(0,0)	(2/3,0) ?
$f_3 = 5.49004(1)$ c d^{-1}	1	(2,-2)	(2,-2)
$f_4 = 5.3343301(3)$ c d^{-1}		(2,-1)	(3,+1)
$f_5 = 10.51340(4)$ c d^{-1}			(2,-1)
$f_6 = 4.24083(5)$ c d^{-1}			(1,+1)

[1] Cugier et al. (1994), [2] Mathias et al. (1994), [3] Aerts (1996).

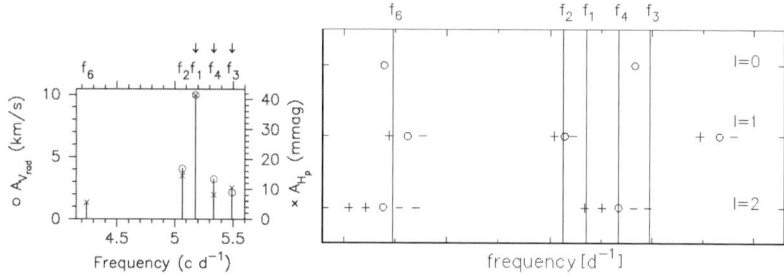

Figure 3. Left: Same as left panel of Figure 1, but for 12 Lacertae. Right: The theoretical model for which the best correspondence is found between the unstable modes (+ for $m > 0$, o for $m = 0$, - for $m < 0$) and the observed modes (full lines). (Figure taken from Dziembowski and Jerzykiewicz, 1999.)

in the spectroscopic and photometric variations, of which $f_5 = f_1 + f_4$ and the three frequencies f_1, f_3 and f_4 form an equidistant triplet (Table II; Figure 3, left panel).

A splitted $\ell = 1$ suggestion is not rejected by the results of the photometric mode identification ([1] in Table II), but the results of the moment method point towards a higher degree mode ([2] and [3] in Table II). Dziembowski and Jerzykiewicz (1999) point out that rotational splitting and nonlinear phase lock are the most plausible models for the observed frequency triplet. The theoretical model for which the best correspondence is found between the unstable frequencies and the observed ones is given in the right panel of Figure 3. In this model, the frequency triplet is caused by a rotationally splitted $\ell = 2$ mode.

2.3. 16 LACERTAE

16 Lacertae (HD 216916, SpT B2IV, $m_V = 5.6$) is another β Cep for which radial velocity variations were already detected in the beginning of last century (Lee, 1910). The brightness variations were detected some 40 years later (Walker, 1951).

Table III
The observed frequencies of 16 Lacertae and their identifications

frequency	ℓ^1	$(\ell,m)^2$
$f_1 = 5.9113184$ c d^{-1}	0	(0,0)
$f_2 = 5.8528991$ c d^{-1}	1/2 ?	(2,0)
$f_3 = 5.5025931$ c d^{-1}	1/2 ?	(1,0) ?

[1] Jerzykiewicz (1993), [2] Aerts et al. (these proceedings).

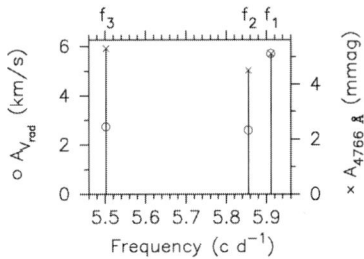

Figure 4. Same as left panel of Figure 1, but for 16 Lacertae.

This star has three well known frequencies of which we list in Table 2.3 the most recent values given by Lehmann et al. (2001) who gathered 942 new spectra with a time base of 475 days and analysed all available spectroscopic data. For this star, a close frequency spacing is found between f_1 and f_2 (Figure 2.3).

The photometric mode identification ([1] in Table 2.3) leads to different ℓ values for f_1 and f_2. However, the results for f_2 and f_3 are ambiguous. Therefore, Dziembowski and Jerzykiewicz (1996) tried to do the mode identification theoretically by considering models without convective overshoot and modes with a degree $\ell \leq 2$. Again, no conclusive results were found for f_2 and f_3 while f_1 could be undoubtfully attributed to a fundamental radial mode. 16 Lac is found to be in the late stage of core-hydrogen burning. Dziembowski and Jerzykiewicz (1996) stress that the exclusion of convective overshoot leads to the conclusion that the interior must rotate faster than the outer layers of the star. If f_3 is due to an $m = 0$ mode, convective overshoot is needed.

The most in depth seismic analysis so far is done by Aerts et al. (these proceedings). They conducted the first spectroscopic mode identification for 16 Lac with the new version of the moment method (Briquet & Aerts, these proceedings) of which the most probable solution is given in Table 2.3. f_1 and f_2 are both zonal modes. Subsequently, Aerts et al. computed a lot of theoretical models and confronted them with the photometric and spectroscopic mode identifications and found only a few possible solutions in terms of age, convective overshoot and metalicity. For these solutions, f_1 and f_2 are attributed to g_1 modes.

2.4. OTHERS

We present here three other promising β Ceps which are currently being studied. The first one is ν Eridani (HD 29248, SpT B2III, $m_V=4.0$), for which 4 frequencies are currently known: $f_1 = 5.76349$ c d^{-1}, $f_2 = 5.6535$ c d^{-1}, $f_3 = 5.6373$ c d^{-1}, $f_4 = 5.6201$ c d^{-1} (Cuypers and Goossens, 1981). f_1, f_2 and f_3 form a non-equidistant triplet (Figure 5, left panel), which suggests a splitted $\ell = 1$ mode. This suggestion is supported by the photometric identification of f_2 (Heynderickx et al., 1994). A photometric and spectroscopic multi-site campaign for ν Eri led by Handler &

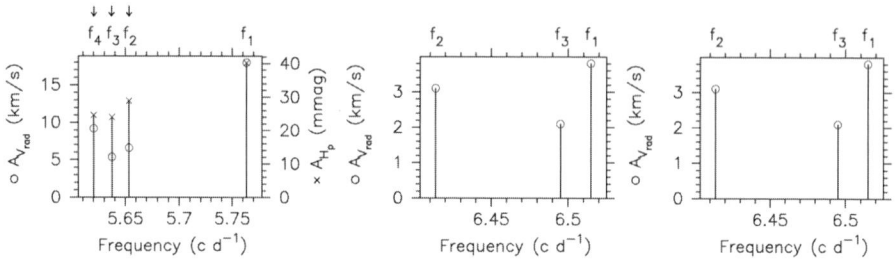

Figure 5. Same as left panel of Figure 1, but for ν Eridani (left), λ Scorpii (middle), and β Centauri (right).

Aerts is scheduled in October 2002 - January 2003 (Handler & Aerts, submitted to *Communications in Asteroseismology*).

λ Scorpii (HD 158926, SpT B2IV, m_V=1.6) is a β Cep with at least 6 pulsation frequencies (f_1 = 4.67942 c d^{-1}, f_2 = 4.41 c d^{-1}, f_3 = 4.12 c d^{-1}, f_4 = 3.89 c d^{-1}, f_5 = 3.57 c d^{-1}, f_6 = 5.37 c d^{-1}). f_1, f_2, f_3, f_4 and f_5 form an equidistant quintuplet (Figure 5, middle panel). However, an ℓ = 2 suggestion is not supported by the mode identification. The most recent results are given by Uytterhoeven et al. (these proceedings).

According to the current ground-based photometric detection limits, β Centauri (HD 122451, SpT B1III, m_V=0.6) is photometrically constant. However, Ausseloos et al. (2002) found at least 3 frequencies in the line profile variations: f_1 = 6.51481 c d^{-1}, f_2 = 6.41356 c d^{-1}, f_3 = 6.49521 c d^{-1} (Figure 5, right panel). This suggests that we are dealing with high degree modes. Mode identification is currently being done.

3. Slowly Pulsating B Stars

Indepth studies of SPBs require a lot of patience. Indeed, their dense theoretical frequency spectrum is reflected in very long beat periods. The asteroseismic studies for SPBs are therefore still in an earlier stage than those of β Ceps. Moreover, there is an urgent need for improved identification techniques. In the following, we concentrate on the most recent results of one bright southern SPB. For results of the other bright SPBs, we refer to De Cat (2001) and Mathias et al. (2001).

ο Velorum (HD 74195, SpT B3IV, m_V = 3.6) is one of the prototypes of the SPBs as given by Waelkens (1991). So far, 4 pulsation frequencies are found for this star (De Cat and Aerts, 2002). Three of them could be members of a frequency multiplet with $\ell \geq 2$ of which not all the components are seen (Figure 6). For the mode identification, the most recent photometric ([1]) and spectroscopic ([2]) results are given in Table IV. Note that the most probable spectroscopic identifications for f_1 and f_2 are consistent with the ones needed for a splitted ℓ = 2 mode. However,

Table IV
The observed frequencies of o Velorum and their identifications

frequency	ℓ^1	$(\ell,m)^2$
$f_1 = 0.35745(9)$ c d^{-1}	1	(2,-1)
$f_2 = 0.35033(9)$ c d^{-1}	1	(3,-1)
$f_3 = 0.34630(9)$ c d^{-1}	1/2 ?	(2,+2)
$f_4 = 0.39864(9)$ c d^{-1}	1/2 ?	(1,-1)

[1] Dupret et al. (these proceedings), [2] Briquet and Aerts (these proceedings).

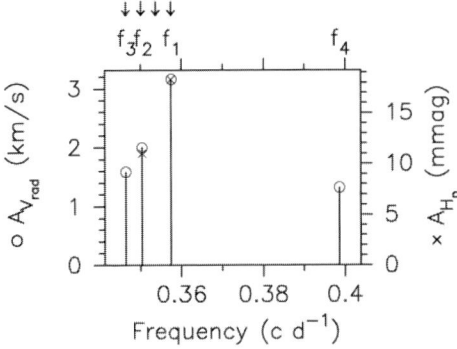

Figure 6. Same as left panel of Figure 1, but for o Velorum

the results of the mode identification are not reliable enough yet to conduct a theoretical modeling.

4. Conclusions and Future Prospects

From an asteroseismic point of view, the most interesting β Ceps are β Cephei, 12 Lacertae, 16 Lacertae, for which the first asteroseismic results still need some fine-tuning. ν Eridani, λ Scorpii, β Centauri are three other interesting multiperiodic β Ceps with close frequency spacings which are currently being studied in detail. IL Velorum, and BU Cir are two fainter, multiperiodic β Ceps for which no detailed line profile study is available yet. o Velorum, V335 Velorum, V869 Centauri are three multiperiodic SPBs for which hints of frequency multiplets are found (De Cat and Aerts, 2002) and for which the mode identification is ongoing. HD 45284, YZ Lep are fainter, multiperiodic SPBs with short beat periods for which a detailed line profile study is lacking.

According to us, the best strategy for future ground-based asteroseismic studies is simultaneous photometric and spectroscopic observational modeling in combination with theoretical modeling as done by Aerts et al. (these proceedings) for 16 Lac. Multicolour photometry is suited to study low degree modes while line profile variations generally reveal more pulsation frequencies and are more suitable to study higher degree modes. Unfortunately, alias problems will remain severe from ground for massive stars, even so for long-term campaigns. However, we are confident that this will be improved significantly with the help of future space missions like COROT and EDDINGTON.

References

Adams, W.S.: 1912, *ApJ* **35**, 163.
Aerts, C.: 1996, *A&A* **314**, 115.
Aerts, C., Mathias, P., Gillet, D. and Waelkens, C.: 1994, *A&A* **286**, 109.
Ausseloos, M., Aerts, C., Uytterhoeven, K., et al.: 2002, *A&A* **384**, 209.
Cugier, H., Dziembowski, W.A. and Pamyatnykh, A.A.: 1994, *A&A* **291**, 143.
Cuypers, J. and Goossens, M.: 1981, *A&AS* **45**, 487.
De Cat, P.: 2001, *Ph.D. Thesis*, Katholieke Universiteit Leuven, Belgium.
De Cat, P.: 2002, in: *Radial and Nonradial Pulsations as Probes of Stellar Physics, A.S.P. Conf. Ser.* **259**, 196.
De Cat, P. and Aerts, C.: 2002, *A&A* **393**, 965.
Dziembowski, W.A. and Jerzykiewicz, M.: 1996, *A&A* **306**, 436.
Dziembowski, W.A. and Jerzykiewicz, M.: 1999, *A&A* **341**, 480.
Frost, E.B.: 1902, *ApJ* **15**, 340.
Guthnick, P.: 1913, *Astronomische Nachrichten* **196**, 357.
Henrichs, H.F., de Jong, J.A., Donati, D.-F., et al.: 2000, in: Y.V. Glagolevskij, I.I. Romanyuk (eds.), *Magnetic Fields of Chemically Peculiar and Related Stars*, p. 57.
Heynderickx, D., Waelkens, C. and Smeyers, P.: 1994, *A&AS* **105**, 447.
Jerzykiewicz, M.: 1993, *Acta Astronomica* **43**, 13.
Lee, O.J.: 1910, *ApJ* **32**, 300.
Lehmann, H., Harmanec, P., Aerts, C., et al.: 2001, *A&A* **367**, 236.
Mathias, P., Aerts, C., Briquet, M., et al.: 2001, *A&A* **379**, 905.
Mathias, P., Aerts, C., De Pauw, M., et al.: 1994, *A&A* **283**, 813.
Shibahashi, H. and Aerts, C.: 2000, *ApJ* **531**, L143.
Shibahashi, H. and Koshikawa, T.: 2002, in: *Radial and Nonradial Pulsations as Probes of Stellar Physics, A.S.P. Conf. Ser.* **259**, 208.
Stebbins, J.: 1917, *Pop. Astr.* **25**, 657.
Telting, J.H., Aerts, C. and Mathias, P.: 1997, *A&A* **322**, 493.
Waelkens, C.: 1991, *A&A* **246**, 453.
Walker, M.F.: 1951, *PASP* **63**, 35.

OLD PULSATORS: WHITE DWARFS AND THEIR IMMEDIATE PRECURSORS

M. SEAN O'BRIEN

Astronomy Department, Yale University, PO Box 208101, New Haven, CT 06520-8101, USA

Abstract. There are now fully six classes of pulsators among white dwarfs and their immediate precursors among central stars of planetary nebulae and on the extended horizontal branch. In this review, we outline those observational and theoretical considerations that link them together and set them apart from other kinds of pulsating stars. We summarize some select astrophysical puzzles to which studies of such pulsators might speak, and we discuss current applications in the fields of atomic, nuclear, and neutrino physics. Finally, we suggest how future observing programs might solve some general problems common not only to the white dwarf and pre-white dwarf pulsators but to many types of variable stars.

1. Introduction

The final stages of stellar evolution include six different classes of pulsating stars. Together they span many orders of magnitude in both temperature and luminosity. Each presents unique challenges, and also unique opportunities to explore stellar structure, evolution, and physics in greater detail than is otherwise possible with non-pulsating stars.

Along the line of evolution from intermediate-mass progenitors, there are first the Planetary Nebula Nucleus Variables (PNNVs) and the slightly more evolved GW Virginis pulsators. With effective temperatures between 80,000 K and 170,000 K, these two classes of pre-white dwarf pulsators inhabit the high-temperature entry point into the cooling track for C/O-core white dwarfs (Figure 1), and some members are among the hottest stars known. They also include the second-richest pulsator after the Sun (GW Vir itself, the first great triumph of asteroseismology; see e.g. Winget et al., 1991; Kawaler and Bradley, 1994), and another fascinating star, PG 0122+200, which should be emitting far more energy in neutrinos than in light (O'Brien and Kawaler, 2000). A campaign is underway to determine the neutrino luminosity of PG 0122+200 via measurement of its rate of period change. If successful, this would mark the first test of lepton theory in dense plasma.

Further down the cooling track are the variable helium-atmosphere white dwarfs (DBVs), just below the DB gap between 29,000 K and 21,000 K. This was the first class of variable stars predicted to exist prior to their discovery (Winget, Robinson and Nather 1982). The prototype, GD 358, was after GW Vir the second significant success story of asteroseismology (Winget et al., 1994); it provided the first asteroseismological measurements of distance and magnetic field strength, and it

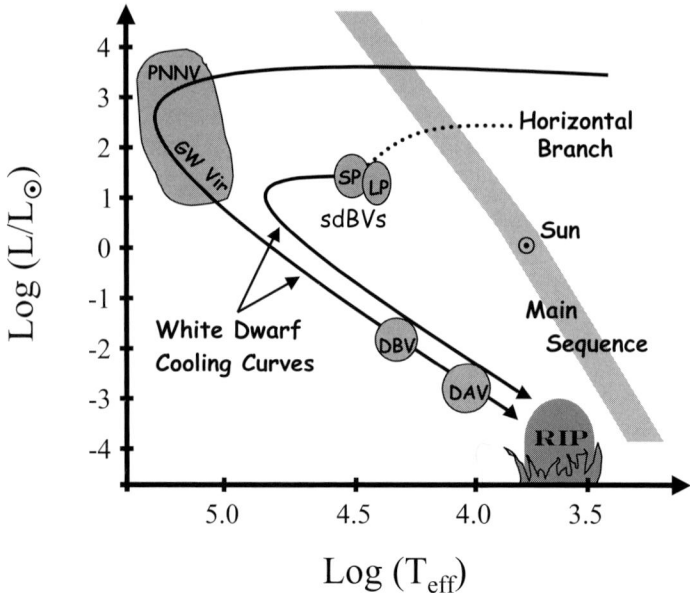

Figure 1. Locations within the H-R diagram of the six different kinds of advanced post-main sequence pulsating stars. Note that all of the pulsators discussed here are evolving down, or about to enter, the white dwarf cooling sequence, on their way to the stellar graveyard.

now serves as a significant test-bed for new mode-identification techniques using high-order cross-frequencies (Vuille et al., 2000; Wu, 2001) and time-series spectroscopy. The DBV stars also show promise as laboratories for nuclear physics: determination the core C/O ratio via asteroseismology would place much stricter constraints on the $^{12}C(\alpha, \gamma)^{16}O$ reaction rate than is possible in earthly laboratories (Metcalfe, Salaris and Winget, 2002).

At the far end of the white-dwarf cooling track are the hydrogen-atmosphere variables, the DAV or ZZ Ceti stars. These are currently the most numerous among the pulsators we will discuss, but also in some ways the most frustrating because individual stars generally show few periods – making mode-identification very difficult. Attempts have been made in recent years to explain their group properties by considering the periods of many stars at once, but more data are needed to prove the worth of this interesting idea. The DAVs have other uses, however: at least one is so massive that theory predicts it should have a crystallized core. Asteroseismology can test and calibrate this forty-year old prediction, but so far results are inconclusive (Nitta et al., 2000).

Finally, two new classes of pulsating 'pre-white dwarfs' were discovered recently: the short-period (p-mode) sdBV (a.k.a. EC 14026) stars (see Koen et al., 1998 for a review), and just this year the long-period (g-mode) sdBVs (Green et al., this proceeding). The sdBVs were the second case of variable stars whose existence was predicted (Charpinet, Fontaine and Brassard, 1996) prior to their discovery (Kilkenny et al., 1997), and they occupy the extreme end of the horizontal branch, just above the 'helium ZAMS'. These stars are not massive enough to make the trip up the AGB and will apparently fall directly onto the cooling track for helium-core white dwarfs. Initial studies of short-period sdBVs are tantalizing but also paradoxical, finding many pulsation modes but little understanding of them. The discovery, properties, and promise of the long-period sdBVs form the subject of a separate review at this conference.

In this review we examine specific applications of studies of old pulsators to problems in physics and astrophysics, and discuss the promise and the problems that confront and unite all these different classes. We also outline future directions that might be taken to fulfill the promise and solve the problems confronting researchers in this still-fledgling field of asteroseismology.

2. Current Applications

2.1. ASTROPHYSICS

A basic question still confronting theorists of stellar evolution is: what kind of a white dwarf will a given star turn into? It is relatively easy in general for stellar evolution codes to make white dwarfs, but it is still not possible to say with certainty which white dwarfs came from which types of precursors. Here the study of pre-white dwarf pulsators can have a great impact. If we can learn the structure and evolution of stars which are in the process of becoming white dwarfs, we have hope of filling in this gap in our theories. So far, we have used asteroseismology to accurately determine the surface layer mass of two stars: the GW Vir star PG 1159-035 (Winget et al., 1991; Kawaler and Bradley, 1994), and the DBV star GD 358 (Winget et al., 1994). Additional work is clearly needed to determine the structure of GW Vir and DBV stars as a group. However, most white dwarf and pre-white dwarf pulsators do not show as rich a pulsation spectrum as either PG 1159 or GD 358, so determining their structure via asteroseismology will require systematic long-term monitoring to pick up new modes as they appear in the light curve. No such programs are currently planned.

There is also the question of the so-called DB gap: helium-atmosphere white dwarfs abound at temperatures above 45,000 K and below 30,000 K, but none are to be found in between. This has come to be understood as an evolutionary problem: either hot DBs cease to look like DBs when they cool to temperatures in the gap, or the hot DBs and the cool DBs have a different origin. One theory (proposed by Fontaine and Wesemael, 1987) requires very small hydrogen abundances in all DBs to explain the gap – the hydrogen is mixed below the surface by convection at high temperatures; when the DB star enters the gap, the convection zone disappears, allowing the H to float to the surface. Clearly knowledge of the H and He abundance in DBV stars will help to test theories of the DB gap, but with only a single star studied so far in any detail, our picture of DB structure remains highly limited.

Next, what are the structural differences between different types of white dwarf? How different are their origins? The DA and DB stars each have their instability strips, which will allow us to answer these questions, but again long-term systematic observation is required to do so, since in general stars show different modes over time.

Finally, asteroseismology of old pulsators will have a great impact on attempts to determine the galactic age via the white dwarf luminosity function (see e.g. Wood & Oswalt, 1998). The rate of evolution of white dwarfs depends primarily on the thickness of their surface layers – a quantity asteroseismology can determine. Long-term monitoring can also tell us the rate of white dwarf evolution directly, via measurement of the rate of secular period change, dP/dt in pulsators. dP/dt has been measured in two or three stars (see e.g. Kepler et al., 2000), but this quantity can depend on other factors than evolution (such as mode stability or the presence of planets or other unseen companions), and we should determine dP/dt for at least several stars from each class of old pulsators before we can claim to trust this method of direct asteroseismological measurement of stellar evolution.

2.2. BASIC PHYSICS

It is not necessary to have a complete picture of white dwarf formation and evolution before we consider using white dwarfs as physics laboratories. That they are relatively simple physical systems (as stars go) is sufficient.

Once the nuclear fires have died and contraction is complete, white dwarf evolution is a simple cooling problem. For the first few million years of their existence, however, they cool not primarily by photon emission as with all other stable stars, but by neutrino emission (O'Brien and Kawaler, 2000). Figure 2 shows the ratio of neutrino luminosity to photon luminosity for two white-dwarf mass sequences. The GW Vir stars occupy the left-hand side of this figure, and among these the coolest and most massive – and therefore most completely dominated by neutrino emission – is PG 0122+200. A multi-year observational campaign is underway to measure dP/dt in PG 0122, attempting for the first time to test standard lepton theory in dense plasma. Even if successful, however, PG 0122 remains an isolated case, highlighting a fundamental problem among all the old pulsators: there are only five known GW Vir stars, and all of them were discovered 'by accident' in a quasar survey. Future surveys should be designed to specifically find faint blue pulsating stars. Without additional pulsators, the era of asteroseismology as a widespread tool in stellar astrophysics will never come.

The DBV and DAV classes each have also their own unique case of a pulsator with important applications to basic physics. Among the DBVs there is GD 358, which shows a rich enough pulsation spectrum that attempts are underway to de-

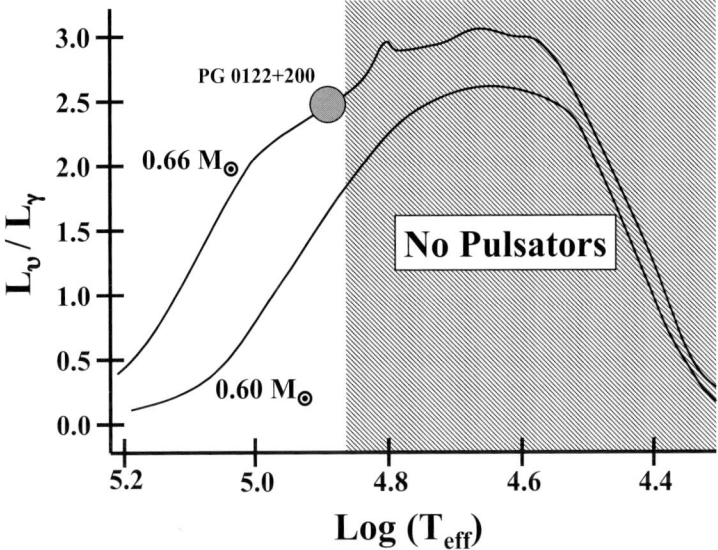

Figure 2. The ratio of neutrino luminosity, L_ν, to the photon luminosity, L_γ, as a function of T_{eff} for white dwarfs of two different masses.

termine the core C/O ratio and therefore to place an important constraint on the famously untested reaction $^{12}C(\alpha, \gamma)^{16}O$ (Metcalfe, Salaris and Winget, 2002). For the DAV stars there is BPM 37093, which is so massive that its core should be crystallized (Winget et al., 1997). A crystallized core would change its pulsation spectrum, which means that detailed study of its pulsation spectrum should yield of determination of the fraction of the core which is so far crystallized (Montgomery and Winget, 1999). Results are inconclusive (Nitta et al., 2000), and BPM 37093 remains the only one of the 29 known DAV stars in which crystallization is suspected.

3. Common Problems and Future Possibilities

Some of the general characteristics of the white dwarf and pre-white dwarf pulsators are given in Table I. With the exception of the long-period sdBVs, all of them pulsate on time scales under half an hour. Considered together, they are hot and of high-gravity, and are best studied in blue or UV light (the spectrum of even the coolest DAV peaks in the UV). They are low-amplitude pulsators, and all except the short-period sdBVs show only non-radial g-modes. They tend to be multiperiodic and variable in amplitude, showing different sets of periods over time – though always, we presume, conforming to the unchanging set of *possible* periods determ-

TABLE I
Properties of the Old Pulsators

Class	Known Pulsators	Visual Magnitudes	B–V	T_{eff} [kK]	Periods [min]	Amplitudes [mag]	Type of Pulsation[†]	Driving Source
sdBV (long period)	13	—	—	25 – 30	40 – 80	~ 0.05	NR g-modes	?
sdBV (short period)	26	12.3 – 15.9	−0.5 – +0.2	33 – 36	2 – 3	0.005 – 0.01	R/NR p-modes	Fe κ-γ
PNNV	10	15.1 – 16.8	~ 0 – −1	130 – 170	10 – 30	0.01 – 0.15	NR g-modes	C/O κ-γ?
GW Vir	5	14.8 – 16.7	−0.4 – −0.6	80 – 140	5 – 15	0.02 – 0.1	NR g-modes	C/O κ-γ?
DBV	8	13.6 – 17.0	~ 0	22 – 29	2 – 15	0.05 – 0.3	NR g-modes	He-II κ-γ
DAV (ZZ Ceti)	29	12.2 – 16.6	0.12 – 0.27	11 – 12	2 – 15	0.01 – 0.3	NR g-modes	Convection

[†]NR = Non-Radial, R = Radial.

ined by each star's individual structure (mass, envelope thickness, rotation rate, etc.). Finally and most vexingly, column two of Table I shows that all of them are rare.

They are not, as a whole, rare in the universe of course. The problem is that the intrinsically bright ones (the PNNV and GW Vir stars) evolve very quickly, and so are truly rare, while countless DBV and DAV stars are simply too faint to observe. The sdBV stars represent in some ways a happy medium, being relatively bright and numerous, but their amplitudes are so low, and their periods so short, that their variability was only discovered within the last few years.

We can then, envision an optimal observing program for old pulsators. The known sample is faint (V is typically 14–17 mag) but not too faint, so we need only a small-to-moderate aperture telescope (or network of telescopes!) and an instrument with quantum efficiency optimized in the blue, capable – because of the short periods – of taking very short exposures with little or no dead-time. Further, we need to observe many individual stars for several nights each to identify multiple pulsation modes, and then continue to monitor them periodically over long periods of time to detect transient pulsations, fill in the mode spectrum, and in some cases determine the rate of evolution through measurement of dP/dt.

Unfortunately, these requirements are precisely at odds with current trends in astronomical instrumentation, observation, and grant application, all of which emphasize very large telescopes equipped with red and IR instruments, looking at many objects for relatively short periods of time, with readout-time essentially irrelevant. What we require, on the other hand, are lots of small telescopes dedicated to time-variable phenomena.

Of course, there are several upcoming satellite projects for asteroseismology, including the French COROT mission, the Canadian MOST mission, and NASA's Kepler. The premise for these missions, however, seems to rest in the idea that helioseismology has shown the way to the new field of asteroseismology, which will truly begin when solar-type oscillations are successfully used to study stellar structure. All three are designed, therefore, to ferret out solar-type oscillations (and detect extra-solar planets), and all are largely irrelevant to studies of old pulsators.

There is another trend within astronomy, however, which holds out great promise for future studies of old pulsators, and indeed for all variable stars. Large surveys like SLOAN generally have a small set of overarching scientific goals, but have become popular partly because they also create large multi-purpose data sets which can then be mined for additional (and in many cases unforeseen) scientific gain. The asteroseismology community needs to create and sustain a dialog with the teams running these various surveys, to make sure that possibilities for discovery of time-variable objects on all possible time-scales are vigorously pursued and followed up.

What is really needed, however, is a ground-based survey dedicated to time-variable objects. Such a survey, undertaken with the express purpose of following a large number of variable objects on many time scales for many years, would be of tremendous service to many fields, including not only asteroseismology but cataclysmic variables, microlensing, and searches for extra-solar planets. A moderate number of robotic telescopes of very modest aperture (1 meter or less), in one location or several, each equipped with inexpensive, off-the-shelf, small-format, blue-sensitive CCDs, would rival the cost of a single space mission or large ground-based telescope, with a specific scientific return which could never be achieved by the large-aperture telescopes currently being erected at a dizzying pace around the world. And unlike the large-telescopes, which seem to be considered relatively 'obsolete' whenever larger apertures are constructed, even very small telescopes engaged in the long-term study of time-variable objects would remain critically useful for decades to come.

References

Charpinet, S., Fontaine, G. and Brassard, P.: 1996, *ApJ* **471**, L103.
Fontaine, G. and Wesemael, F.: 1987, in: A.G.D. Philip, D.S. Hayes and J. Liebert (eds.), *Proc. IAU Colloq. 95: The Second Conference on Faint Blue Stars*, Davis (Schenectady), 319.
Kawaler, S.D. and Bradley, P.A.: 1994, *ApJ* **427**, 415.
Kepler, S.O. et al.: 2000, *ApJ* **534**, L185.
Kilkenny, D., Koen, C., O'Donoghue, D. and Stobie, R.S.: 1997, *MNRAS* **285**, 640.
Koen, C., O'Donoghue, D., Kilkenny, D. and Stobie, R.S.: 1998, in: F.-L. Deubner, J. Christensen-Dalsgaard and D. Kurtz (eds.), *Proc. IAU Symp. 185: New Eyes to See Inside the Sun and Stars*, Kluwer Academic Publishers (Dordrecht), 361.
Metcalfe, T.S., Salaris, M. and Winget, D.E.: 2002, *ApJ* **573**, 803.
Montgomery, M.H. and Winget, D.E.: 1999, *ApJ* **526**, 976.
Nitta, A., Kanaan, A., Kepler, S.O., Koester, D., Montgomery, M.H. and Winget, D.E.: 2000, *Baltic Astron.* **9**, 97.
O'Brien, M.S. and Kawaler, S.D.: 2000, *ApJ* **539**, 372.
Vuille et al.: 2000, *ApJ* **314**, 689.
Winget, D.E., Robinson, E.L. and Nather, R.E.: 1982, *ApJ* **262**, L11.
Winget, D.E. et al.: 1991, *ApJ* **378**, 376.
Winget, D.E. et al.: 1994, *ApJ* **430**, 839.
Winget, D.E., Kepler, S.O., Kanaan, A., Montgomery, M.H. and Giovannini, O.: 1997, *ApJ* **487**, L191.
Wood, M.A. and Oswalt, T.D.: 1998, *ApJ* **497**, 870.
Wu, Y.: 2001, *MNRAS* **323**, 248.

δ SCUTI PULSATIONS IN PRE-MAIN SEQUENCE STARS

CLAUDE CATALA
Observatoire de Paris, LESIA, France

Abstract. Intermediate mass pre-main sequence stars (1.5 to 5 $M_\odot$) cross the instability strip on their way to the main sequence. They are therefore expected to be pulsating in a similar way as the δ Scuti stars. In this review, I present the status of observational studies of pulsations in these stars, and comment on prospects for future investigations of these pulsations from the ground and from space.

1. Introduction

Pre-main sequence (PMS) stars with masses above 1.5 $M_\odot$ are known as Herbig Ae/Be stars (Herbig, 1960; Strom et al., 1972). They are usually found within star formation regions, like their lower mass counterparts the T Tauri stars; they show variable emission lines in their spectra, in particular at Hα, which is often seen as a signature of variable structured stellar winds and/or extended active atmospheres; they also exhibit strong infrared excesses, due to the presence of circumstellar (CS) dust, presumably in CS accretion disks.

Due to this dusty environment and active atmospheres and winds, they exhibit strong photometric and spectroscopic variability, on various time scales: variable dust obscuration produces photometric variations as high as several magnitudes, with time scales between months and years (van den Ancker et al., 1998), and rotational modulation, due to magnetic activity, yields spectroscopic and photometric variability on time scales of hours to days (Catala et al., 1999).

The Herbig Ae/Be stars raise several important questions, which still need to be answered. First, their PMS nature needs to be confirmed, as their location in the HR diagram above the main sequence (Strom et al., 1972; van den Ancker et al., 1997) leaves us with the ambiguity that they could be either PMS or post-MS objects. Once their PMS nature is ascertained, the Herbig Ae/Be stars could be used to constrain our modelling of PMS evolution and of the coupling of the stars with their CS environment, involving e.g. magnetic processes, accretion/ejection processes, exchanges of angular momentum, etc... In order to study these problems, we need to gather much information on these stars, in particular concerning their internal structure.

Asteroseismology of the Herbig Ae/Be stars can in principle allow us to test our modelling of PMS evolution, by providing a probe of their internal structure and rotation, and also carries the potential to detect evolutionary frequency changes, which can be a measure of PMS evolution time scales (Kurtz and Catala, 2001).

PMS stars with masses above 1.5 $M_\odot$ are expected to cross the instability strip on their way to the main sequence, spending typically 5 to 10% of their PMS phase within it (Marconi and Palla, 1998). This is sufficiently long for a significant number of Herbig Ae stars to be present in this strip, and therefore presumably to exhibit δ Scuti pulsations.

However, other sources of variability in these stars may prevail over pulsational variability: spectroscopic variations due to stellar winds and magnetic activity, or photometric variations due to variable dust obscuration or to magnetic activity indeed occur with higher amplitudes than the expected δ Scuti-type variations, and this difficulty is probably at the origin of the lack of observational results concerning pulsations in these stars. On the other hand, the time scales for these high amplitude variations are in principle separated from that of δ Scuti-type pulsations: keplerian rotation of the CS envelope, presumably responsible for variable dust obscuration, occurs on time scales of months, and the starÕs rotation, at the origin of the variability due to surface activity is typically of the order of one to several days, while p-modes in these stars are expected with periods of minutes to hours. It is therefore likely that δ Scuti pulsations can be found in Herbig Ae stars.

In this paper, we review the observational results obtained so far concerning δ Scuti pulsations in intermediate mass PMS stars, and discuss the prospect for future work in this field. Section 2 presents results and prospects for photometric observations in this field, while Section 3 deals with spectroscopic data. Future prospects for space-based photometric investigation of δ Scuti pulsations in Herbig Ae stars are discussed in Section 4. Finally, Section 5 gives a general conclusion.

2. Photometric Study of δ Scuti Pulsations in PMS Stars

Breger (1972) was the first to report short-term photometric variations resembling δ Scuti pulsations in two PMS stars belonging to the young cluster NGC 2264: V 588 Mon and V 589 Mon. Although these results are based on data obtained over a few hours on three nights separated by several months, δ Scuti pulsations are clearly suggested in these photometric variations, with frequencies of 9.09 and 8.06 c/d, respectively. Only monoperiodicities were evidenced, with considerable ambiguity on their reality and large uncertainties on their frequencies, due to the limited amount of data. Besides, the membership of these two stars to the cluster was not totally doubtless (Breger, private communication), and no further observation of them was attempted since then.

Years later it was shown from photometric monitoring that δ Scuti pulsations in the Herbig Ae star HR 5999 can be distinguished from the higher amplitude, longer time scale variations due to variable dust obscuration or surface activity (Kurtz and Marang, 1995). Only one frequency, at 4.8 c/d, was found in these data. Photometric data of HR 5999 from 9 years earlier were recently re-analyzed (Kurtz and Catala, 2001), and shown to exhibit the same mode. This indicates that pulsation modes in these stars can be stable over periods of at least a decade, and that PMS stars, and in particular HR 5999, are excellent candidates for searching for mode frequency changes due to evolution of the star's internal structure, as already suggested (Breger and Pamyatnykh, 1998). PMS evolutionary time scales are indeed short enough to yield relative variations of pulsation periods of the order of $\dot{P}/P = 10^{-6}$ yr^{-1}, which translate into drifts of about 0.4 hrs in 10 years for the timing of maxima in the light curve, and should be observable. Further photometric observations of this star and other Herbig Ae stars, leading to a regular monitoring over a decade or so, should be undertaken, and could lead to a direct detection of signatures of PMS stellar evolution and measurements of its time scale.

Photometric monitoring of PMS stars has recently led to the discovery of δ Scuti pulsations in several of them. These results are summarized in Table I. Although still rather short, the list of PMS stars detected as δ Scuti pulsators has increased noticeably in the last few years, after the existence and location of a PMS instability strip was predicted, on the basis of stability analysis of the first three radial overtones (Marconi and Palla, 1998). Most of the stars presented in Table I indeed fall within this instability strip, but many further observations of δ Scuti pulsations in PMS stars are necessary to define observationally the boundaries of this instability strip.

Due to the poor duty cycle and short time span of the observations performed so far, only monoperiodicities were detected in most cases, and the few reported multiperiodicities are still very uncertain. Besides, the strong 1 day alias present in these observations and their short duration make the frequency measurements unreliable and uncertain. This is illustrated by Figure 1, where the power spectrum and spectral window of data on HR 5999 are shown.

Finally, even if the measured frequencies indeed represent the actual stellar pulsation modes, these modes are not identified and can correspond to radial or nonradial pulsations.

The photometric studies of δ Scuti pulsations in PMS stars are therefore still in a preliminary stage, and can hardly be exploited asteroseismologically, although various attempts have been made in this direction (see Marconi, these proceedings). It is very clear that data of much better quality are needed, in particular data spanning at least a couple of weeks, with improved duty cycles. Multi-site campaigns and/or space-based observations are necessary to reach this goal.

3. Spectroscopic Study of δ Scuti Pulsations in PMS Stars

Spectroscopic monitoring of PMS stars can also provide information on their δ Scuti pulsations. Radial velocity curves can be used to study low-degree modes, with a sensitivity to the various modes which differs from that of photometric observations. Radial velocity curves therefore provide a tool of investigation for

TABLE I

Summary of photometric investigations of pulsations in PMS stars

star	ν_1 (c/d)	ν_2 (c/d)	Sp. type	references
V 351 Ori	15.5	11.9?	A7	(Marconi et al., 2000) (Marconi et al., 2001) (Balona et al., 2002)
V 346 Ori	34.2	21.2	A5	(Pinheiro et al., 2002)
HD 104237	33.3	36.6?	A5	(Kurtz and Müller, 1999) (Donati et al., 1997)
HR 5999	4.81		A7	(Kurtz and Marang, 1995) (Kurtz and Catala, 2001)
HP 57	12.73	15.52		(Pigulski et al., 2000)
BL 50	13.91	9.89		(Pigulski et al., 2000)
V 588 Mon	9.1		A7	(Breger, 1972)
V 589 Mon	8.1		F2	(Breger, 1972)
HD 142666	21.4		A8	(Kurtz and Müller, 2001)
HD 35929	5.1		A5	(Marconi et al., 2000)

pulsations which is independent from and complementary to photometry, and can be used for instance to confirm the frequencies found in photometry.

Only two papers reporting δ Scuti pulsations from radial velocity curves of PMS stars can be found in the literature. First, observations made at the Anglo-Australian Telescope with the UCLES spectrograph revealed the probable presence of a mode with an amplitude of 1.3 kms^{-1} and a period of 37 hrs in the A5 Herbig star HD 104237 (Donati et al., 1997). This mode (see Table I) was subsequently observed in photometry (Kurtz and Müller, 1999). In addition, a recent multisite spectroscopic campaign was organized on HD 104237, using a network of 2m class telescopes equipped with echelle spectrographs, with a total duration of three weeks. Although these data are still being analyzed, it seems that they confirm the presence of the same mode, and show at least 2 additional modes (Böhm, private communication).

Another Herbig Ae star, V 351 Ori, was also observed extensively both in photometry and in high resolution spectroscopy from the South African Astronomical Observatory (Balona et al., 2002). A previous photometric monitoring had been performed on this star (Marconi et al., 2000; Marconi et al., 2001), resulting in

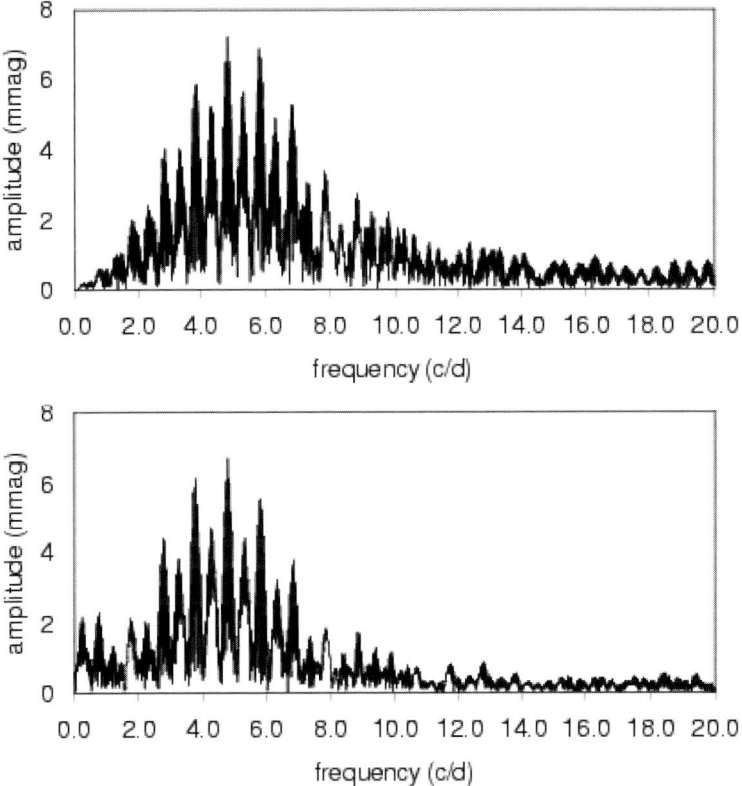

Figure 1. Amplitude spectrum (top) and spectral window (bottom) for the Kurtz and Marang (1995) V data on HR 5999. The spectral window was generated with an artificial, noise-free sine wave sampled at the times of observation, with the frequency and amplitude of the highest peak in the amplitude spectrum. Note that although the 4.8 c/d peak was chosen as most probable, the next peak near 5.8 c/d could have been selected instead with a nearly equal confidence. The figure is from Kurtz and Catala (2001)

a list of 6 potentially detected frequencies, the most important two of them being presented in Table I. The radial velocity curve from these new observations (Balona et al., 2002) indicates three frequencies, two of which being identical to those found in the previous analysis by Marconi et al. The third frequency present in Balona et al's data does not correspond to any of the 6 frequencies published by Marconi et al. These new results therefore bring a nice independent confirmation of the two frequencies common to both analysis, but beyond this positive comparison, they also illustrate the difficulties of exploiting data with a poor duty cycle, and the unreliability of frequencies derived with such data.

For stars that rotate fast enough that the rotationally broadened photospheric lines are well resolved, line profile variations induced by nonradial pulsations can be detected. They give access to higher degree modes than photometric and radial velocity monitoring, typically up to $\ell = 10$, and can also provide constraints on mode identification, each mode having its own signature in the line profile variations (see e.g. Telting, Balona, De Ridder, these proceedings).

An early study of the Herbig Ae star HD 163296 showed complex short-term line profile variations, that were tentatively interpreted as due to noradial pulsation (Baade and Stahl, 1989). This interpretation was however questioned and the observed variations shown to be consistent with rotational modulation of the stellar surface and a stellar wind (Catala et al., 1989).

The high resolution monitoring of V 351 Ori presented by Balona et al. (2002) shows series of bumps moving from blue to red across the line profiles, clearly indicating a high-degree pulsation in this star. A frequency of 20 c/d and a degree $\ell = 8$ are suggested by these data.

The results obtained so far on pulsations in PMS stars using spectroscopic techniques, although still very few, are promising, and this kind of research should certainly be continued in the future. As in the case of photometric investigation, the most important difficulty rests with the poor duty cycle leading to very strong 1d alias, and most of the efforts to spend in coming years in this area should be directed toward the organization of multi-site spectroscopic campaigns.

4. Space-Based Investigation of Pulsations in PMS Stars

Space-based observations are definitely the best strategy for asteroseismology, providing both a very good spectral window and a very low level of photometric noise. Among the four ongoing asteroseismology space projects, three have the possibility of observing PMS stars as part of their programme: COROT, MONS, and EDDINGTON.

COROT (Baglin et al., 2002) is a small space photometry mission, for asteroseismology and exoplanet search. COROT (for **CO**nvection, **RO**tation and planetary **T**ransits) is being developed by CNES, the french space agency, with a significant participation from other european countries, Brasil and ESA, and includes a 27cm CCD photometer placed on a low-earth orbit, covering a 3 degree field. It will be launched in 2005, and will reach a noise level as low as 0.7 (resp. 2.8) parts per million (ppm) in Fourier space in 5 days for 6th (resp. 9th) magnitude stars, which is remarkably better than what is achievable from the ground. The duty cycle of the observations will be better than 96%, which also is by far more efficient than multisite groundbased campaigns. COROT will observe a few fields for 5 months continuously, and also a few other fields for 10 to 20 days, in order to perform a seismological exploration of the HR diagramme. There will be a possibility to observe a few PMS stars in this exploratory phase of the mission, in particular in the young cluster NGC 2264, which falls in the continuous viewing zone of the instrument. This will give us the first opportunity to detect and measure low amplitude modes, with a good frequency resolution and most of all no alias.

MONS (for **M**easuring **O**scillations in **N**earby **S**tars) is a danish space project, with the main goal of studying solar-like oscillations in about 20 bright stars (Christensen-Dalsgaard, 2002). The main instrument is a 32cm photometric telescope, making two-colour measurements. In addition a field monitor, and star trackers on the platform, will be used to simultaneously study other targets. Including some PMS stars as part of these additional observations is being considered. The project is still awaiting definitive approval and funding, and if funded, will be scheduled for a launch in 2006.

Finally, EDDINGTON (Favata, 2002) is a recently approved ESA mission for asteroseismology and exoplanet search. The instrument, a set of four 60cm Schmidt telescopes, equipped with CCD mosaics, will be placed in orbit at Lagrangian point L2, after a launch foreseen in the 2007-2008 time frame. It will cover a 6 degree field, and reach a noise level of 1.1ppm in 30 days for 11th magnitude stars. The duty cycle will be better than 90%. EDDINGTON will give us the possibility to observe a statistically significant sample of PMS stars. Like COROT, EDDINGTON will detect low amplitude modes, and allow us to measure them with a good frequency resolution, and with no alias.

5. Conclusion

The study of δ Scuti pulsations in the PMS Herbig Ae/Be stars, although still very preliminary, bears the promise of very significant advances in the field of stellar evolution and stellar formation. Two types of major ground-based observational efforts must be undertaken in order to progress in this area: (i) regular monitoring of some of the known PMS pulsators on time scales of at least a decade, in order to detect and measure evolutionary frequency changes; (ii) multisite photometric and spectroscopic campaigns, in order to limit the aliasing in the derived power spectra and obtain reliable and precise measurements of the frequencies.

A special attention should be directed to the few known PMS eclipsing binaries, whose masses and radii are well-known. In particular, one of them, RS Cha, has both components inside or very near the predicted instability strip, and, if a seismological analysis was available, would provide powerful tests of the stellar evolution models in the PMS phase.

Finally, future space based asteroseismology missions, such as COROT, MONS and EDDINGTON, have the potential to provide very high quality seismic data for PMS stars.

References

Baade, D. and Stahl, O.: 1989, *A&A* **209**, 268.
Baglin, A., Auvergne, M., Barge, P. et al.: 2002, in: F. Favata, I.W. Roxburgh and D. Galadi (eds.), *Proceedings of the First Eddington Workshop on Stellar Structure and Habitable Planet Finding*, *ESA SP*-**485**, 17.
Balona, L.A., Koen, C. and van Wyk, F.: 2002, *MNRAS* **333**, 923.
Breger, M.: 1972, *ApJ* **171**, 539.
Breger, M. and Pamyatnykh, A.A.: 1998, *A&A* **332**, 958.
Catala, C., Simon, T., Praderie, F. et al.: 1989, *A&A* **221**, 273.
Catala, C., Donati, J.F., Böhm T. et al.: 1999, *A&A* **345**, 884.
Christensen-Dalsgaard, J.: 2002, in: F. Favata, I.W. Roxburgh and D. Galadi (eds.), *Proceedings of the First Eddington Workshop on Stellar Structure and Habitable Planet Finding*, *ESA SP*-**485**, 25.
Donati, J.F., Semel, M., Carter, B.D. et al.: 1997, *MNRAS* **291**, 658.
Favata, F.: 2002, in: F. Favata, I.W. Roxburgh and D. Galadi (eds.), *Proceedings of the First Eddington Workshop on Stellar Structure and Habitable Planet Finding*, *ESA SP*-**485**, 3.
Herbig, G.H.: 1960, *ApJS* **4**, 337.
Kurtz, D.W. and Marang, F.: 1995, *MNRAS* **276**, 191.
Kurtz, D.W. and Müller, M.: 1999, *MNRAS* **310**, 1071.
Kurtz, D.W. and Müller, M.: 2001, *MNRAS* **325**, 1341.
Kurtz, D.W. and Catala, C.: 2001, *A&A* **369**, 981.
Marconi, M. and Palla, F.: 1998, *A&A* **507**, L141.
Marconi, M., Ripepi, V., Alcala, J.M. et al.: 2000, *A&A* **355**, L35.
Marconi, M., Ripepi, V., Bernabei, S. et al.: 2001, *A&A* **372**, L21.
Pigulski, A., Kolaczkowski, Z. and Kopacki, G.: 2000, *AcA* **50**, 113.
Pinheiro, F., Folha, D., Monteiro, M.J. et al.: 2002, *A.S.P. Conf. Ser.* **259**, 352.
Strom, S.E., Strom, K.M., Yost, J. et al.: 1972, *ApJ* **173**, L65.
van den Ancker, M.E., Thé, P.S., Tjin, A. et al.: 1997, *A&A* **324**, L33.
van den Ancker, M.E., de Winter, D., Tjin, A. and Djie, H.R.E.: 1998, *A&A* **330**, 145.

SOLAR-LIKE OSCILLATIONS IN SEMIREGULAR VARIABLES

TIMOTHY R. BEDDING

School of Physics, University of Sydney 2006, Australia

Abstract. Power spectra of the light curves of semiregular variables, based on visual magnitude estimates spanning many decades, show clear evidence for stochastic excitation. This supports the suggestion by Christensen-Dalsgaard et al. (2001) that oscillations in these stars are solar-like, i.e., stochastically excited by convection, with mode lifetimes ranging from years to decades.

Keywords: stars: AGB and post-AGB – stars: oscillations

Oscillating red giants with high luminosity – the long period variables – are conventionally divided into Miras and semiregulars. Both can be monitored visually, thanks to the extreme temperature sensitivity of the TiO absorption bands that dominate the visible spectrum. For some stars, visual magnitude estimates by amateur astronomers span many decades.

Mira variables have large amplitudes and are very regular, reflecting the nature of the driving process, which is self-excitation via opacity variations. Semiregulars (SRs), on the other hand, have lower amplitudes, less regularity and often show two or three periods (Bedding and Zijlstra, 1998; Kiss et al., 1999; Wood et al., 1999). In these stars, it seems plausible that there is a substantial contribution from convection to the excitation and damping. Indeed, Christensen-Dalsgaard et al. (2001) have suggested that the amplitude variability seen in SRs is consistent with the pulsations being solar-like, i.e., stochastically excited by convection. The subject of Mira-like versus solar-like excitation has also been discussed in the context of K giants by Dziembowski et al. (2001).

Figures 1 and 2 show power spectra of visual observations for some of the best-studied long period variables. The first star, R Leo, is a typical Mira (period 310 d) and shows a narrow peak in the power spectrum. This indicates that the pulsation is stable in both period and phase.

The other stars are SRs, and all show strong evidence for stochastic excitation. The power from each oscillation mode is split into a series of peaks under a narrow envelope. This structure is typical of a stochastically excited oscillator and is strikingly similar to close-up views of individual peaks in the power spectrum of the Sun (Toutain and Fröhlich, 1992). We can estimate the mode lifetime from the width of the envelope. This seems to range from a few decades (L_2 Pup) down to only a few years (e.g., X Her). For most doubly-periodic stars, the mode lifetime is similar for both periods.

RR CrB appears to show two closely space modes (periods 54 and 60 d), which at first sight seem hard to understand as consecutive low-order radial modes. How-

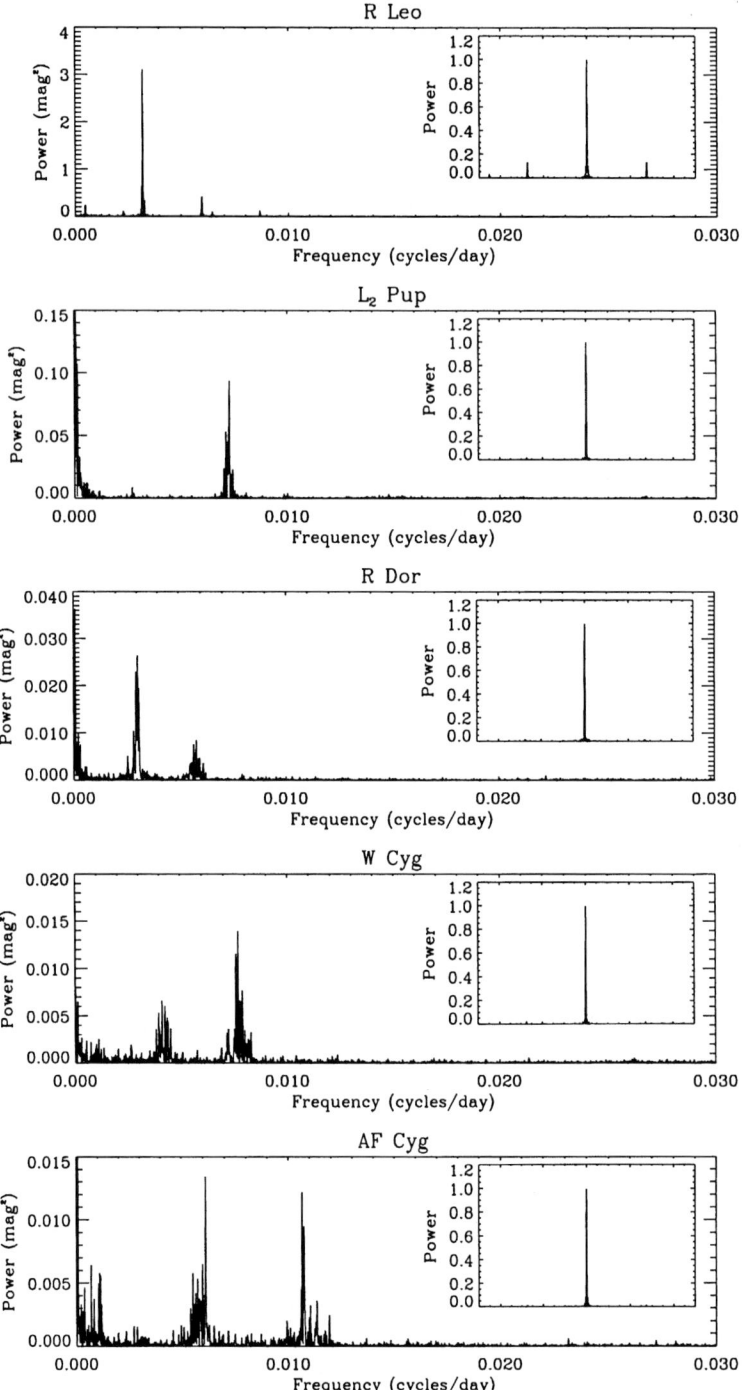

Figure 1. Power spectra of visual observations of a Mira and four semiregulars. In each case, the inset shows the spectral window.

SOLAR-LIKE OSCILLATIONS IN RED GIANTS 63

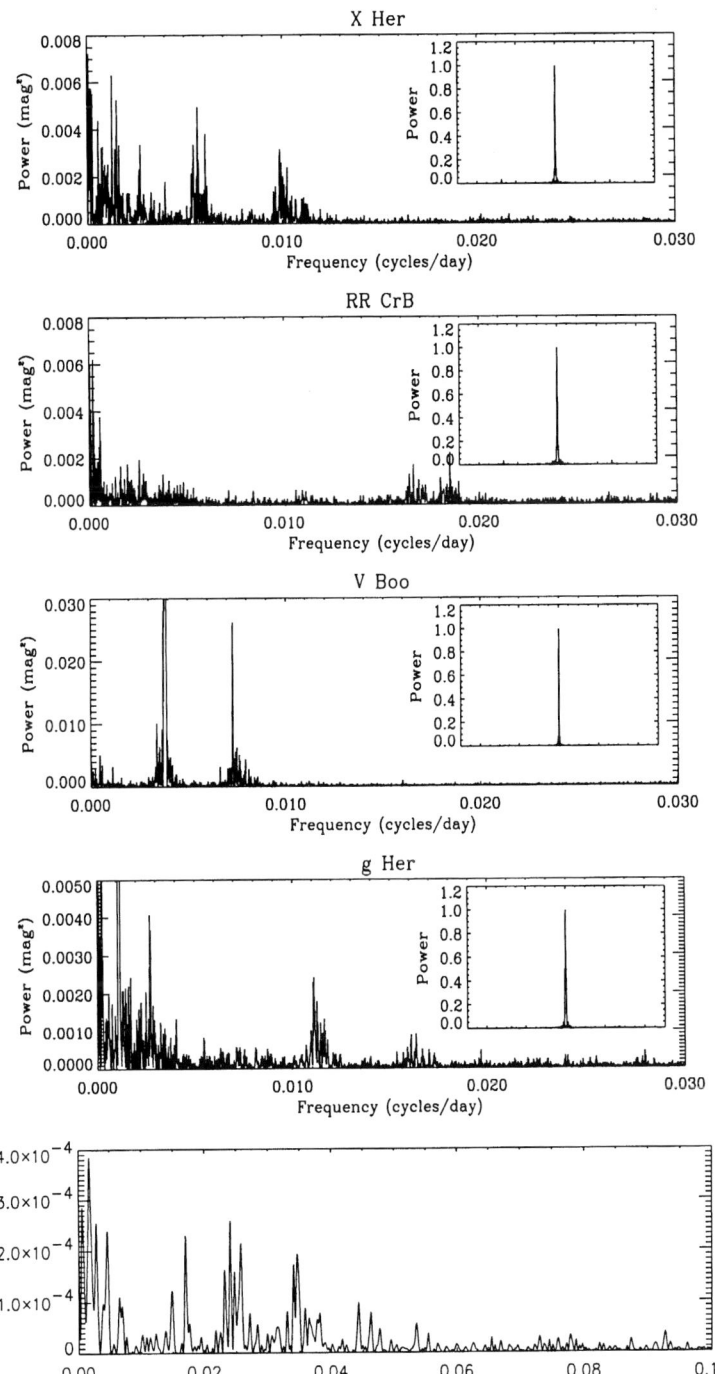

Figure 2. Same as Figure 1, for four more semiregulars, plus one red giant from MACHO observations of the LMC (bottom panel; note change of horizontal scale).

ever, observations of semiregulars both locally (Kiss et al., 1999) and in the LMC (Wood et al., 1999) also show some stars with period ratios close to 1.1, and models by Wood et al. (1999) indicate plausible identifications with low-order modes.

V Boo is an unusual star, with a Mira-like mode (258 d) whose amplitude has decreased steadily over the past 90 years, plus a shorter-period mode (137 d) that has remained relatively constant in amplitude (Szatmáry et al., 1996; Bedding et al., 1998; Kiss et al., 1999). As expected, the power spectrum at the long period is very strong (the peak is way off scale, at $0.1\,\text{mag}^2$), while the spectrum around the short period shows a low broad hump. However, there is also a narrow peak just left of centre in the latter. This peak does not coincide with the harmonic of the longer period, and apparently indicates a coherent long-lived component to an otherwise stochastically excited oscillation.

g Her (= 30 Her) has two pulsation modes with solar-like envelopes (90 and 60 d), but also a much longer period (890 d) that is coherent (the peak is off the top of the graph, at $0.033\,\text{mag}^2$). The latter is a typical example of a long secondary period, often seen in SRs and probably due to binarity (Wood et al., 1999; Huber et al., these Proceedings).

Finally, the bottom panel of Figure 2 shows the power spectrum of a red giant in the LMC, based on seven years of data from the MACHO database. We see evidence for perhaps as many as five equally-spaced modes, all with similar envelopes. There is clearly much to be learned from data such as these.

References

Bedding, T.R. and Zijlstra, A.A.: 1998, *ApJ* **506**, L47.
Bedding, T.R., Zijlstra, A.A., Jones, A. and Foster, G.: 1998, *MNRAS* **301**, 1073.
Christensen-Dalsgaard, J., Kjeldsen, H. and Mattei, J.A.: 2001, *ApJ* **562**, L141.
Dziembowski, W.A., Gough, D.O., Houdek, G. and Sienkiewicz, R.: 2001, *MNRAS* **328**, 601.
Kiss, L.L., Szatmáry, K., Cadmus, R.R. and Mattei, J.A.: 1999, *A&A* **346**, 542.
Szatmáry, K., Gál, J. and Kiss, L.L.: 1996, *A&A* **308**, 791.
Toutain, T. and Fröhlich, C.: 1992, *A&A* **257**, 287.
Wood, P.R., Alcock, C., Allsman, R.A. et al.: 1999, in: T. Le Bertre, A. Lebre and C. Waelken (eds.), *IAU Symp. 191: Asymptotic Giant Branch Stars*, 151.

DISCOVERY OF A NEW CLASS OF PULSATING STARS: GRAVITY-MODE PULSATORS AMONG SUBDWARF B STARS

ELIZABETH M. GREEN, KEITH CALLERAME, IVO R. SEITENZAHL,
BROOKE A. WHITE, ELAINA A. HYDE and MELISSA GIOVANNI
Steward Observatory, University of Arizona, Tucson, AZ 85721 USA

MIKE REED
Physics, Astronomy, & Material Science, SW Missouri State University, Springfield, MO 65804 USA

GILLES FONTAINE
Dépt. de Physique, Université de Montréal, CP6128, Station Centreville, Montréal, QC H3C 3J7, Canada

ROY ØSTENSEN
Isaac Newton Group of Telescopes, 37800 Santa Cruz de La Palma, Canary Islands, Spain

Abstract. During the course of an ongoing CCD monitoring program to investigate low-level light variations in subdwarf B (sdB) stars, we serendipitously discovered a new class of multimode pulsators with periods of the order of an hour. These periods are a factor of ten longer than those of previously known multimode sdB pulsators (EC 14026 stars), implying the new pulsations are due to gravity modes rather than pressure modes. The iron opacity instability that drives the short period EC 14026 stars is effective in hot sdB's. The long period pulsators are found only among cooler sdB stars, where they are surprisingly common. The mechanism responsible for exciting the deeper g-modes in cool sdB's is currently unknown, but the temperature and gravity range in which these stars occur must be an important clue. We present the first observational results for this new class of pulsating sdB stars, and discuss some possible implications.

Keywords: pulsating sdB stars, gravity modes

1. Observational Characteristics

The prototype long period sdB variable, PG 1716+426, happens to be in a post-common envelope binary with an orbital period of 1.77736 days. Its variability was discovered when it was included in a pilot program to monitor the light curves of 15 binaries with large radial velocity variations, looking for evidence of eclipses, ellipsoidal effects and reflection effects. A first analysis of 81 hours of photometric monitoring of PG 1716+426 (13 nights in 2 months) indicates the presence of at least 3–5 pulsation modes with periods between 0.8 and 1.4 hours.

Once the pulsations in PG 1716+426 were proved likely to be real, we broadened our photometry program to include an unbiased sample of both binary and non-binary sdB stars. In the past year, we have identified 17 sdB stars with probable pulsation periods of about an hour, out of a sample of 61 sdB stars monitored

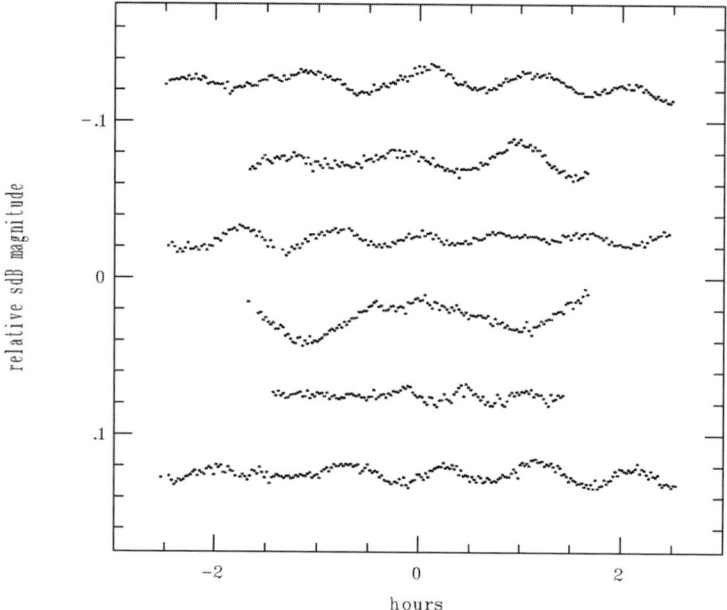

Figure 1. Typical light curves for the largest amplitude long period pulsators. The time between crests varies from about 35 mins to 120 mins.

for 2 to 7 hours each. The apparent pulsation periods are a factor of 10 longer than typical periods of previously known multimode sdB pulsators, *i.e.* the EC 14026 stars (Kilkenny et al., 1997; Charpinet et al., 1996, 1997; Koen et al., 1999; Charpinet, 2001). Due to such lengthy periods, extensive, well-coordinated multi-site campaigns would be needed for fairly complete mode characterizations. The κ mechanism responsible for the 80–500 sec acoustic variations (p-modes) in EC 14026 stars is incapable of exciting the high radial order gravity modes needed to produce such long periods, leaving the g-mode driving mechanism unknown.

Initially, long period sdB pulsators were all found in binaries with orbital periods of a few hours to several days, but this was likely due to chance. In our kinematically unbiased sample of 52 sdB's with precise radial velocity data, 29 stars (56%) occur in short period binaries. 14 of the 52 are long period pulsators, and 5 of these are almost certainly single stars. Therefore, the binary fraction of the long period pulsators is not significantly different from that of sdB stars in general.

The peak-to-peak amplitudes in the light curves of long period sdB pulsators range from nearly 0.05 mag to only a few millimags, less than 25% of the amplitude of PG 1605+072 (admittedly the most extreme example of an EC 14026 star). The strongest new pulsators clearly show the irregular amplitudes and periods characteristic of multimode pulsations (Figure 1). However, an equal number of

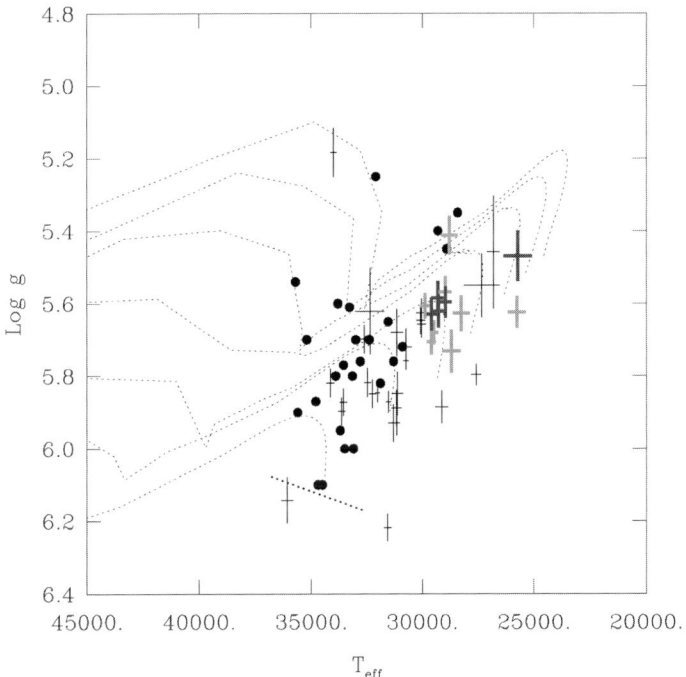

Figure 2. The log g vs. T_{eff} diagram for various pulsating sdB stars superposed on Dorman's EHB evolutionary tracks (light dotted lines) for models with a range of H envelope masses. The thick 2σ error bars represent the long period pulsators; thin error bars show the non-pulsating sdB stars in our sample. Filled circles are the known EC 14026 sdB pulsators. The thicker dotted line represents the zero-age He-burning main sequence (*i.e.* the zero hydrogen envelope limit to the EHB).

long period sdB stars pulsate so weakly they are barely detectable even with the stability, precision, and good sky coverage possible with CCD's.

The new pulsators all have $25,000 < T_{eff} < 30,000$K and $5.4 < \log g < 5.7$ (Figure 2). There is little or no overlap with the hotter EC 14026 stars. [Note, however, that the derived atmospheric parameters for the short and long period pulsators are not necessarily on a uniform scale. Temperatures and gravities for the plotted long period pulsators were determined from a homogeneous sample of high S/N, 1Å resolution MMT spectra. The data for the known EC 14026 stars (Charpinet, 2001) come from a variety of inhomogeneous sources.]

The fraction of sdB stars that are long period pulsators is quite high: about 75% of sdB stars cooler than 30,000 K, or 25–30% of all sdB stars. This contrasts once again with the EC 14026 stars, which comprise $< 5\%$ of all sdB stars (Billères et al., submitted to *ApJ*).

2. Discussion

The limited temperature range of the long period sdB pulsators in Figure 2 implies the pulsation mechanism is intrinsic to the star, rather than an external driver such as tidal excitation. (In any case, the available statistics do not support binarism as the primary explanation.) Cooler sdB stars have larger mass hydrogen envelopes. Evidently, larger envelopes are required to sustain long period pulsations. This seems puzzling at first, since even the coolest sdB has an almost negligible envelope compared to the majority of normal core He-burning stars. Closer investigation of the stars with $T_{\text{eff}} < 30{,}000$K shows that the two with highest gravity – nearest to the zero-age extended horizontal branch – do not pulsate, and only very small amplitude long period pulsators are found slightly above the zero-age EHB. The strongest long period pulsators occur significantly higher in the diagram, just at the point where the sdB is beginning to exhaust its helium core and evolve away from the EHB (after adjustment for the small apparent offset between the theoretical tracks and the data points). In a normal horizontal branch star, this is where the hydrogen shell is revving up due to heating from a nascent helium shell, as the star heads toward the asymptotic giant branch. SdB stars have such tiny envelopes that shell burning is quickly damped, as they begin contracting to white dwarfs. Still, the present data appear consistent with a driving mechanism involving the hydrogen shell, small and weak though it may be.

Acknowledgements

This work was supported by NSF grants AST-9731655 and AST-0098699.

References

Charpinet, S., Fontaine, G., Brassard, P. and Dorman, B.: 1996, *ApJ* **471**, L106.
Charpinet, S., Fontaine, G., Brassard, P., Chayer, P., Rogers, F.J., Iglesias, C.A. and Dorman, B.: 1997, *ApJ* **483**, L123.
Charpinet, S.: 2001, *AN* **322**, 387.
Kilkenny, D., Koen, C., O'Donoghue, D. and Stobie, R.S.: 1997, *MNRAS* **285**, 640.
Koen, C., O'Donoghue, D., Kilkenny, D., Stobie, R.S. and Saffer, R.A.: 1999, *MNRAS* **306**, 213.

RECENT OBSERVATIONS OF PMS δ SCUTI STARS

VINCENZO RIPEPI

Osservatorio Astronomico di Capodimonte, Naples, Italy; E-mail: ripepi@na.astro.it

Abstract. We present time series observations of the intermediate mass Pre-Main Sequence star H254 belonging to the young star cluster IC 348 and of the Herbig Ae star V351 Ori.

Both these stars present light variation on short time scale (a few hours) typical of the δ Scuti pulsation. The new data are briefly described together with the plan for future observational campaigns on PMS δ Scuti stars.

Keywords: stars: δ Scuti – stars: pre-main sequence

1. Introduction

The existence of δ Scuti type pulsation during the pre-main sequence (PMS) evolutionary phase was postulated by Breger (1972) on the basis of time series observations of two stars, namely V588 Mon and V589 Mon, members of the young cluster NGC 2264. Later observations (see Marconi et al., 2002 and references therein) confirmed Breger's hypothesis. These evidences stimulated the first theoretical investigation of the PMS δ Scuti instability strip, based on nonlinear convective hydrodynamical models (Marconi and Palla, 1998). As a result, the topology of the PMS instability strip for the first three radial modes was identified and a list of possible PMS pulsating candidates was provided on the basis of spectral types. Several of these candidates were tested for pulsation and many resulted actually PMS δ Scuti variables (see Marconi et al., 2002 for details). The comparison between empirical and theoretical periods and period-ratios is expected to provide important constraints on stellar mass, whereas the comparison of the position in the color-magnitude diagram suggested by observations and/or the pulsational analysis with PMS and post-MS tracks may allow to constrain the evolutionary state. However, in order to derive precise stellar parameters from the comparison between theory and observations, we need to measure the frequencies of pulsation with good accuracy. To this aim we have undertaken an observational campaign centred on the two stars H254 (member of the cluster IC348) and V351 Ori. In the following we shall summarize our observational results for these two stars, whereas the comparison with pulsation models will be reported in a separate paper of this volume (Marconi and Palla, these proceedings).

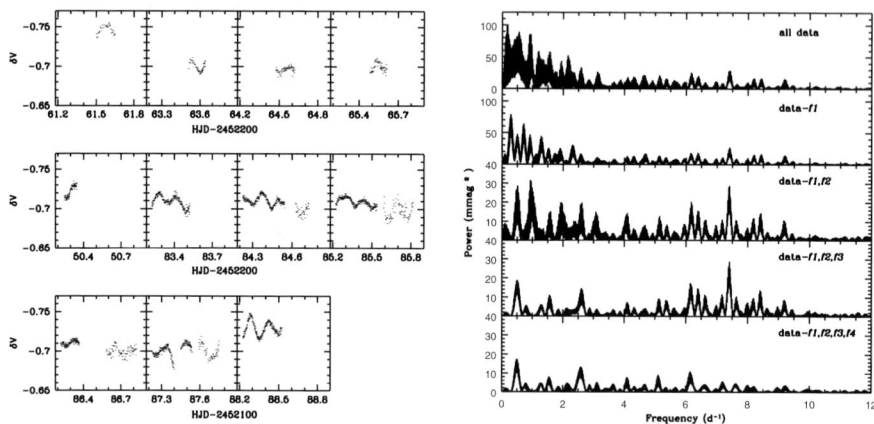

Figure 1. Left: Differential light curve for H254; Right: Periodograms of V photometry for H254. The sequence from top to bottom shows the change of the periodogram pre-whitened by f_1, f_2, f_3, and f_4, respectively.

2. Observations on H254 in IC 348

IC 348 is a young cluster located near the edge of the Perseus molecular cloud complex at a distance of about 320 pc. On the basis of observations by Luhman et al. (1998) and Herbig (1998) it is possible to place cluster members in the HR diagram and, in turn, to verify if there are stars falling inside the theoretical instability strip by Marconi and Palla (1998). In this way we were able to identify three candidates: H83, H254, H261. In order to test the variability of these stars, we obtained time series CCD observations of IC 348 at the 1.5m Loiano telescope (Italy) during 4 nights in Sept. 2001. As a result, only H254 showed variability on a few hours time scale. Therefore we re-observed H254 during a two site observing campaign (6 nights in Jan. 2002, telescope/instrument used: i) 1.5 m Loiano Telescope equipped with a three channel photometer; ii) San Pedro Martir, Mexico, uvby photometer). The resulting photometry is shown in Figure 2 (left side), whereas the Fourier analysis of the data is shown in Fig. 2 (left side). Apart from three frequencies which take into account the long term behaviour ($f_1 = 0.157$ c/d, $f_2 = 0.283$ c/d and $f_3 = 0.931$ c/d in the figure), there is one frequency ($f_4 = 7.406$ c/d, amplitude 5.4 mmag), corresponding to a period of around 3.2 h, which is characteristic of δ Scuti pulsation. As already remarked the scientific interpretation of this result will be presented elsewhere in this volume.

3. Observations on V351 Ori

The importance of V351 Ori as a target for PMS δ Scuti studies has already been stressed in various papers (Marconi et al., 2000, 2001; Balona et al., 2002; Ripepi

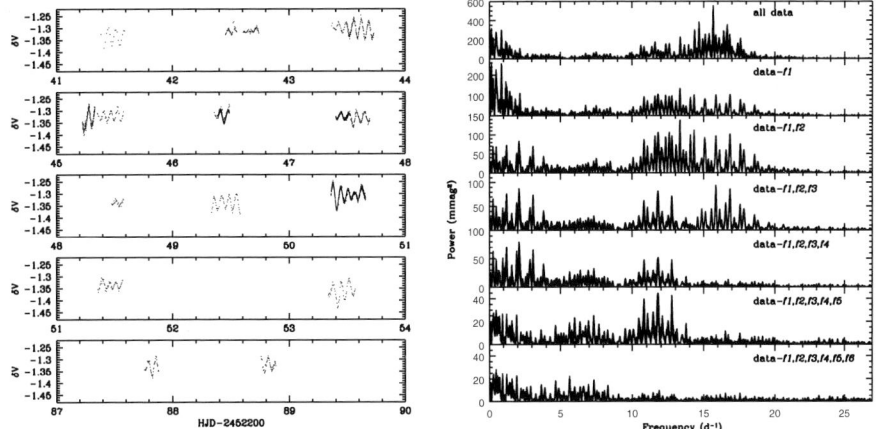

Figure 2. Left: Light curve of V351 Ori. The sense of δV is Variable-Comparison star. Right: Frequency analysis for the whole set of data.

et al., 2002). In this contribution we shall only briefly describe the multisite campaign that we carried out during 2001 (for details see Ripepi et al., 2002). Four observatories were involved in the observational campaign*. The observations were supplemented by contemporaneous data from SAAO (see Balona et al., 2002).

The multisite campaign on V351 Ori was performed in the Johnson V filter during around 2 months from Nov. 2001 to Jan. 2002. The bulk of the observations was obtained in the period Nov. 27-Dec. 9. The resulting sequence of light variations is shown in Figure 2 (left side), together with the result of the Fourier analysis (right side). The measured frequencies are also reported in Table I in comparison with results obtained in recent papers (Marconi et al., 2001; Balona et al., 2002). An inspection of the table reveals that at least four frequencies fall in the domain of δ Scuti pulsation. We also note that these frequencies are in reasonable agreement with the results obtained in previous papers on V351 Ori (Marconi et al., 2001; Balona et al., 2002).

4. Conclusions

The observational results on the two PMS δ Scuti stars H254 in IC348 and V351 Ori have been presented. V351 Ori is up to now the best studied variable of this class. Many more observations (on both photometric and spectroscopic side) are needed in order to increase our knowledges on this class of variables. In particular we need: 1) to enlarge the number of variables observed (up to now only 12 stars have at least one measured periodicity, against ~ 600 'normal' δ Scuti stars;

* 1.0 m ING-JKT; 85 cm BAO; 1.5 m Loiano; 1.5 m San Pedro Martir – SPM

TABLE I

Frequencies, amplitudes and phases derived from the Fourier analysis of the data. For comparison we also report the frequencies found in previous works (a '(?)' mean uncertain correspondence between our and Balona et al. (2002) frequencies)

	Frequency (d^{-1})	Amplitude (mmag)	Phase	Marconi et al. (2001)	Balona et al. (2002)
f_1	15.687	19.5	0.838	15.49	15.675
f_2	0.145	14.4	0.432	–	0.053 (?)
f_3	13.337	11.8	0.224	–	14.335
f_4	16.868	8.7	0.434	16.27	–
f_5	2.099	10.1	0.349	–	1.90 (?)
f_6	11.780	7.3	0.944	11.89	11.877

2) to observe in more detail the already known variables (multisite campaigns, spectroscopy, radial velocity measurements, etc.).

In the future we plan both to re-observe the already known PMS δ Scuti stars with poor data and to find new variables of this class (many candidates are known, both in the north and in the south).

References

Balona, L. A., Koen, C. and Van Wyk, F.: 2002, *MNRAS* **333**, 923.
Breger, M.: 1972, *ApJ* **171**, 539.
Herbig, G.H.: 1998, *ApJ* **497**, 736.
Luhman, K.L., Rieke, G.H., Lada, C.J. and Lada, E.A.: 1998, *ApJ* **508**, 347.
Marconi, M. and Palla, F.: 1998, *ApJ* **507**, L141.
Marconi, M., Ripepi, V., Alcalá, J. M., Covino, E., Palla, F. and Terranegra, L.: 2000, *A&A* **355**, L35.
Marconi, M., Ripepi, V., Bernabei, S., Palla, F., Alcalá, J. M., Covino, E. and Terranegra L.: 2001, *A&A* **372**, L21.
Marconi, M., Palla, F. and Ripepi, V.: 2002, *Comm. in Asteroseismology* **141**, 13.
Ripepi, V., Marconi, M., Bernabei, S., Palla, F., Pinheiro, F.J.G., Folha, D.F.M., Terranegra, L., Arellano Ferro, A., Jiang, X.J., Alcalá, J.M. and Oswalt, T.D.: 2002, *A&A*, submitted.

MULTICOLOR PHOTOMETRY FOR MODE IDENTIFICATION

M.D. REED
Department of Physics, Astronomy, & Material Science, SW Missouri State University, 901 S. National, Springfield, MO 65804, USA

Abstract. The goal of asteroseismology is to discern the physical conditions of stars by comparing observed pulsations with models. To obtain this goal, the observed pulsation periods and the spherical harmonics (n, ℓ, and m) need to match the theoretical model. Typically the most difficult part in this process is the identification of the pulsation modes in the observations. Multicolour photometry is one method that has proven useful for identifying pulsation modes. By observing stars through various wavebands, and comparing the amplitudes and phases, it is possible to determine the spherical harmonics. This contribution will emphasize the work of Watson (1988), which has since been applied to many different types of variable stars including δ Scuti (Garrido et al., 1990), γ Doradus (Breger et al., 1997), β Cepheid (Cugier et al., 1994), and EC 14026 (Koen, 1998) stars. I will also discuss the technique of summing spectra (especially UV) into various wavebands which are then used to identify modes as pioneered by Robinson, Kepler, and Nather (1982) and applied to white dwarf stars (Kepler et al., 2000).

Keywords: multicolor photometry

1. Introduction

Asteroseismology is the art of matching observed pulsations to theoretical models to infer the internal or evolutionary properties of stars. The information we learn about the physical conditions of stars does not come from the stars themselves, but from the models. To ensure that the models realistically represent the stars under study, we must connect the models to the observations. To make this connection, we need to identify the spherical harmonics (n, ℓ, and m) of the observed pulsations. However, identifying the spherical harmonics has been difficult.

One successful method for identifying pulsation modes is through period spacing. Nowhere has this succeeded so well as in PG 1159, a pre-white dwarf star. Figure 1 shows a small frequency range in PG 1159 that serves to illustrate the method. In the figure, three groups of pulsations are easily seen. Two of the groups show obvious triplets, with the third being a probable triplet. The period spacing between the groups is nearly constant and indicates that the groups have constant ℓ values and successive n values. In the asymptotic limit, $n \gg \ell$, such spacing results has the following form:

$$\Pi_{n,\ell} \approx \frac{\Pi_o}{\sqrt{\ell(\ell+1)}} (n + \epsilon) \quad \text{for g modes} \tag{1}$$

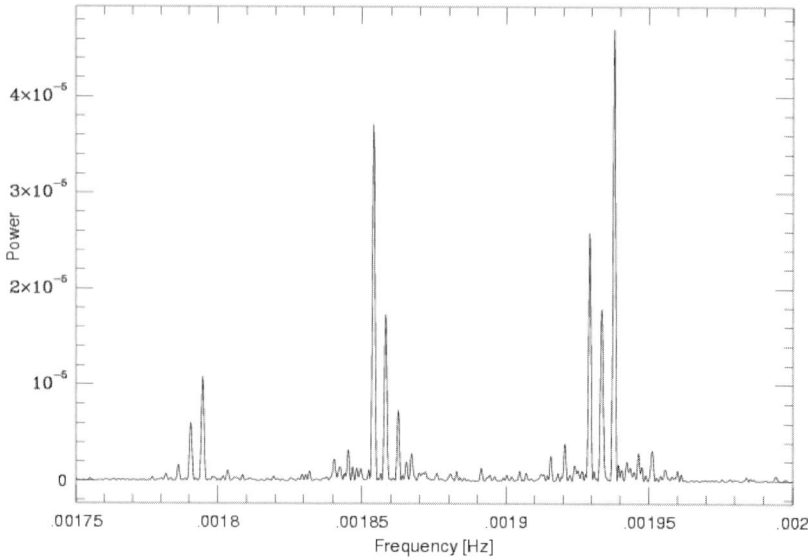

Figure 1. Portion of the temporal spectrum of PG 1159 (Winget et al., 1991).

$$\nu_{n,\ell} \approx \nu_o \left(n + \frac{\ell}{2} + \epsilon \right) \quad \text{for } p \text{ modes.} \tag{2}$$

However, it is very seldom that stars cooperate with observers to show such spacings. Thus we need to develop other methods for identifying pulsation modes. With recent advancements in CCD technology and observing techniques, time-series-spectroscopy is becoming quicker and easier. Though callibration is still an issue, techniques such as line profile fitting (see Balona, these proceedings, or DeRidder et al., these proceedings, for a review of the moment method), equivalent width fitting (Baldry et al., 1999), and spectral summing (Kepler et al., 2000) should prove useful for mode identification.

The tool that will be discussed in this paper (and that of Dupret et al., these proceedings) is multicolor photometry. The basic tenet of multicolor photometry is to compare observed color ratios and phase differences between two (or more) different filters to theoretical values. In the next section, I will outline the theoretical background as developed by Watson (1988; hereafter W88) to determine color amplitue ratios and phase differences. In §3 I will review multicolor observations of several different variable star classes which are compared to theoretical values; some successfully, others less so. Note that because of space constraints, I will not be able to show plots of matches for all classes. In §4 I will conclude with some remarks on apparent prerequisites for successful mode identification using multicolor photometry and discuss how multicolor photometry may be able to help with other theoretical problems to increase our understanding of the internal processes of stars.

2. Theoretical Background

Pulsations in stars viewed as brightness variations are caused by temperature and area (radial) variations. Thus the pulsations have a dependence on temperature, geometry, and limb darkening (as a measure of *observed* effective temperature), as well as several of their derivatives. An analytic expression for flux variations from non-radial pulsations was given in Dziembowski (1977). It was subsequently modified by Buta and Smith (1979) and applied to several different variable classes (as discussed in §3). The formulation I will follow is that of W88.

As written in W88, a linearized equation for flux variation (in magnitudes) for nonradial pulsations with indices ℓ and m is:

$$\Delta m(\lambda, t) = -1.086\epsilon \, P_{\ell,|m|}(\mu_o) \left[(T_1 + T_2)\cos(\omega t + \psi_T) + (T_3 + T_4 + T_5)\cos(\omega t) \right], \tag{3}$$

where ϵ is an amplitude parameter, $P_{\ell,m}$ are the Legendre polynomials and μ_o is the cosine of the pulsation axis to the observer. Since ϵ and $P_{\ell,|m|}(\mu_o)$ are wavelength-independent factors, they can be eliminated by taking the ratio of Δm for different wavebands (known as amplitude ratios). With ωt describing the pulsations themselves, only the T_n terms (and the ψ_T phase term) need to be understood to determine the usefulness of multicolor photometry.

The T_1 and T_3 terms indicate the local surface temperature and geometry changes. The T_2 and T_5 terms indicate the variation in limb darkening with local surface temperature and gravity (via gas pressure). T_2 and T_5 generally are not important for modes with $\ell < 3$ (which are difficult to observe due to geometric cancellation) and are often ignored. The T_4 term accounts for local surface pressure changes. Expressions for these terms are as follows:

$$T_1 = b_{\ell\lambda} \, \alpha_T(\lambda) \, B \tag{4}$$

$$T_2 = \frac{\partial b_{\ell\lambda}}{\partial \log T_{\text{eff}}} \frac{B}{2.3026} \tag{5}$$

$$T_3 = b_{\ell\lambda}(2+\ell)(1-\ell) \tag{6}$$

$$T_4 = -b_{\ell\lambda} \, p^\star C \, \alpha_g(\lambda) \tag{7}$$

$$T_5 = -\frac{\partial b_{\ell\lambda}}{\partial \log g} \frac{p^\star C}{2.3026} \tag{8}$$

2.1. FACTORS WITHIN THE T_n TERMS

Terms T_1, T_3, and T_4 have a direct dependence on $b_{\ell\lambda}$, the weighted limb-darkening integral defined as:

$$b_{\ell\lambda} = \int_0^1 h_\lambda(\mu)\, \mu\, P_\ell(\mu)\, d\mu, \qquad (9)$$

where $h_\lambda(\mu)$ is the limb-darkening function (Wade & Rucinski, 1985). Terms T_2 and T_5 tend to be small for $\ell < 3$ because of the dependence on the derivative of the limb-darkening integral.

The T_4 and T_5 terms are proportional to $p^\star C$. $p^\star$ describes the variation of the atmospheric gas presure with surface gravity in flux-emitting layers and C contains physical parameters of the star; mostly inside the pulsation constant Q. They are defined as:

$$p^\star \approx \left(\frac{\partial \log g}{\partial \log p_g} \right)_{\tau=1} \qquad (10)$$

$$C = \left(4 + \frac{3\pi}{G\rho_\odot Q^2} \right) - \ell(\ell+1)\left(\frac{G\rho_\odot Q^2}{3\pi} \right), \qquad (11)$$

with $Q \approx -6.456 + \log P_i + 0.5 \log g + 0.1 \log M_{Bol} + \log T_{\text{eff}}$, where P_i are the observed pulsation periods.

The B dependence of T_1 and T_2 is a measure of the local fractional temperature amplitude to the local fractional radius amplitude and can be written as:

$$B = R\left(1 - \Gamma_2^{-1}\right) C \qquad (12)$$

where R and ψ_T (Eq. 3) measure the departure from adiabicity. If the pulsations are assumed to be adiabatic, then R and ψ_T equal 1 and π respectively.

$\alpha_T(\lambda)$ and $\alpha_g(\lambda)$ in terms T_1 and T_4 are the local flux derivatives with respect to T_{eff} and g.

The T_n terms contain a total of 13 (not entirely independent) variables which need to be determined in order to identify pulsation modes using multicolor photometry. These variables must come from a number of different sources, each with their strengths and weaknesses. Two variables (ϵ, $P_{\ell|m|}$) have already been eliminated by using amplitude ratios. Observations supply the pulsation constant, Q, which contains 4 variables (P_i, M_{Bol}, $\log g$, and T_{eff}), but the remaining factors must come from theory. Atmospheric models are the most important; providing 4 variables including $b_{\ell\lambda}$, p_g, α_T and α_g, as well as the $b_{\ell\lambda}$ derivatives with respect to $\log g$ and T_{eff}. Other models (pulsation, interior) or theory must supply the remaining 3 variables: Γ_2, R, and ψ_T. Though theory has advanced significantly in the last few decades, it is still typically the limiting factor in the effectiveness of multicolor photometry. Assumptions and simplifications, particularly concerning departures from adiabicity, limit the resolution of the models to separate different ℓ modes and introduce systematics that offset theory from the observations.

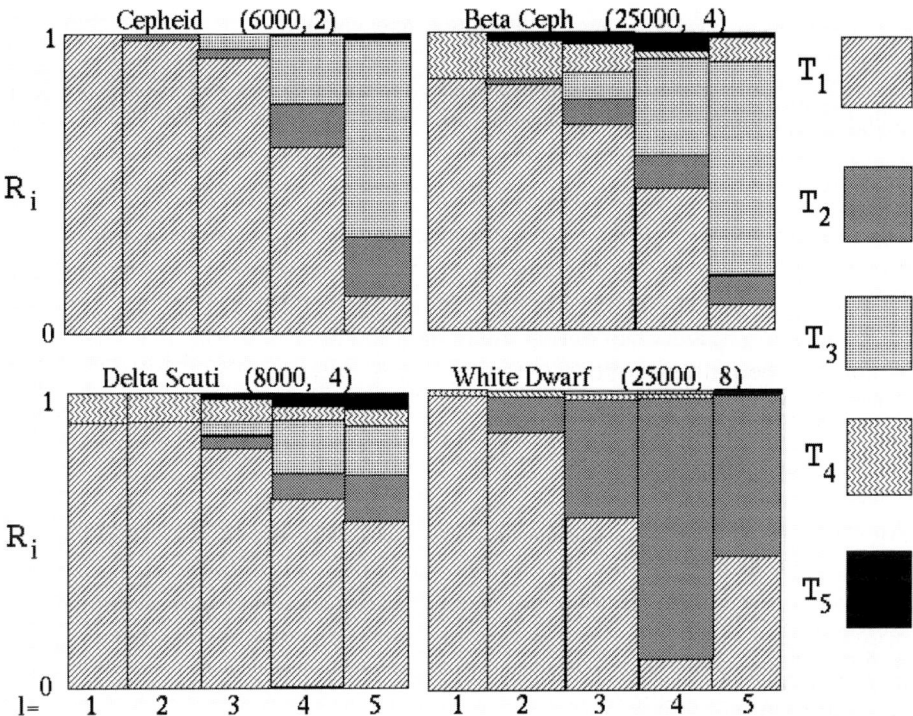

Figure 2. Relative importance (R_i) of T_n terms for different ℓ modes in various pulsation classes. T_{eff} and $\log g$ are given in parentheses. (Adapted from W88.)

2.2. RELATIVE IMPORTANCE OF THE T_n TERMS

The relative importance of the T_n terms will vary between pulsation classes. This is of course due to the differing physical characteristics of the stars themselves. This is shown schematically in Figure 2. As indicated in the figure, the T_1 term dominates for most classes for $\ell < 3$. For classes with Q values close to their radial fundamental (mostly classes above the main sequence) and for $\ell > 2$, the T_3 term has a significant contribution (except for White Dwarfs).

Though the *relative* importance of each T_n term does not indicate whether multicolor photometry will be useful, stars with a significant T_3 contribution tend to have better model matches.

2.3. PREDICTIONS

As every T_n term has limb darkening contained within it, multicolor photometry will be useful when limb darkening is considerably different between wavebands considered. Additionally, stars with pulsation constants (Q) close to the radial fundamental period will have larger geometry terms, aiding mode identification via multicolor photometry.

3. Multicolor Observations

The real test for the usefulness of multicolor photometry is comparing observations to theoretical models. For observations, Eq. 3 becomes:

$$\Delta m = A_\lambda \cos(\omega t + \phi_\lambda). \tag{13}$$

Thus for observations in any 2 wavebands, one can compare the amplitude ratio $(A_{\lambda 1}/A_{\lambda 2})$ and the phase difference $(\phi_1 - \phi_2)$ to theoretical values (from Eq. 3). Most observers work in the amplitude ratio–phase difference plane. This plane makes it easy to deduce not only *if* multicolor photometry works, but when it does, to identify the pulsation modes. I will now review recent multicolor photometry work.

3.1. CEPHEID VARIABLES

For short period Cepheids, the Q value is about 0.04d, which is close to the radial fundamental or first overtone values. As such, multicolor photometry could yield additional information. Balona and Stobie (1979) and W88 both examined multicolor data of Cepheids. Their results are consistent with Cepheids being radial (l=0) pulsators. Thus for Cepheids, multicolor photometry is a good test to ensure that pulsations are radial.

3.2. β CEPHEID VARIABLE

For β Cepheids, the pulsation constant Q is close to the radial fundamental or first overtone and models suggest a clear separation in the amplitude ratio-phase difference plane (W88). Figure 3 shows a good match between observations and theory (left). Figure 3 also shows the results of Cuiger, Dziembowski, and Pamyatniykh (1994; herafter CDP94), who examined how various filter sets (right) can effect the results. They also examined the dependence of models on metallicity and opacities. Their results show that filter colors need to be carefully chosen to avoid colors where different ℓ values can overlap in the amplitude ratio-phase difference plane. They also show how dependent mode identification is on the models. Incorrect metallicity and shortcomings in opacities can create large differences in how amplitude ratios and phases differences are interpreted. Even allowing for problems such as filter selection and model dependence, CDP94 were able to identify the spherical harmonics for the dominant mode of 36 β Cepheids.

3.3. δ SCUTI STARS

δ Scuti stars have proven to be more difficult than β Cepheids. Their Q value is close to their radial fundamental, which means that they should be quite good for multicolor studies. In the left panel of Figure 4, W88 show how mode identifications and physical parameters can be determined by comparing a good model

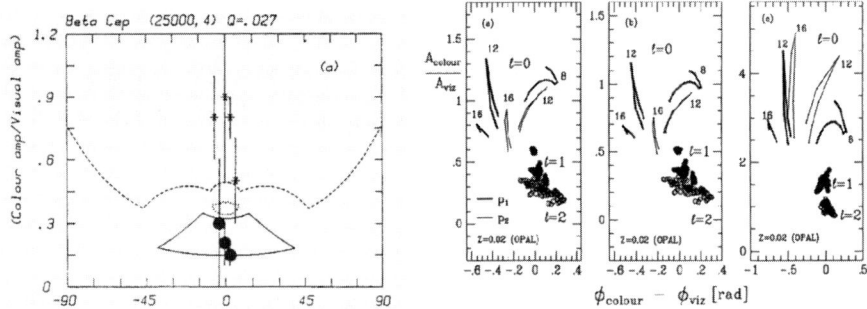

Figure 3. Left: Comparison of observed amplitude ratios and phase differences in β Cepheid stars with modes determined using line profile fitting (asterisks are radial modes, dots are nonradial modes) to calculated mode areas. Dashed, dotted, and solid regions correspond to $\ell=0$, 1, and 2. (From W88). Right: Comparison of calculated amplitude ratios and phase differences for β Cepheid stars using different filter systems. Left is Geneva, middle is Walraven, and right is UV filters. For $\ell=0$ modes, thick lines are for fundamental, and thin line is for first overtone modes. (From CDP94.)

match (top) with a bad one (bottom). Garrido, Garcia-Lobo, and Rodriguez (1990; hereafter GGR90) compared Stromgren observations with models for 15 δ Scuti stars. Their results, like CDP94, indicate the importance of proper filter selection. In the right panel of Figure 4, GGR90 results show that the best mode separation occurs for Stromgren (b-y)-(y) color combination (top) while other combinations (u-v shown in bottom panel) have severely overlapping ℓ regions. The work of GGR90 also indicate that departures from adiabicity play a major role in multicolor photometry for δ Scuti stars. Such mixed results seem common for δ Scuti stars. In Figure 5 results are shown from Handler et al. (2000). They clearly resolved the pulsations in the δ Scuti star XX Pyx, but could not get a decent model fit within spectroscopic constraints for $\log g$ and T_{eff} (left panel). In another star, Bi Cam, Breger et al. (2002) also resolved the pulsations. Their results are shown in the right panel of Figure 5 and they claim to identify 7 of the 20 observed pulsation modes based only on phase. Though this claim is somewhat suspect, they do some very important work by comparing models with varying metallicity and convection parameters. Their work indicates just how model-dependent the mode identifications are. With similar observational results, Koen et al. (2002) were able to identify 2 and constrain 5 other modes in the δ Scuti star QQ Tel.

Not shown are some excellent results from Balona (2001) for the δ Scuti star 1 Mon. Balona (2001) used a different method in that he *only* plots amplitude ratios or phase difference versus wavelength. In such a manner, he can directly compare how each filter matches his calculated values.

3.4. γ DORADUS STARS

With the same spectral type as δ Scuti stars, it would seem that γ Doradus stars would have similar results using multicolor photometry. The pulsation constant

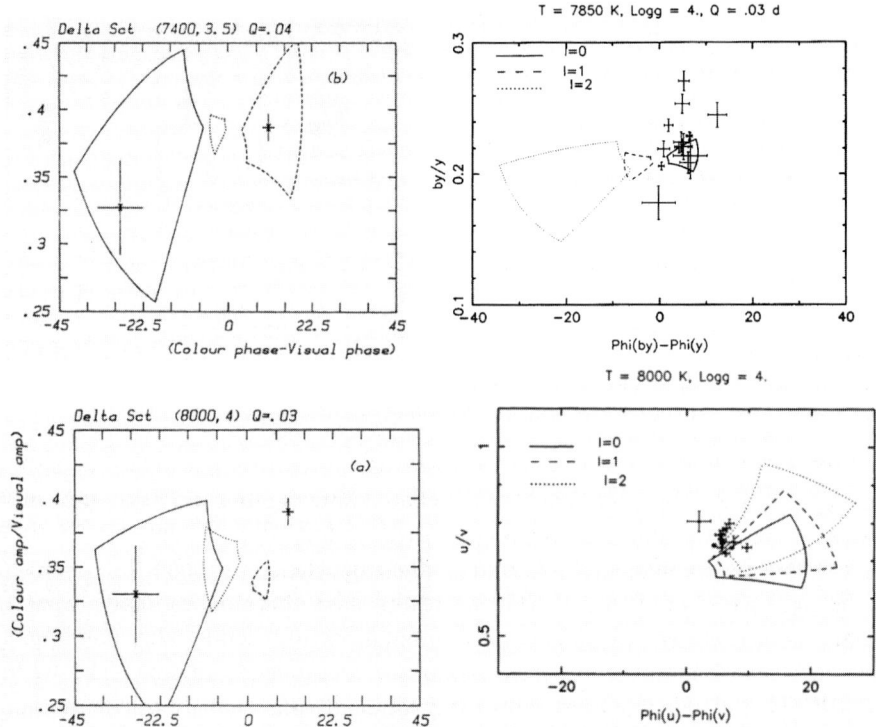

Figure 4. Comparison of successful mode identifications (top) with unsuccessful cases (bottom) for δ Scuti stars. The left panels show the dependence on model log g and T_{eff} (from W88) and the right panels show different color combinations in the Stromgren system (from GGR90).

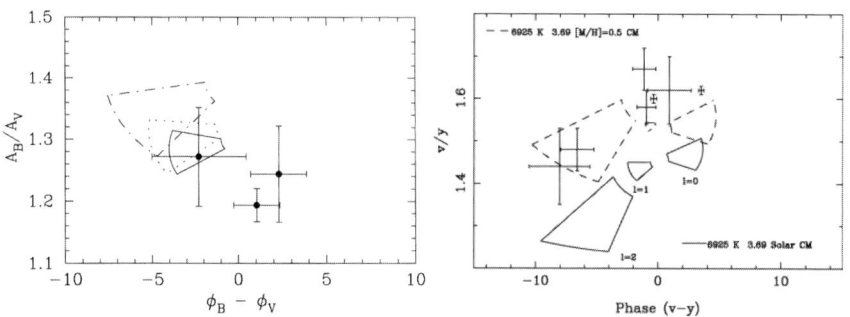

Figure 5. Recent results for δ Scuti stars show mixed results. Handler et al. (2000) could not identify modes using multicolor photometry (left) while Breger et al. (2002) could, but only using the phases (right).

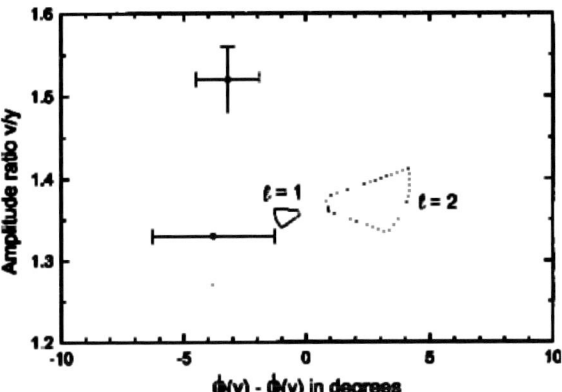

Figure 6. Comparison of amplitude ratios and phase differences for a γ Doradus star (from Breger et al., 1997). The observations support an ℓ=1 identification.

(Q) for γ Dor stars is ≈0.3 days, considerably longer than the radial fundamental mode. This is indicative of the g-mode nature of the pulsations. Figure 6 shows the results of the only multicolor work published to date (Breger et al. 1997). Though the errorbars are fairly large, they are consistent with ℓ=1 modes. Figure 6 shows that multicolor photometry holds some promise for mode identifications in γ Dor stars.

3.5. RAPIDLY OSCILLATING AP STARS

For roAp stars, the Q value (0.0015d) is much smaller than the radial fundamental period and W88 was unable to put any corresponding meaning to the amplitude ratio-phase difference plot. This is not surprising considering that Ap stars have significant surface inhomogeneities due to their large magnetic fields.

3.6. EC 14026 STARS

The only multicolor work on EC 14026 stars has been done by Koen (1998, 2002). He has obtained multicolor photometry for two EC14026 stars. Though there is some correspondence between the various colors and spherical harmonic, ℓ, there is no clear correlation. This is expected in the optical as sdB stars have rather small limb darkening coefficients. Additionally, EC 14026 stars have shown large variability in pulsation amplitudes, making it difficult to interpret amplitude ratios.

3.7. WHITE DWARF STARS

For white dwarfs, the Q value (5d) is much longer than the radial fundamental. As in the γ Dor case, this is indicative of their g-mode nature. Robinson et al. (1982) examined multicolor variations for the ZZ Ceti (DAV) stars. They determined that the pulsation phases must be the same for all (visible) colors and the pulsation

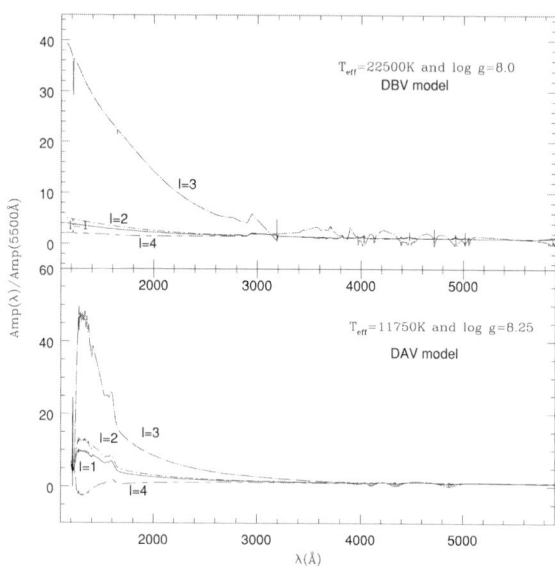

Figure 7. The wavelength dependence of color ratios for white dwarf stars in the UV (from Robinson et al., 1995).

amplitudes are the same for all (visible) filters (once temperature effects are corrected for different filters). Thus (visible) multicolor photometry will not separate ℓ values, but merely confirm the g-mode nature of the pulsations.

Robinson et al. (1995) revised their 1982 calculations to observations in the ultraviolet. Their results are shown in Figure 7 and show that in the ultraviolet, limb darkening is sufficient (along with a temperature dependence of the flux) to separate spherical harmonics, ℓ. Kepler et al. (2000) used HST to observe DAV stars using STIS while obtaining simultaneous ground observations. Their results are shown in Figure 8 and are consistent with both pulsations being $\ell=1$ modes. However, this method really has not been sufficiently callibrated. To properly determine how useful the method will be, observations need to be calibrated with a pulsator where multiple ℓ modes have previously been identified. Such observations were obtained in 2002 (O'Brien, private communication) and we look forward to the results.

4. Conclusions

We have reviewed the theoretical work of W88 on multicolor photometry. His work indicates that stars with substantial limb darkening or radial modes (and thus large geometry terms) are good prospects for mode identification via multicolor photometry. We then reviewed the observational work to date. The observations seem to confirm the theory in that stars with lower $\log g$ tended to have modes which could

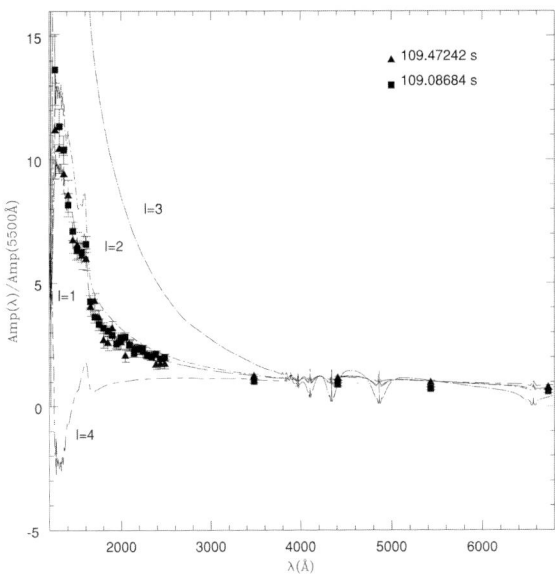

Figure 8. Comparison of amplitude ratios in the UV for observations with models for a DAV white dwarf. (From Kepler et al., 2000.)

be identified through multicolor photometer (Cepheids, β Cepheids, and to some extent δ Scuti stars). We found some cases where multicolor photometry seems like it should work, but only does occasionally (likely due to incomplete model physics), and some cases where it fails altogether, at least in the optical (roAp, EC 14026, and white dwarf stars).

Though the theory behind multicolor photometry is sound, in practice, models are quite often not up to the task. This serves to remind us just how far we have to go in our knowledge of stars. They are not a solved problem. It also shows that the tool to complete that knowledge is quite likely asteroseismology. By using combinations of tools, such as multicolor photometry and time-series spectroscopy, observers may be able to provide sufficient constraints for theorists to determine the impact of such physics as departures from adiabicity, chemical stratification, semiconvection, and convective overshoot.

We hope that future observations will include multicolor photometry of γ Dor and δ Scuti stars to search for cases where it provides clues to mode identification. Time-series UV spectroscopy of EC 14026 stars and white dwarfs is likely another method to identify modes. Of course we also eagerly watch the continued develop of time-series spectroscopic methods. Yet in the long run, identifications using multiple means are necessary to calibrate the various methods (such as optical spectroscopic studies of line profiles or velocity analysis). Once properly calibrated, such work will allow us a finer probe of the interior of stars.

Acknowledgements

I would like to thank the local organizing committee for funds to support my trip to the conference.

References

Baldry, I.K. et al.: 1999, *MNRAS* **302**, 381.
Balona, L.A. et al.: 2001, *MNRAS* **321**, 239.
Balona, L.A. and Stobie, R.S.: 1979, *MNRAS* **189**, 649.
Breger, M. et al.: 2002, *MNRAS* **329**, 531.
Breger, M. et al.: 1997, *A&A* **324**, 566.
Buta, R.J. and Smith, M.A.: 1979, *ApJ* **232**, 213.
Cugier, H., Dziembowski, W.A. and Pamyatnykh, A.A.: 1994, *A&A* **291**, 143.
Dziembowski, W.: 1977, *AcA* **27**, 3, 203.
Garrido, R., Garcia-Lobo, E. and Rodriguez, E.: 1990, *A&A* **234**, 262.
Hander, G. et al.: 2000, *MNRAS* **318**, 511.
Kepler, So.O. et al.: 2000, *ApJ* **539**, 379.
Koen, C. et al.: 2002, *MNRAS* **330**, 567.
Koen, C.: 1998, *MNRAS* **300**, 567.
Robinson, E.L. et al.: 1995, *ApJ* **438**, 908.
Robinson, E.L., Kepler, S.O. and Nather, R.E.: 1982, *ApJ* **259**, 219.
Wade, R.A. and Rucinski, S.M.: 1985, *A&AS* **60**, 471.
Watson, R.D.: 1988, *Phys. Space Sc.* **140**, 255.
Winget, D.E. et al.: 1991, *ApJ* **378**, 326.

HIGH-RESOLUTION SPECTROSCOPY FOR PULSATION-MODE IDENTIFICATION

JOHN TELTING

Nordic Optical Telescope, Apartado 474, E-38700 Santa Cruz de La Palma, Spain
E-mail: jht@not.iac.es

Abstract. Asteroseismology relies on accurate mode identification. High-resolution spectroscopy allows to detect such crucial information as the pulsational degree ℓ, the azimuthal number m, and pulsation amplitudes, directly from time series of observations. The advantage of high-resolution spectroscopy over standard photometric techniques is that not only pulsational temperature variations can be detected, but also the pulsational velocity field, yielding valuable extra information. In this paper I review the mode-identification methods that have been developed over the last decades, with emphasis on the application to hot main-sequence stars.

Keywords: spectroscopy, time series, line profiles, pulsations, mode identification

1. Introduction

The study of stellar pulsations provides a promising tool to further refine our knowledge of the internal structure of stars. Provided that one can conclusively derive pulsational parameters from the observations, asteroseismology can constrain stellar structure and evolution models.

Ledoux (1951) suggested that the observed variability in the width of the spectral lines of β CMa (a prototype β Cephei variable) is due to non-radial pulsations. Osaki (1971) calculated line profiles of non-radially pulsating stars to compare them with observations. With the work of Smith (1977), who used a similar model to fit line-profile variations, the field of pulsation-mode identification came off the ground.

All over the Hertzsprung-Russell diagram there are types of stars that pulsate in non-radial modes. The mode-identification (mode-ID) methods discussed in this paper are typically applied to early-type stars, such as δ Scuti and β Cephei variables. These stars have apparent pulsation periods of about one to a few hours, and with the present observing facilities harmonic degrees up to $\ell \sim 20$ can be detected spectroscopically (Doppler imaging). More recently, applications to longer period g-mode pulsators such as γ Doradus stars and Slowly Pulsating B-stars (SPB) were presented in the literature (see next section).

Stellar pulsations are reflected as line-profile variations in absorption lines formed in the photosphere. In rotating stars, where the Doppler imaging principle applies, the line-profile variations due to non-radial pulsations appear as blue-to-

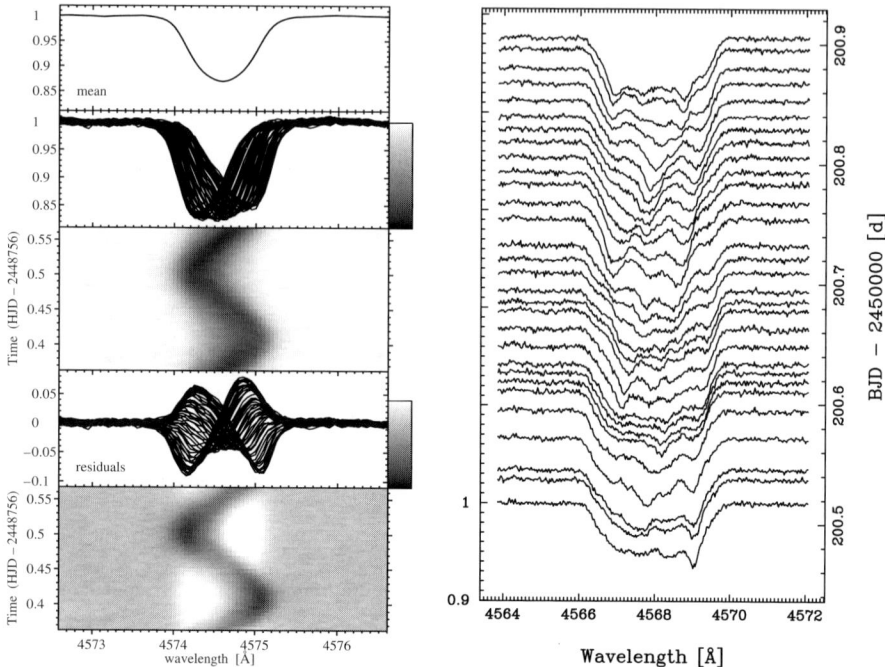

Figure 1. Examples of pulsational line-profile variations. Left: The dominant radial mode in β Cep, a slow rotator with $v \sin i \sim 25$km/s (Aerts et al., 1994; Telting et al., 1997). Right: Intermediate-degree ($2 \lesssim \ell \lesssim 5$) multimode pulsations in the rotating ($v \sin i \sim 115$km/s) β Cep star ϵ Cen (Schrijvers et al., submitted).

red moving bumps (Vogt and Penrod, 1983). In general, the line-profile variations are caused by the stellar pulsational velocity field, and by the photospheric pulsational temperature variations. The latter give rise to local changes in brightness, and to local changes of the equivalent width of the intrinsic stellar line profile (e.g. Buta and Smith, 1979; Lee et al., 1992). Effects induced by local surface gravity variations are usually neglected. In rotating stars, the pulsational velocity field cannot be described by a single spherical harmonic. In the case of slow rotation toroidal terms are induced due to the Coriolis force (e.g. Aerts and Waelkens, 1993); when the pulsation frequency in the corotating frame, $\omega_{\rm cor}$, is comparable to the rotation frequency, Ω, a whole series of spherical harmonic functions is needed to describe an eigenfunction (Lee and Saio, 1990; Townsend, 1997).

A further complication in the interpretation of line-profile variations is that the inclination angle of the star, i, is usually not known. Other parameters that play a role in the appearance of the line-profile variations are: the limb-darkening coefficient, the intrinsic profile width, the rotational velocity, the ratio of horizontal to vertical pulsational motion, the amplitude of the local brightness variations, the amplitude of the local equivalent width variations, and the non-adiabatic phase lag between velocity and temperature variations. Not all of these are free parameters: some parameters depend on others, and some can be modelled with spectral synthesis codes and non-adiabatic pulsation codes.

To get reliable values for the parameters that one wants to obtain from high-resolution spectroscopy, the degree ℓ, the azimuthal number m, and the pulsational velocity amplitude, one has to perform a careful analysis on data with high signal-to-noise ratio (S/N). Depending on the line-profile shape, the degree of the mode, and the identification method used, a spectral resolution of R = 20000–60000 is necessary. Doppler imaging and profile-fitting methods used for high-ℓ modes require the highest resolution in that range. In the ideal data set one-day aliasing does not play a role, and all the intrinsic frequencies as well as the harmonic, sum and beat frequencies, that are caused by the non-linear mapping of the stellar variations onto the line profiles, are unambiguously resolved. In reality, most data sets are far from ideal, but experience has shown that from the best data we are able to estimate values of ℓ and $|m|$ with an accuracy of ± 1 or better. In the following sections I will review some of the most used spectral analysis and mode-ID techniques. Excellent reviews on spectroscopic mode-ID were given by Aerts and Eyer (2000), Balona (2000a), and Mantegazza (2000).

2. Some Recent Applications

In the recently published literature, spectroscopic mode-IDs have been reported for different kinds of pulsating stars. Examples are listed below.

- *β Cep stars:* Evidence for pulsations in double-lined spectroscopic binaries was presented for κ Sco (Uytterhoeven et al., 2001) and ψ^2 Ori (Telting et al., 2001). The runaway star and prototypical line-profile variable ζ Oph was studied by Jankov et al. (2000), and another prototypical line-profile variable ϵ Per by Saio et al. (2000). Schrijvers & Telting (2002) identified modes in the close binary ν Cen.
- *Be stars:* Besides variability caused by circumstellar matter, pulsational line-profile variations have been detected for ω Ori (e.g. Neiner et al., 2002). There is still a debate whether line-profile variations in μ Cen are due to $\ell=2$ and $\ell=3$ gravity modes (Rivinius et al., 2001), or due to circumstellar activity (Balona et al., 2001a).
- *SPB stars:* Spectroscopic campaigns are ongoing, in order to obtain sufficient data for mode-ID. First results were published by Mathias et al. (2001) showing line-profile variations in candidate SPBs.
- *δ Scuti stars:* The importance of simultaneous photometric and high-resolution spectroscopic measurements for reducing the ambiguities in mode-ID was demonstrated by Koen et al. (2002; QQ Tel), Balona et al. (2001b: 1 Mon), Mantegazza et al. (2001: BV Cir; 2000: X Cae). Mode-IDs based on spectroscopic measurements only were presented by Chadid et al. (2001: 20 CVn) and Balona (2000b: o^1 Eri).

- *roAp stars:* Kochukhov and Ryabchikova (2001) used 1.5 hours of spectroscopic observations to identify non-radial modes in γ Equ.
- *γ Dor stars:* Aerts and Kaye (2001) spectroscopically identified the non-radial mode in the mono-periodic star HD207223.
- *Solar-like oscillations:* Pressure-mode oscillations have been detected with amplitudes in the order of 0.5 m/s in β Hyi (Carrier et al., 2001; Bedding et al., 2001) and α Cen A (Bouchy and Carrier, 2001).
- *sdB stars:* Using cross-correlation of Balmer lines, Woolf et al. (2002) investigated if the moments of correlation profiles of PG1605+072 can be used for mode-ID. They found that variations in the higher moments (variance, skewness) of these broad profiles are difficult to detect, whereas the radial velocity as derived from the line centers varies with an amplitude in the order of 10km/s (see also O'Toole et al., 2000).

3. Combining Different Spectral Lines

Spectroscopic mode-ID requires data of very high quality. The line-profile intensity variations caused by stellar pulsation are often on the order of 1% of the continuum or smaller. In order to detect and monitor such variations, continuum S/N values higher than 300 per wavelength bin are required. Another problem in the analysis of line-profile variations is line blending. The variations on the stellar surface are mapped onto the full stretch of the line profile; if lines are blended, the resulting profile variations can be very confusing and difficult to analyse. For many types of stars unblended absorption lines are hard to find. To overcome the problems of limited S/N and line-blending, techniques have been developed to merge the information of all stellar absorption lines of the recorded spectrum into a single representative profile.

Kennelly et al. (1998) introduced the method of Doppler deconvolution (DD) in their analysis of the δ Scuti star τ Peg. They use a synthetic spectrum to represent the intrinsic stellar spectrum, which is then convolved with a time-dependent broadening profile. For each observed spectrum the broadening profile represents the average Doppler broadening of all (blended) absorption lines. It accounts for Doppler shifts induced by stellar rotation and pulsation, and serves as a high-S/N line profile for mode-ID (see Figure 2). A similar result can be obtained by cross correlating the spectra of a data set. Here, the time-variable cross-correlation function (CCF) serves as an unblended profile with high S/N, which can be used for mode-ID.

Both methods assume that all profiles in the spectrum show the same temporal behaviour, which means that the differences between the equivalent-width variations of the individual lines are neglected. It has not been extensively tested whether a mode-ID based on DD broadening profiles or CCF profiles gives the same results as for single unprocessed unblended line profiles of excellent quality.

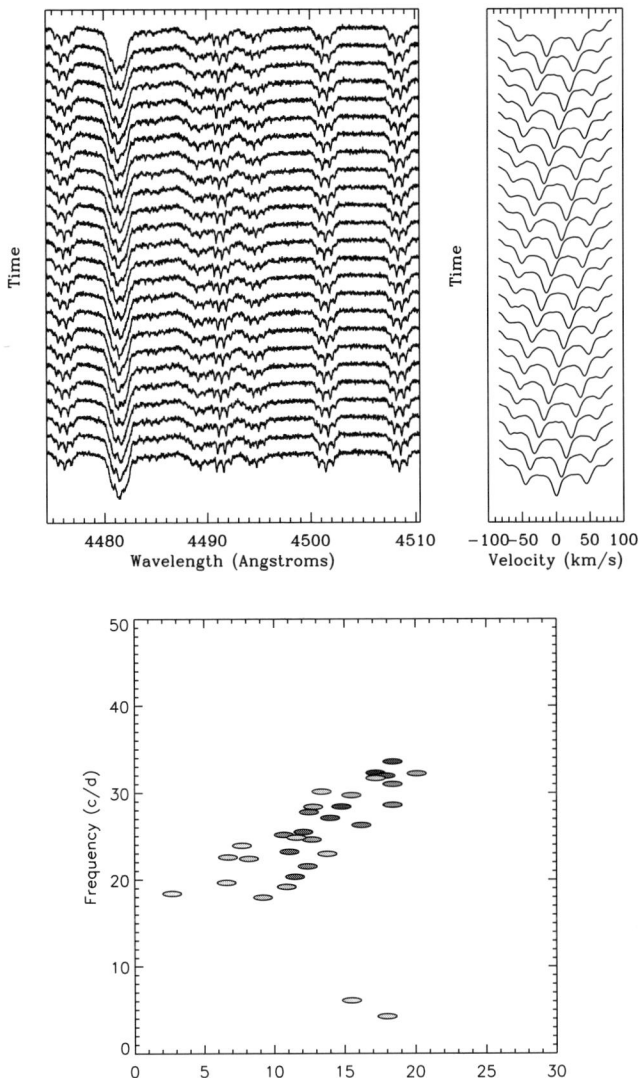

Figure 2. Broadening functions are derived from blended line spectra (Doppler Deconvolution), and are subsequently subject to a two-dimensional transform into frequency and ℓ space (Fourier Doppler Imaging), revealing many non-radial modes in the δ Scuti star τ Peg. Taken from Kennelly et al. (1998).

Chadid et al. (2001) showed that the CCF profiles, obtained by correlating with a template star, are significantly wider than an unblended profile in the original spectra. For this reason they had to adapt the mode-ID method they used (the moment method, see below), which shows that the DD and CCF methods should be used with care.

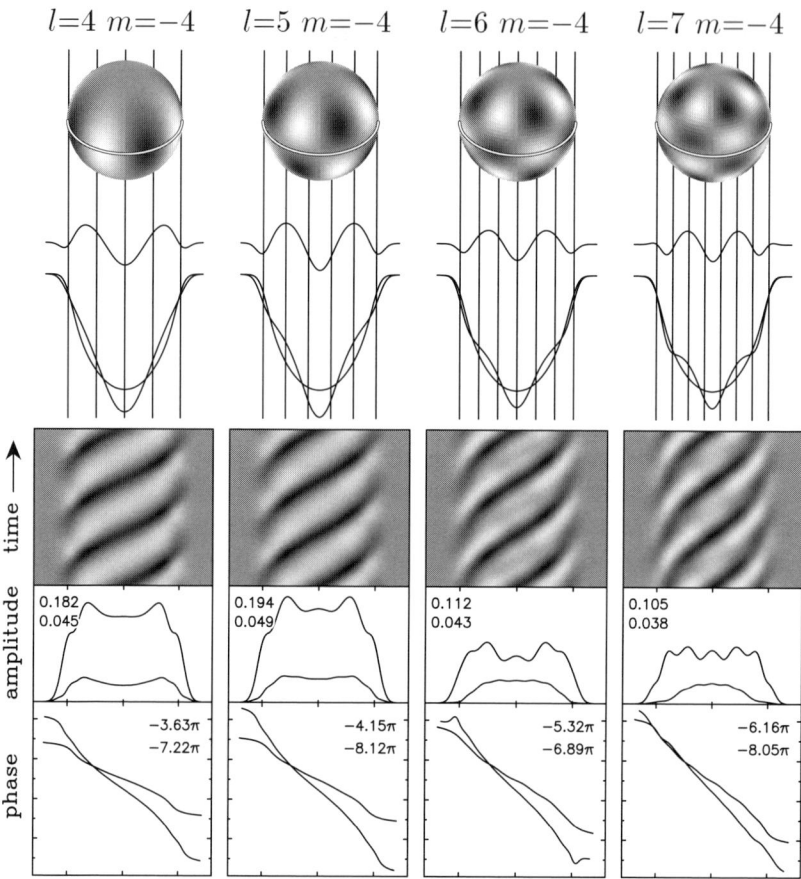

Figure 3. The IPS method. The variations are Fourier transformed for each position in the line profile, forming amplitude and phase diagrams. For high rotational broadening (Doppler imaging) there is a direct relation between phase and ℓ (Telting and Schrijvers, 1997). For increasing ℓ the slope of the phase diagram steepens, whereas the slope of the harmonic phase diagram, which is related to $|m|$, stays constant.

4. Frequency Considerations

In general, it is of no use trying to identify modes from line-profile variations, if the full intrinsic frequency spectrum is not known. In this paper I will not discuss the difficult task of determining the intrinsic frequencies responsible for the line-profile variations. Once the frequencies are known, or once good candidate frequencies have been derived, mode-ID can be attempted for each frequency individually. It is important to realize, however, that the non-linear mapping of the stellar variations onto the line profiles, introduces beating between signals at different frequencies. Even for linear pulsations, such beating will be present in the data set of line profiles (see e.g. Telting and Schrijvers, 1997, their Figure 11). Because of such beating, mode-ID in multiperiodic stars can in some cases not be done for each frequency individually: one has to take into account the effects of the simultaneous precence of other modes, independent of the identification method that is used.

For multiperiodic stars the radial velocity variations are often used for pulsation frequency determinations, as the pulsation velocities sum linear into the radial velocity curve. However, when temperature effects (pulsational variations in brightness and equivalent width) become important the pulsational velocities do not add linearly into the radial velocity curve anymore, such that sum and beat frequencies become apparent, even for linear (sinusoidal) pulsations.

The three most used methods for obtaining mode-ID from line-profile variations are discussed below. It should be noted that none of the methods give an immediate answer to the sign of the azimuthal number m, which determines if a mode is prograde ($m<0$) or retrograde ($m>0$) with respect to the stellar rotation. In many cases the sign can be derived by considering the relation between observed and corotating pulsation frequency and the rotation frequency $\pm\omega_{obs} = \omega_{cor} - m\Omega$, where the minus sign in the lefthand term comes in for retrograde modes that appear prograde in the observers' frame.

5. Mode-Identification Methods

5.1. MOMENT METHOD

In this method the statistical moments of the line profile $P(v)$ are analysed as a function of time (Balona, 1986). Using the definition of the moments given by Aerts (1996), M_1 is equal to the centroid or radial velocity, the second moment relates to the variance of the profile as $M_2 = var(P(v)) + M_1^2$, and the third moment relates to the skewness of the profile as $M_3 = skew(P(v)) + 3M_1 var(P(v)) + M_1^3 = skew(P(v)) + 3M_1 M_2 - 2M_1^3$. The observed moments are compared with theoretically calculated moment variations, using an elaborate weighing function.

The moment method is more sensitive to zonal ($m = 0$) modes than to sectoral modes ($|m|=\ell$). Moment variations are only detectable for $\ell \lesssim 4$. As this method does not rely on Doppler imaging, it can be used for stars with very low rotation rates. The method derives all pulsation parameters, including ℓ and $|m|$ with an accuracy of ± 1. As the method typically returns a few modes that fit the moment variations equally well, confrontation of the results with the observed line-profile variations provides an often necessary confirmation.

A new version of the moment method, optimized for multiperiodic stars, is presented by Briquet and Aerts (these proceedings). Currently, the moment method does not account for effects of the Coriolis force on the shape of the eigenfunctions, nor for brightness and equivalent width variations. Review papers considering the moment method can be found in Aerts and Eyer (2000) and De Ridder (these proceedings).

5.2. DOPPLER IMAGING

The Doppler-imaging (DI) method is possibly the most popular of the mode-ID methods, as some of its applications require no modelling at all in order to derive the modal degree ℓ for mono- and multi-periodic stars. The DI method comes in many flavours, and is complementary to the moment method, as it is sensitive to sectoral modes ($|m|=\ell$) in particular. Whereas the moment method analyses variations in velocity space, the DI methods analyse intensity variations as a function of position in the line profile. DI can detect values of ℓ as high as $\sim$20 if the rotational broadening is large enough.

5.2.1. *Intensity Period Search (IPS): amplitude and phase diagrams*

With this method, introduced by Gies and Kullavanijaya (1988) and further developed by Telting and Schrijvers (1997), the variations are separated in frequency for each position in the line profile. For each frequency the amplitude and phase of the variations are plotted as a function of position in the profile. The full set of amplitude and phase diagrams form a complete description of the profile variations in Fourier space.

The power of the method lies in the fact that whereas the difficult-to-interpret amplitude diagrams contain information about many stellar and pulsational parameters, the phase diagrams contain mostly information about ℓ and $|m|$ (see Figure 3). Telting and Schrijvers showed from Monte-Carlo simulations that it is easy to derive ℓ with accuracy ± 1 from the phase diagram without having to model the pulsations, and that the phase diagram of the first harmonic is related to $|m|$. The linear relationship between ℓ and the phase difference of the blue and red line wings was explained by Hao (1998) and Montgomery (2002).

Telting et al. (1997) have shown that the amplitude and phase diagrams can also be used for mode-ID for stars that do not have rotationally broadened profiles, by fitting pulsation models to the diagrams for each frequency present in the data (see Figure 4).

5.2.2. *Fourier Doppler imaging*

Kennelly et al. (1992, 1998) have taken the technique of Gies and Kullavanijaya (IPS) one step further: the temporal variations and the line profiles themselves are transformed into frequency and ℓ space by means of a two-dimensional Fourier transform (see Figure 2). The advantage over the IPS method is that modes having the same frequency can still be resolved in ℓ space if $\Delta\ell \gtrsim 4$.

Unlike the IPS method, the Fourier Doppler imaging (FDI) method has never been calibrated in detail. Kennelly et al. (1998) have shown by simulations that in the case of the δ Scuti star τ Peg low values of ℓ are overestimated by 2, and that for high values of ℓ FDI overestimates ℓ by about 1. For the general case, as proper calibrations are not available yet, I estimate the accuracy of the retrieval of ℓ as ± 2 for FDI.

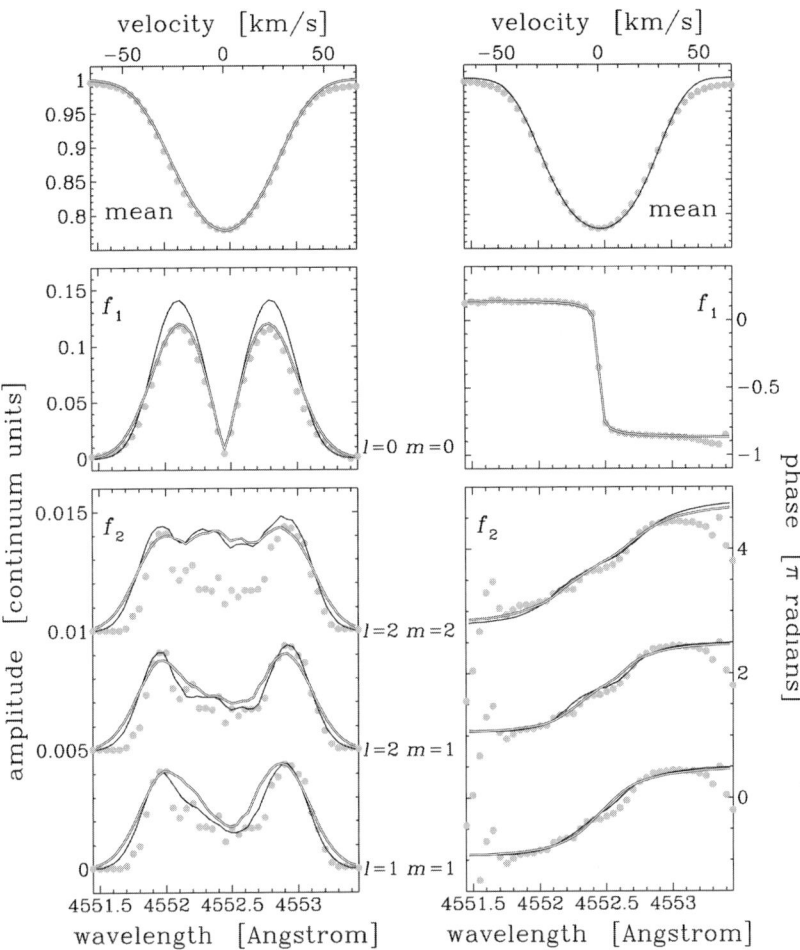

Figure 4. Fits to IPS amplitude and phase diagrams provide mode identification, even for slow rotators (Telting et al., 1997): for the non-radial mode in β Cep these fits point to $m=1$, $\ell=1$ or $\ell=2$, consistent with the results of the moment method.

5.2.3. *Doppler imaging: image reconstruction, surface mapping*

Berdyugina et al. (2000) showed that surface structure modelling through Doppler image reconstruction is a useful aid in mode-ID. Although the modelling of the profile variations only accounts for surface temperature variations, and neglects the major contribution of the pulsational velocity field, the method results in surface maps that can readily be interpreted in terms of ℓ and $|m|$ (see Figure 5). Note that for surface mapping the modelling does not assume the surface-feature distribution to be in the form of spherical harmonics or any other presupposed shape.

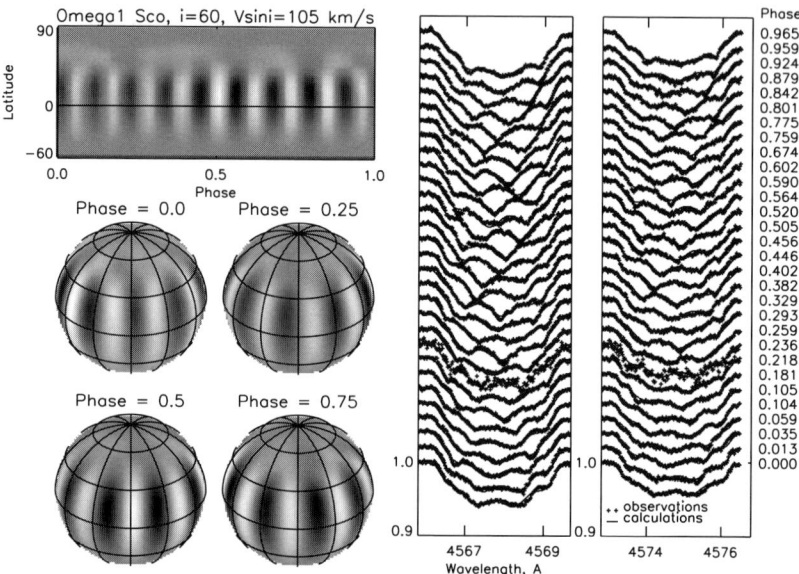

Figure 5. Doppler image reconstruction for the β Cephei pulsations in ω^1 Sco (Berdyugina et al., 2000). In this surface mapping method, line profiles are modelled with the temperature in many surface grid points as free parameter.

5.3. LINE-PROFILE FITS

Line-profile fits provide the ultimate test to confirm the results of the moment and Doppler imaging methods. For the profile fits one generates spectra using a model that accounts for the pulsational velocity field and temperature perturbations on the stellar surface. The complexity of the profile-fitting method lies in the fact that besides all pulsational parameters also many stellar parameters relating to the actual shape of the line profile have to be tweaked, whereas this is not necessarily the case for the moment and Doppler imaging methods. The profile fits provide values for all important pulsation parameters.

As profile fitting can be computationally intensive, the observed line profiles are often folded in only several phase bins for each apparent frequency, such that the variations at other frequencies average out. Then only few profiles have to be fitted instead of few hundreds or more. Phase binning is equivalent to extending the exposure times of the spectra to a significant part of the pulsation period. This gives rise to phase smearing and will have an impact on the results of the mode-ID, and on the ambiguity of the results.

Review papers on line-profile fits can be found in Mantegazza (2000), Balona (2000a, these proceedings). Balona has found that in many cases the accuracy of the derived values of ℓ and $|m|$ is not better than ± 1.

5.4. OTHER METHODS

- *Spectroscopic nonadiabatic observables* (Cugier and Daszynska, 2001): this model-dependent method can complement conventional techniques that do not specifically account for nonadiabatic effects.
- *Radial velocity and line-bisector analysis* (Hatzes and Cochran, 2000): applied primarily to pulsations in giants, this method can be useful for stars were higher velocity moments cannot be accurately determined.
- *Equivalent width analysis* (Viskum et al., 1998; Balona, 2000c): Balona concluded that in general the information contained in equivalent width variations of Balmer lines is similar to that of multi-colour photometry.
- *Spatial wavelet analysis* (Townsend, 1999): similar to FDI, but does not take temporal variations into account. May be used to determine the degree ℓ of a dominant pulsation mode from very few spectra.
- *Merged line profiles* (Hao, 2000): offers a one-dimensional form of FDI, which enables to derive estimates for ℓ from very few spectra.
- *Interferometric Fourier Doppler imaging* (Jankov et al., 2001): this extension to the Doppler imaging technique adds significantly to the largely azimuthal information obtained through high-resolution spectroscopy of rotating stars, as the interferometric measurements will provide independent latitudinal information as well.

6. Concluding Remarks

Mode identification requires data of very high quality. Only from the best data we can identify ℓ and $|m|$ with accuracy ± 1 or better. As some mode-ID methods are sensitive to different modes than others, complementary use of different methods gives the most conclusive results.

It has been shown that for mode-ID of δ Scuti stars it is very valuable to get simultaneous multi-colour photometry and high-resolution spectroscopy. However, for δ Scuti stars and other types of stars there are many spectroscopically detected frequencies that are due to modes with such high ℓ values ($\ell \gtrsim 3$) that they have no counterpart in ground-based photometric data. Especially in the case of (rotating) β Cep stars it is necessary to obtain high-resolution spectroscopic data together with space based photometry for accurate frequency determination. Once the exact frequency spectrum is known, spectroscopic mode-ID will be easier as aliasing is not a problem anymore.

References

Aerts, C.: 1996, *A&A* **314**, 115.
Aerts, C., Mathias, P., Gillet, D. and Waelkens, C.: 1994, *A&A* **286**, 109.

Aerts, C. and Waelkens, C.: 1993, *A&A* **273**, 135.
Aerts, C. and Eyer, L.: 2000, *A.S.P. Conf. Ser.* **210**, 113.
Aerts, C. and Kaye, A.B.: 2001, *ApJ* **553**, 814.
Balona, L.A.: 1986, *MNRAS* **219**, 111.
Balona, L.A.: 2000a, *A.S.P. Conf. Ser.* **210**, 170.
Balona, L.A.: 2000b, *MNRAS* **318**, 289.
Balona, L.A.: 2000c, *MNRAS* **319**, 606.
Balona, L.A. et al.: 2001a, *MNRAS* **324**, 1041.
Balona, L.A., Bartlett, B. and Caldwell, J.A.R.: 2001b, *MNRAS* **321**, 239.
Bedding, T.R., Butler, R.P. and Kjeldsen, H.: 2001, *ApJ* **549**, L105.
Berdyugina, S. et al.:2000, *A.S.P. Conf. Ser.* **214**, 268.
Bouchy, F. and Carrier, F.: 2001, *A&A* **374**, L5.
Buta, R.J. and Smith, M.A.: 1979, *ApJ* **232**, 213.
Carrier, F., Bouchy, F., Kienzle, F. et al.: 2001, *A&A* **378**, 142.
Chadid, M., De Ridder, J., Aerts, C. and Mathias, P.: 2001, *A&A* **375**, 113.
Cugier, H. and Daszynska, J.: 2001, *A&A* **377**, 113.
Gies, D.R. and Kullavanijaya, A.: 1988, *ApJ* **326**, 813.
Hao, J.: 1998, *ApJ* **500**, 440.
Hao, J.: 2000, *Ap&SS* **271**, 145.
Hatzes, A.P. and Cochran, W.D.: 2000, *AJ* **120**, 979.
Jankov, S., Janot-Pacheco, E. and Leister, N.V.: 2000, *ApJ* **540**, 535.
Jankov, S. et al.: 2001, *A&A* **377**, 721.
Kennelly, E.J., Walker, G.A.H. and Merryfield, W.J.: 1992, *ApJ* **400**, L71.
Kennelly, E.J., Brown, T.M., Kotak, R. et al.: 1998, *ApJ* **495**, 440.
Kochukhov, O. and Ryabchikova, T.: 2001, *A&A* **374**, 615.
Koen, C., Balona, L., van Wyk, F. et al.: 2002, *MNRAS* **330**, 567.
Ledoux, P.: 1951, *ApJ* **114**, 373.
Lee, U. and Saio, H.: 1990, *ApJ* **349**, 570.
Lee, U., Jeffery, C.S. and Saio, H.: 1992, *MNRAS* **254**, 185.
Mantegazza, L.: 2000, *A.S.P. Conf. Ser.* **210**, 138.
Mantegazza, L., Zerbi, F.M. and Sacchi, A.: 2000, *A&A* **354**, 112.
Mantegazza, L., Poretti, E. and Zerbi, F.M.: 2001, *A&A* **366**, 547.
Mathias, P., Aerts, C., Briquet, M. et al.: 2001, *A&A* **379**, 905.
Montgomery, M.H.: 2002, *A.S.P. Conf. Ser.* **259**, 216.
Neiner, C., Hubert, A.-M., Floquet, M. et al.: 2002, *A&A* **388**, 899.
O'Toole, S.J., Bedding, T.R. and Kjeldsen, H.: 2000, *ApJ* **537**, L53.
Osaki, Y.: 1971, *PASJ* **23**, 485.
Rivinius, Th., Baade, D., Stefl, S. et al.: 2001, *A&A* **369**, 1058.
Saio, H., Kambe, E. and Lee, U.: 2000, *ApJ* **543**, 359.
Schrijvers, C. and Telting, J.H.: 2002, *A&A* (in press).
Smith, M.A.: 1977, *ApJ* **215**, 574.
Telting, J.H. and Schrijvers, C.: 1997, *A&A* **317**, 723.
Telting, J.H., Aerts, C. and Mathias, P.: 1997, *A&A* **322**, 493.
Telting, J.H., Abbott, J.B. and Schrijvers, C.: 2001, *A&A* **377**, 104.
Townsend, R.H.D.: 1997, *MNRAS* **284**, 839.
Townsend, R.H.D.: 1999, *MNRAS* **310**, 851.
Uytterhoeven, K., Aerts, C., De Cat, P. et al.: 2001, *A&A* **371**, 1035.
Viskum, M., Kjeldsen, H. and Bedding, T.R.: 1998, *A&A* **335**, 549.
Vogt, S.S. and Penrod, G.D.: 1983, *ApJ* **275**, 661.
Woolf, V.M., Jeffery, C.S. and Pollacco, D.L.: 2002, *MNRAS* **329**, 497.

THEORETICAL CLUES FOR MODE IDENTIFICATION - INSTABILITY RANGES AND ROTATIONAL SPLITTING PATTERNS

A.A. PAMYATNYKH

Copernicus Astronomical Center, Bartycka 18, 00-716 Warsaw, Poland
Institute of Astronomy, Russian Academy of Sciences, Pyatnitskaya Str. 48,
109017 Moscow, Russia
Institute of Astronomy, University of Vienna, Türkenschanzstr. 17,
A-1180, Vienna, Austria

Abstract. Frequency spectra of unstable modes for typical models of β Cephei, SPB and δ Scuti variables are presented. Comparison of theoretical and observed instability ranges allows to constrain possible parameters of the stellar models. As an example, models of θ^2 Tau are considered. The main uncertainties in the determination of the theoretical frequency ranges for δ Sct variables are due to unsatisfactory treatment of convection. The structure of rotationally split multiplets for the low-order modes excited in δ Sct stars is discussed. The rotational coupling between close modes of spherical harmonic degree, ℓ, differing by 2, can significantly disturb the frequency spectrum.

Keywords: stellar oscillations, instability ranges, rotational splitting

1. Introduction

In this review I will discuss some properties of the theoretical frequency spectra of different main-sequence variables like β Cep, SPB and δ Sct stars. The main goal is to show typical regularities and non-regularities in these spectra, if we take into account effects of stellar rotation. The numerical results were obtained in Wojtek Dziembowski's group in Warsaw and in Mike Breger's group in Vienna. The most important effects of rotation on the frequencies were studied during the last ten years by Wojtek Dziembowski, Philip Goode, Marie-Jo Goupil, Hideyuki Saio and their collaborators who created corresponding codes for linear analysis of radial and nonradial oscillations (Dziembowski and Goode, 1992; Soufi et al., 1998; Goupil et al., 2000; Saio, 2002). Basic theoretical aspects of stellar pulsation, including effects of rotation on the oscillation frequencies, were summarized recently by Christensen-Dalsgaard and Dziembowski (2000).

This paper is a continuation of reviews on pulsational instability domains in the upper part of the main sequence – namely, on β Cep and SPB instability domains (Pamyatnykh, 1999; Paper I), and on the δ Sct instability domain (Pamyatnykh, 2000; Paper II). In previous papers we considered the position of unstable models in the HR diagram and studied the effects of variations of different stellar parameters (chemical composition, opacity, convection, overshooting from the convective core, rotation) on the position of the instability domains in the HR diagram.

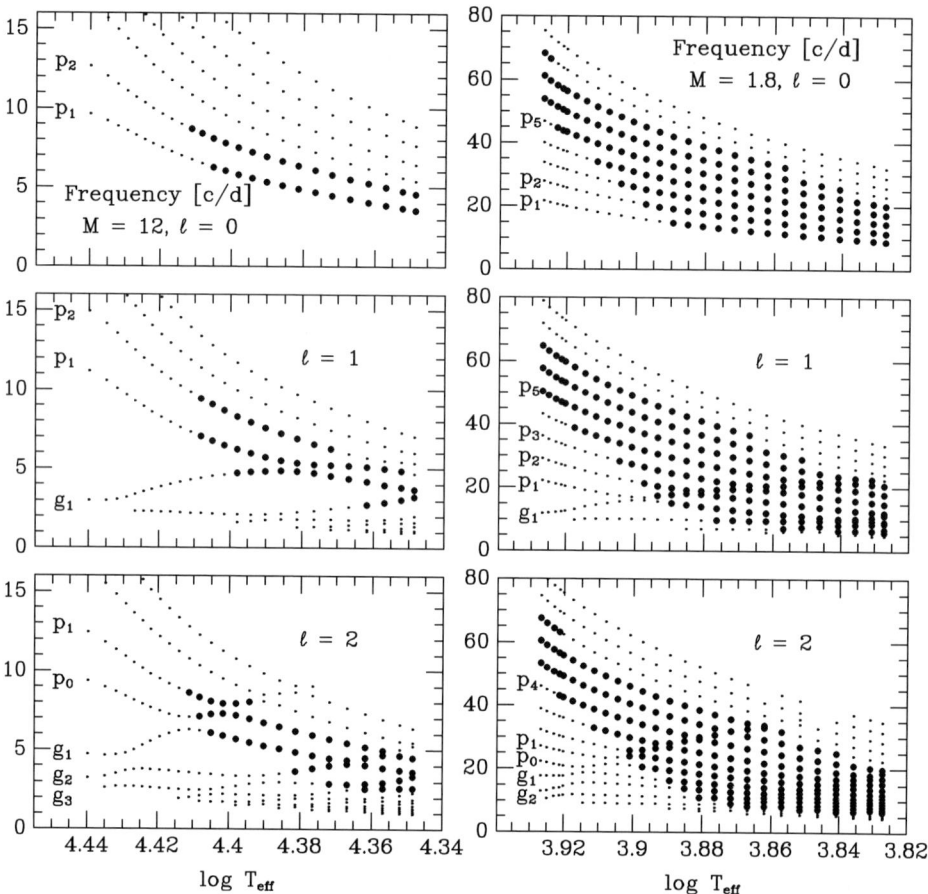

Figure 1. Frequencies of low-order p- and g-modes with low degree, ℓ, for models of 12 and 1.8 $M_\odot$ in the main sequence evolutionary phase. In each panel, the leftmost and rightmost points correspond to the ZAMS and TAMS models, respectively. The large dots mark unstable modes. According to Pamyatnykh (2000).

In Section 2 we briefly discuss the theoretical frequency spectra of unstable modes for β Cep, SPB and δ Sct variables. In Section 3 we give an example of model constraints from the comparison of observed and theoretical frequency ranges. In Section 4 we show and discuss the structure of rotationally split multiplets in typical models of a δ Sct-type variable, and in the last section we outline some problems.

2. Structure of the Frequency Spectra of Low-Degree Modes

Some properties of the pulsations within the β Cep and δ Sct instability domains are given in Figure 1, where the frequency oscillation spectra for stellar models of

12 $M_\odot$ and 1.8 $M_\odot$ during their evolution from the ZAMS to the TAMS are plotted. The oscillations of both types of variables are similar in many aspects because in both cases low-order acoustic and gravity modes are excited by the same classical κ–mechanism. The main difference is that the oscillations of the β Cep and δ Sct variables rely on different opacity bumps (see Paper II). For radial modes ($\ell = 0$) we see an almost equidistant frequency separation between consecutive modes. The complicated patterns of nonradial modes are caused by evolutionary changes in the stellar interiors, in the region surrounding the convective core. Due to these changes, the p- and g-modes are not separated in frequency already in mid- or early-MS evolution, and the phenomenon of 'avoided crossing' between p- and g-modes takes place (Aizenman et al., 1977). This results in a mixed character of the low-order nonradial modes: they are similar to pure acoustic modes in the outer stellar layers and to pure gravity modes in the interior. Both in the β Cep and δ Sct star models we find unstable low-order, low-degree p-, g- and mixed modes. The frequency range of the unstable modes in the 1.8 $M_\odot$ models is more extended than that in the 12 $M_\odot$ models. This is in agreement with the fact that the observed frequency range of the unstable modes is wider in δ Sct than in β Cep stars. For a more detailed discussion of these frequency spectra see Paper II.

In Figure 2 the periods of unstable low-degree modes of an evolutionary sequence of typical SPB models are plotted. This figure is similar to Figure 8 in Dziembowski et al. (1993) and to Figure 4 in Dziembowski (1995), but the newest data on stellar opacity were used (OPAL data of 1996, see Paper I for details). The excited oscillations here are high-order gravity modes which can be accurately described by the asymptotic theory. In each model of a given effective temperature a large number of modes is excited simultaneously. These modes are nearly equidistant in period. For a more detailed discussion of theoretical spectra of the SPB models see Dziembowski et al. (1993) and Paper I.

3. Model Constraints from Comparison of Theoretical and Observed Frequency Ranges

Detailed quantitative fitting of observed frequencies of a multiperiodic variable with the theoretical frequency spectrum of an appropriate stellar model seems to be still an open issue. We don't know any successful example of such a fitting (see, for example, Goupil and Talon, 2002; where problems of asteroseismology of δ Sct stars are discussed). Also, models usually predict much larger number of unstable modes than it is observed in an individual star. For example, a model of the evolved δ Sct-type variable 4 CVn predicts approximately 500 unstable modes of low degree, $\ell < 3$, which is 25 times larger than the number of observed modes (at least 17 frequencies in the range 4.7–9.7 c/d, see the short discussion in Paper II). The mechanism of the modal selection is still unknown.

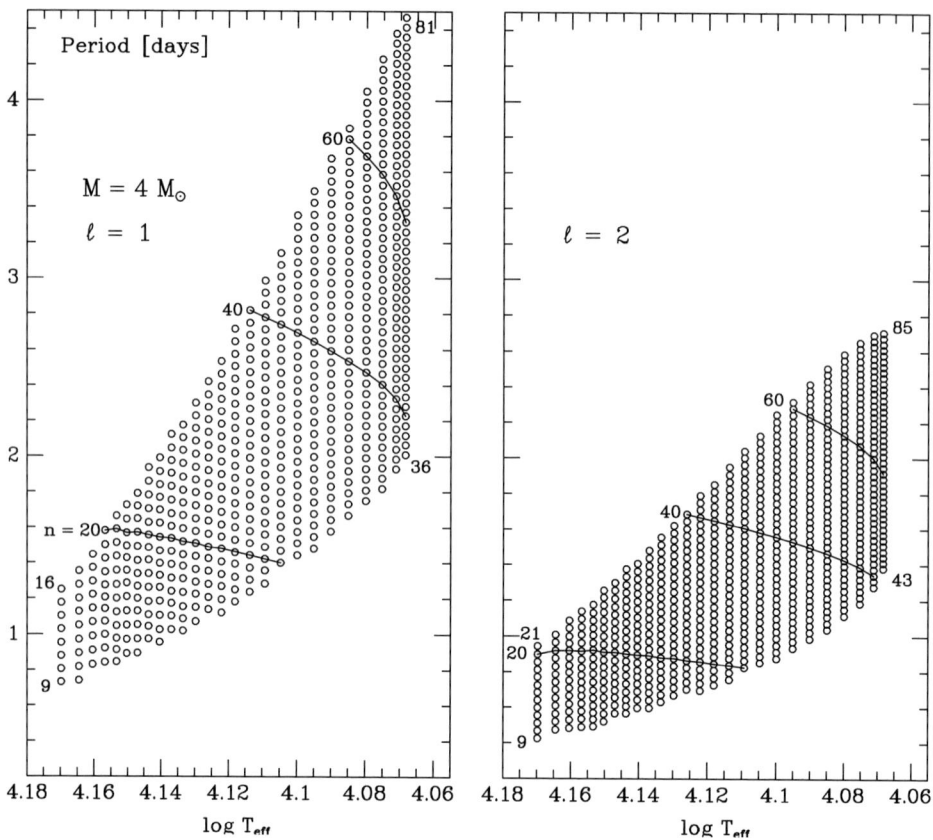

Figure 2. Periods of unstable high-order gravity modes of $\ell = 1$ and 2 in the sequence of 4 $M_\odot$ models during evolution on the main sequence. Leftmost points in each panel correspond to the ZAMS model, whereas rightmost points correspond to the TAMS model. Numbers close to some modes give corresponding radial orders. A few solid lines connect modes of the same radial order in the sequence of the models.

However, we can try to compare the frequency range as a whole with the corresponding theoretical ranges of unstable modes and to obtain some constraints on model parameters. Figure 3 shows the results of such a comparison for the primary component of θ^2 Tau, a δ Sct-type variable, in which 11 frequencies in the 10.8 to 14.6 c/d range are detected (see Breger et al., 2002; for more details). The normalized growth rates of radial and nonradial modes are plotted against frequency for nine models of different mass and effective temperature. Only axisymmetric modes ($m = 0$) are shown. The independence of the growth rate on the spherical harmonic degree, ℓ, is a typical feature of modes excited by the κ-mechanism. The best fit between the theoretical and observed frequency ranges is achieved for models with $T_{\text{eff}} \approx 7800$ K (or slightly higher), in agreement with photometric calibrations. The instability range spans two or three radial orders in the range p_4 to p_6 for radial modes.

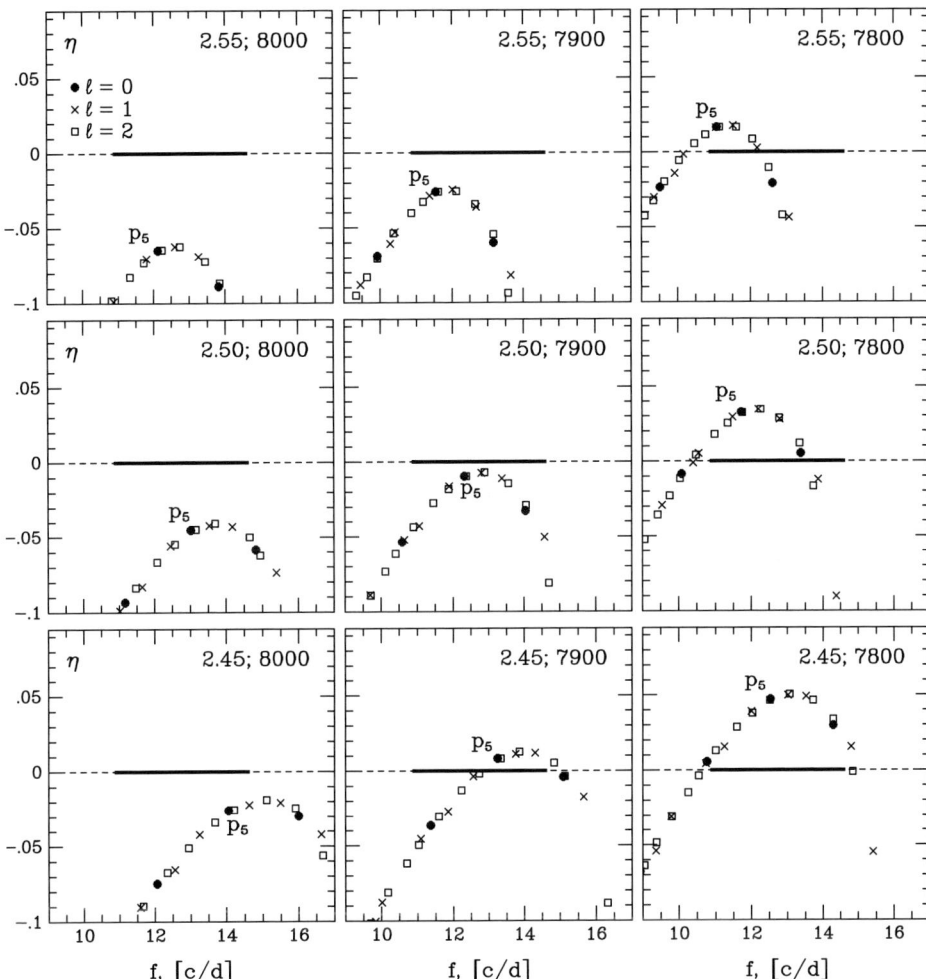

Figure 3. Normalized growth rates, η, plotted against frequency, f, in test models of the primary component of θ^2 Tau. Positive values of η correspond to unstable modes. The values of M (in solar units) and T_{eff} are given in each panel. The symbol p_5 is plotted near the corresponding radial overtone. The thick horizontal line shows the range of frequencies (10.865 to 14.615 cd^{-1}) observed in the primary component of θ^2 Tau. From Breger et al. (2002).

The main uncertainty of such a study is the unsatisfactory description of convection in the stellar envelope and its interaction with pulsations. We used the standard mixing-length theory and the assumption of frozen-in convection. This simple assumption is probably incorrect in the hydrogen convection zone. Therefore, the reality of the additional excitation in the hydrogen zone must be examined by using a nonlocal time-dependent treatment of convection. The first promising results in this direction (see Michel et al., 1999; Houdek, 2000) exist. Note that the total driving in a δ Sct star (with the main contribution due to the κ-mechanism operating in the second helium ionization zone) only slightly exceeds the total damping. Therefore even small contributions to the driving are important.

We note also the nonlocal results by Kupka and Montgomery (2002), who point out the necessity to use very different values of the mixing-length parameter in the hydrogen and helium convection zones if we still use local mixing-length theory. According to nonlocal studies of A-star envelopes, it is necessary to use a small value of this parameter in the hydrogen zone and a significantly higher value in the deeper helium zone. We performed some tests and did find a possibility to fit observations with slightly hotter models. However, we probably can exclude modes involving the second radial overtone (mode p_3) and lower-order modes, as well as all hot models of the primary with $T_{\text{eff}} > 8000$ K. The reason is that all these models are stable in the observed frequency range.

4. Rotational Splitting and Rotational Mode Coupling

To illustrate the effects of rotation on stellar oscillation frequencies, I will follow Christensen-Dalsgaard and Dziembowski (2000) and Goupil et al. (2000). The third order expression for a rotationally split frequency may be written in the form:

$$\nu_m = \nu_0 + m(1 - C_{n\ell})\frac{\Omega}{2\pi} + \frac{\Omega^2}{2\pi \nu_0}(D_0 + m^2 D_1) + m\frac{\Omega^3}{\nu_0^2}T$$

where subscript m denotes the azimuthal order of a mode. The subscripts (n, ℓ) which denote, respectively, the radial order and the degree of the mode, have been omitted. The temporal dependence of the oscillations is assumed to be $\exp(-i\omega t)$, so that prograde modes correspond to $m > 0$. The frequency ν_0 includes effects of the horizontally averaged centrifugal force in the equilibrium model. (In the computations of stellar evolution we assumed solid-body rotation and conservation of global angular momentum during evolution.) The Ledoux constant, C, determines the usual equidistant splitting valid in the limit of slow rotation. The term $m\Omega$ stands for transformation of the co-rotating coordinate system to the inertial coordinate system of the observer. The second and third order coefficients D_0, D_1 and T are determined by a perturbation method and take into account non-spherically symmetric distortion due to the centrifugal force and second and third order Coriolis effects. The quadratic terms destroy the symmetry of the multiplet and also predict a frequency shift for the radial modes and nonradial axisymmetric modes. The cubic term affects the value of $(\nu_m - \nu_{-m})/m$, which can be used to determine the rotational velocity, and may result in fictitious variation of the rotation velocity with depth, as shown by Goupil et al. (2000).

Moreover, an important additional correction to the frequency, which is not taken into account in Eq. (1), arises when rotation couples close modes of spherical harmonic degree, ℓ, differing by 2 and of the same azimuthal order, m. We will illustrate how this effect can be significant at typical velocities of rotation and for

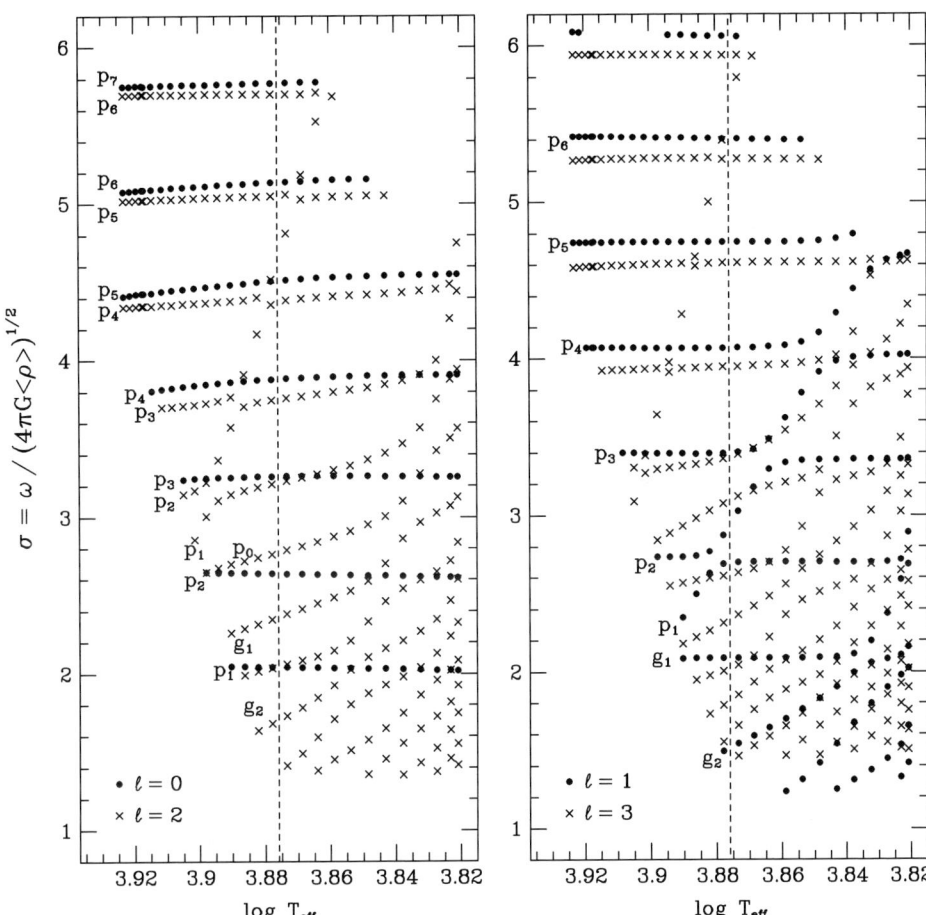

Figure 4. Evolution of frequency spectra of unstable modes with $\ell = 0, 2$ (left panel) and with $\ell = 1$, 3 (right panel) in a sequence of δ Sct models with mass of 1.8 $M_\odot$. For simplicity, only modes of $\ell = 1$ are identified in the right panel. Dotted vertical line corresponds to the model of $T_{\mathrm{eff}} = 7515$ K. Effects of rotation on frequency spectrum of this model are presented in Figure 5.

typical models of main sequence stars. A detailed discussion of the occurrence of rotational mode coupling is given by Daszyńska-Daszkiewicz et al. (2002) and we will follow this description.

In Figure 4 we show the behaviour of frequencies of unstable low-degree modes in a typical δ Sct-type model of 1.8 $M_\odot$ which evolves from the ZAMS to the TAMS. This plot is similar to Figure 1 in Daszyńska-Daszkiewicz et al. (these Proceedings) for a β Cep-type model of 12 $M_\odot$. However, the range of unstable frequencies is now much larger; it extends up to 7 radial modes. Both avoided crossing phenomenon and presence of close frequencies of modes with spherical harmonic degree, ℓ, differing by 2, are seen very clearly. The modes are designated according to the avoided-crossing principle, i.e. each mode preserves its initial designation on the ZAMS besides the fact that in the course of the evolution this mode can significantly change its properties. As mentioned by Daszyńska-Daszkiewicz et al. (2002), the physical nature of the modes – that is, the relative proportion of the contribution from acoustic and gravity propagation zones to the mode's energy – is reflected in the slope of the σ (log T_{eff}) function. Slow and rapid rises correspond to dominant acoustic and gravity mode characters, respectively.

The rotational mode coupling is important when the frequency distance between the modes becomes comparable to the rotational frequency. For our model of 1.8 $M_\odot$ we assumed uniform (solid-body) rotation with equatorial velocity of 100 km/s on the ZAMS, and we have also assumed conservation of the global angular momentum during evolution. With these assumptions the rotational velocity on the TAMS is equal to 81 km/s. For these velocities we find that the dimensionless rotational frequency is $\sigma_{\text{rot}} \equiv \Omega/\sqrt{4\pi G <\rho>} \approx 0.12 - 0.13$. It is clear from Figure 4 that the coupling must be significant in many cases.

Figure 5 illustrates all the main effects of rotation on the frequency spectrum of unstable modes in the 1.8 $M_\odot$ model with T_{eff} = 7515 K and V_{rot} = 92 km/s. The gravitational or acoustic character of the modes involved can be easily understood from Figure 4. The two upper panels in Figure 5 show the effects of rotation on low-order and higher-order modes in an extended scale. As noted by Daszyńska-Daszkiewicz et al. (2002), the coupling strength depends on mode properties. Coupling between acoustic modes is stronger than that involving one or more gravity modes. This is so because the effect of the centrifugal distortion is only important in the acoustic cavity and it increases with the mode frequency. Such a tendency is seen very clearly in the two upper panels, where the coupling is much more pronounced for higher frequencies. Also, due to centrifugal distortion, the asymmetry of the rotational splitting is larger for higher frequencies, as can be seen in the lower panel for dipole modes.

The Ledoux constant is small for acoustic modes, therefore linear splitting for these modes is due to the transformation from the co-rotating coordinate system to the inertial coordinate system of the observer. In contrast, the Ledoux constant is about 0.5 for dipole gravity modes, therefore linear splitting can be approximately two times smaller for gravity modes than for acoustic modes. Such a case takes place in our model for low-order dipole modes, $\ell = 1$, see lower panel.

The rotational splitting results in the widening of the frequency instability range. This effect is relatively more important for β Cep than for δ Sct stars due to typically higher rotational velocities and due to a smaller frequency range of unstable modes (see Figure 1).

In Figure 6 we show the effect of the rotational coupling on the period ratio of the two lowest radial modes in the same evolutionary sequence of the 1.8 $M_\odot$ models. Due to rotational coupling between radial modes and closest quadrupole modes, very large and nonregular perturbations to the period ratio occur. If rotation is fast enough, this effect must be taken into account. The effect of rotational coupling on the period ratio is, probably, not important for well-studied high-

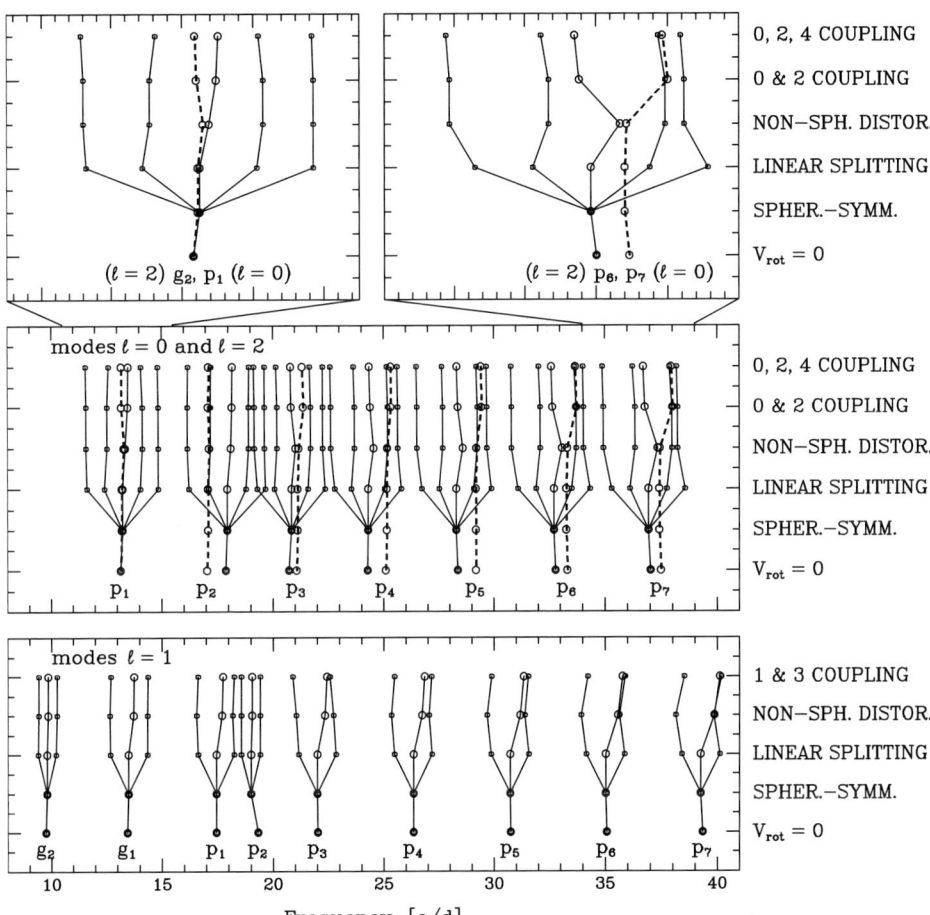

Figure 5. Effects of rotation on the frequencies of close pairs of $\ell = 0$ and 2 (upper and middle panel) and $\ell = 1$ and 3 modes (lower panel) for the 1.8 $M_\odot$ model with $T_{\rm eff}$ = 7515 K and $V_{\rm rot}$ = 92 km/s. At each step upward a new effect is added. In an analogy with Figure 11 from Christensen-Dalsgaard and Dziembowski (2000).

amplitude δ Sct-type variables like AI Vel which rotate slowly (see Petersen and Christensen-Dalsgaard, 1996). This may not be true for one of the best studied δ Sct variables, FG Vir, which has $v \sin i$ of about 20 km/s, but there are indications from spectroscopy in favour of a low inclination angle of the rotation axis, therefore true rotation velocity may be around 80 km/s (Mantegazza and Poretti, 2002).

5. Some Conclusions

We have seen that simple regular patterns in the theoretical frequency spectra are significantly disturbed – both in the spacing between modes of consecutive order

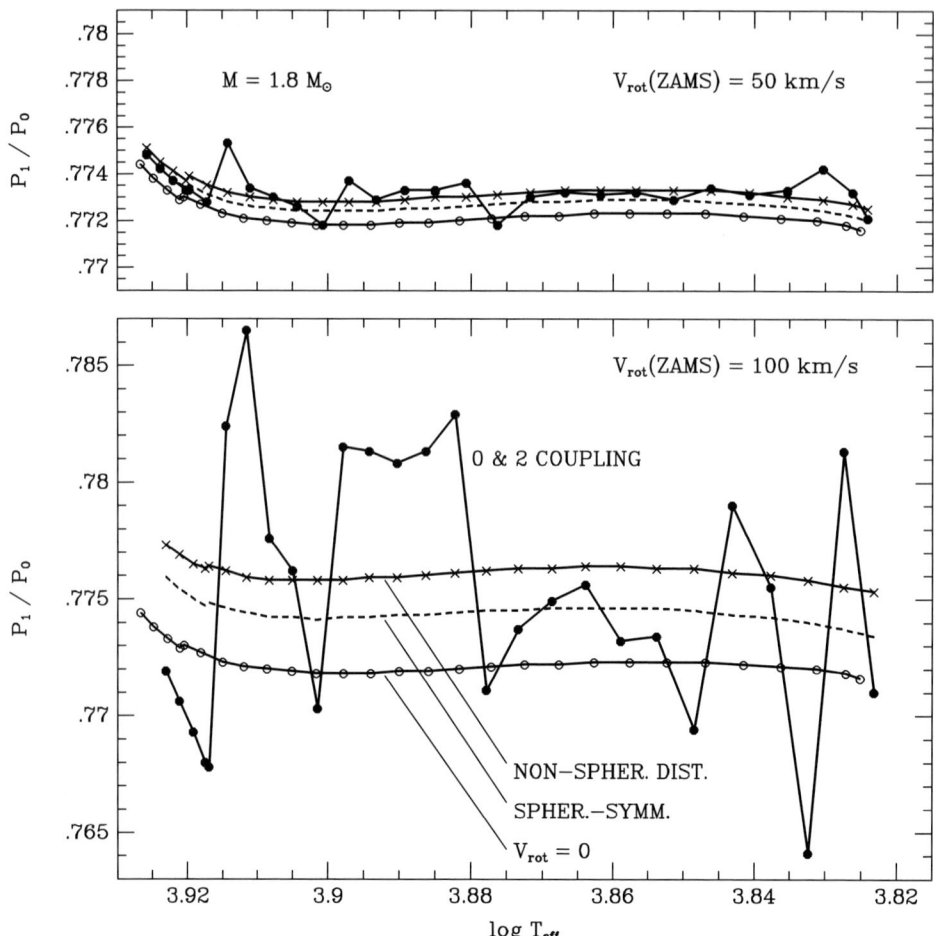

Figure 6. Variation of the period ratio of the radial first overtone to the fundamental mode during evolution of the 1.8 $M_\odot$ model from the ZAMS to the TAMS. According to Goupil et al. (in preparation).

and in the rotational splitting. For the radial order of modes, the equidistant pattern is disturbed by avoided crossing phenomenon, whereas the rotational splitting is asymmetric due to second order and higher order effects.

However, if we have a rich observed frequency spectrum it is still expedient to search for statistically significant equidistant spacings to infer information on stellar mean density (as was suggested, for example, by Handler et al., 1997; for the XX Pyx) and/or on stellar rotation.

The important effect of rotation is the coupling between close frequency modes of spherical harmonic degree, ℓ, differing by 2, and of the same azimuthal order, m. Such a coupling may occur at typical rotation velocities of stars in the upper part of the main sequence. It must be a rather often phenomenon because close frequencies of relevant modes occur over wide ranges of the frequencies and effective temperatures of the models in the instability domains.

The rotational coupling has also strong influence on photometric diagnostic diagrams (amplitude ratio versus phase difference in two passbands of multicolour photometry), as discussed by Daszyńska-Daszkiewicz in these Proceedings (the more detailed results of this study are given by Daszyńska-Daszkiewicz et al., 2002).

Acknowledgements

All numerical results presented here were obtained in the Wojtek Dziembowski's group in Warsaw and in the Mike Breger's group in Vienna. I am indebted to M.-J. Goupil for useful discussions during the Conference. It is my pleasure to thank the organizers of the Conference for the very nice scientific atmosphere and for the hospitality. The work was supported by Polish KBN grant No. 5 P03D 012 20.

References

Aizenman, M.L., Smeyers, P. and Weigert, A.: 1977, *A&A* **58**, 41.
Breger, M., Pamyatnykh, A.A., Zima, W. et al.: 2002, *MNRAS* **336**, 249.
Christensen-Dalsgaard, J. and Dziembowski, W.A.: 2000, in: C. Ibanoglu (ed.), *Variable Stars as Essential Astrophysical Tools*, Kluwer Academic Publishers, NATO Science Series, Series C, Mathematical and Physical Sciences, vol. **544**, 1.
Daszyńska-Daszkiewicz, J., Dziembowski, W.A., Pamyatnykh, A.A. and Goupil, M.-J.: 2002, *A&A* **392**, 151.
Dziembowski, W.A.: 1995, in: R.K. Ulrich, E.J. Rhodes Jr. and W. Däppen (eds.), *GONG'94: Helio- and Astero-Seismology*, A.S.P. Conf. Ser. **76**, 586.
Dziembowski, W.A. and Goode, P.R.: 1992, *ApJ* **394**, 670.
Dziembowski, W.A., Moskalik, P. and Pamyatnykh, A.A.: 1993, *MNRAS* **265**, 588.
Goupil, M.-J., Dziembowski, W.A., Pamyatnykh, A.A. and Talon, S.: 2000, in: M. Breger and M.H. Montgomery (eds.), *Delta Scuti and Related Stars*, A.S.P. Conf. Ser. **210**, 267.
Goupil, M.-J. and Talon, S.: 2002, in: C. Aerts, T.R. Bedding and J. Christensen-Dalsgaard (eds.), *Radial and Nonradial Pulsations as Probes of Stellar Physics*, A.S.P. Conf. Ser. **259**, 306.
Handler, G., Pikall, H., O'Donoghue, D. et al.: 1997, *MNRAS* **286**, 303.
Houdek, G.A.: 2000, in: M. Breger and M.H. Montgomery (eds.), *Delta Scuti and Related Stars*, A.S.P. Conf. Ser. **210**, 454.
Kupka, F. and Montgomery, M.H.: 2002, *MNRAS* **330**, L6.
Mantegazza, L. and Poretti, E.: 2002, *A&A* (in press).
Michel, E., Hernández, M.M., Houdek, G. et al.: 1999, *A&A* **342**, 153.
Pamyatnykh, A.A.: 1999, *Acta Astr.* **49**, 119 (Paper I).
Pamyatnykh, A.A.: 2000, in: M. Breger and M.H. Montgomery (eds.), *Delta Scuti and Related Stars*, A.S.P. Conf. Ser. **210**, 215 (Paper II).
Petersen, J.O. and Christensen-Dalsgaard, J.: 1996, *A&A* **312**, 463.
Saio, H.: 2002, in: C. Aerts, T.R. Bedding and J. Christensen-Dalsgaard (eds.), *Radial and Nonradial Pulsations as Probes of Stellar Physics*, A.S.P. Conf. Ser. **259**, 177.
Soufi, F., Goupil, M.-J. and Dziembowski, W.A.: 1998, *A&A* **334**, 911.

PEAK BAGGING FOR SOLAR-LIKE STARS

THIERRY APPOURCHAUX

*Research and Science Support Department, European Space Agency,
Keplerlaan 1, 2200 AG, Noordwijk, The Netherlands*

Abstract. The identification of the low-degree p modes in other stars is the challenge of future asteroseismology space missions such as COROT, MONS, MOST or Eddington. The identification is based on a priori knowledge of the characteristics of the modes. We shall review the most common assumptions needed for the identification such as basic stellar structure, visibilities, rotational splittings or linewidths. We shall describe a few tools needed for facilitating the identification. As soon as modes are properly identified, the peakbagging of the mode characteristics can be done using Maximum Likelihood Estimation. We give examples of the whole process using solar data and hare-and-hound exercises performed in the frame work of the COROT project.

Keywords: Sun, stars, seismology

1. Introduction

In the very near future, there shall be a fleet of space missions aiming at understanding the internal structure of the stars: MOST*, COROT**, MONS*** and Eddington[‡]. All these missions will observe global oscillations of the stars by measuring tiny light fluctuations; they are due to the perturbation of the star surface by the oscillations. The detection and identification of these modes of oscillation is the challenge of all these missions. This challenge is for most stars extremely difficult (e.g. Cepheids) but easier for solar-like stars. For these latter, the Sun has been and is a great aid and example.

Hereafter after having defined what is meant by solar-like stars, I explain how the current identification can be based on that of the Sun case. When the identification is achieved, peak bagging[‡‡] can be performed; the theory of which is explained here. The practice of mode identification and peak bagging is developed through

---

* Microvariability and Oscillations of Stars, a Canadian mission to be launched in April 2003 (Matthews et al., 2000).

** COnvection and ROTation, a CNES mission to be launched in Nov 2005 (Baglin and The COROT Team, 1998).

*** Measuring Oscillations in Nearby Stars, a Danish mission under study to be launched in 2005 (Kjeldsen et al., 2000).

[‡] A mission part of the 'Cosmic Vision' programme of ESA, to be launched in 2008 (Favata et al., 2000).

[‡‡] *Peak bagging* is a term coined by Jesper Schou who is a keen mountain climber. Such individuals record climbed peaks in a book and refer to it as the peak bagging list...

the use of hare-and-hound exercise. The results of such exercises carried out by the COROT Data Reduction Group are presented here.

2. What is a Solar-Like Star?

The definition of a solar-like star depends on whom you talk to. There are basically 3 'definitions' driven by:
1. *the stellar structure*: the solar-like star is similar to the Sun, i.e. with an outer convection zone and a radiative zone. Stars with a mass smaller than 1.2 $M_\odot$ satisfy the criterion as they do not have a convective core yet (Iben and Ehrman, 1962; Cox and Giuli, 1968).
2. *the excitation process of the oscillations*: the process is similar to that of the Sun, i.e. due to turbulent convection (Houdek et al., 1999). It occurs in stars with mass smaller than 2 $M_\odot$ with an outer convection zone even if it is very shallow (Houdek et al., 1999); the efficiency of the process is driven by the turbulent Mach number M_t (Houdek et al., 1999) that is maximum for 1.6 $M_\odot$. Above roughly 1.5 $M_\odot$ the stars have overstable modes; they enter the Cepheids instability strip (Houdek et al., 1999).
3. *the structure of the oscillation spectrum*: a regular spacing of the modes must be observed as predicted by asymptotic frequencies (Tassoul, 1980). The regular spacing provides the basis for a diagnostic tool: the *echelle diagramme* devised by Grec (1981) for the Sun.

I prefer the last definition because the structure of the ridges in the echelle diagramme of a solar-like star looks like those of the Sun. The ridge structure can simply be derived from the asymtotic frequency expression given by Tassoul (1980):

$$\nu_{n,l} \approx (n + \frac{l}{2} + \epsilon)\Delta\nu_0 - \frac{Al(l+1)}{n + \frac{l}{2} + \epsilon} \tag{1}$$

where n is the order of the modes, l is the degree, $\Delta\nu_0$ is the large frequency separation (or acoustic diameter of the star); and A and ϵ are constant involving integrals over the stellar model. From Eq (1), it can be noted that the even modes ($l = 0 - 2$) and odd modes ($l = 1 - 3$) have nearly the same frequency; they are only separated by the small frequency separation, i.e. δ_{02} or δ_{13} related to the second term in Eq (1). In the solar case, $\delta_{02} = 10\mu$Hz for $l = 0 - 2$ and $\delta_{13} = 13\mu$Hz for $l = 1 - 3$. It is this difference that allowed helioseimologists to identify properly the degree of the ridges. It outlines that the experience gained in helioseismology will be of considerable help for mode identification.

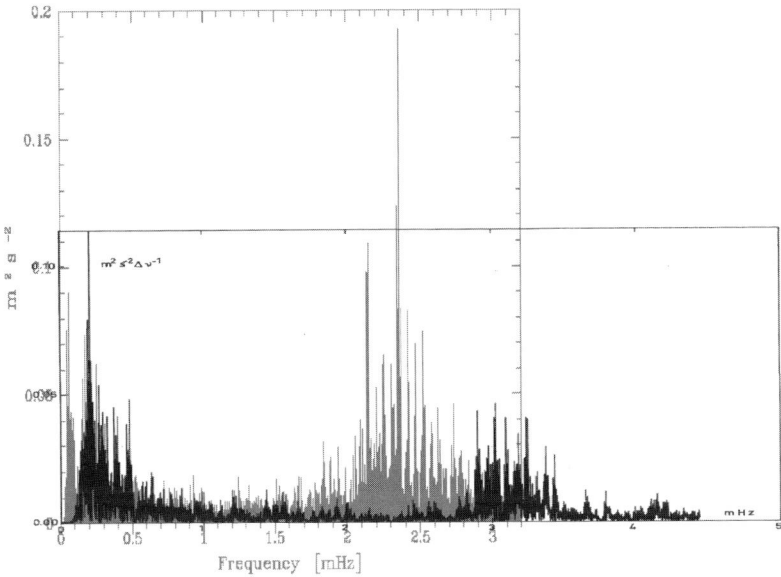

Figure 1. The power spectra of the Sun observed by Grec and Fossat (1977) (Black) and of α Cen observed by Bouchy and Carrier (2002) (Grey). Thanks to modern technology, they were put to the same scale showing that the stellar observations of today match the solar observations of the 70's. Both spectra show what can be obtained for a few days of observation stellar radial velocities. The frequency at which the modes are maximum depends upon the mass of the star (Houdek et al, 1999).

3. Past and Present Mode-Identification Methodology

3.1. IN THE EARLY AGE OF HELIOSEISMOLOGY

The first detection and identification of the solar oscillations as global modes (low degree) is attributed to Claverie et al. (1979). They identified the regular spacing based on the prediction made for low degree, low order mode frequencies by Iben and Mahaffy (1976). They had enough foresight for finding that the detected modes were higher order modes of about $n > 20$. Unfortunately, they could not identify the degree of the modes because the length of the time series did not allow for separating the odd degrees ($l = 1 - 3$) from the even degree ($l = 0 - 2$) (See above). Such an identification was performed by Grec et al. (1980) using the echelle diagramme and the properties of the small frequency separation. The length of the time series was such that they had to collapse the power along the ridges in order to increase the signal-to-noise ratio (Fossat et al., 1981). At this stage, such a technique allowed to resolve the rotational splitting (Grec et al., 1983). The identification of the order n was made by identifying the f-mode ridge in the (l, ν) diagramme obtained by making images of the Sun (Deubner, 1975).

3.2. IN THE XXIST CENTURY AND BEYOND

From helioseismology, we can identify three steps in the mode identification, all using the echelle diagramme:

1. determination of the large frequency separation (i.e. $\Delta \nu_0$)

2. degree identification and determination of the small frequency separation (i.e. δ_{02}, δ_{13})

3. rotational splitting and star inclination estimation

These steps are intimately linked to the length of the time series and the signal-to-noise ratio. A typical large frequency separation for a solar-like star ranges from 40 μHz to 170 μHz which is easily resolved by observing a star over a few hours (Audard and Provost, 1994). The small frequency separation is more difficult to retrieve as it ranges typically from 3 to 13 μHz (Audard and Provost, 1994), and requires a few days before resolving it. The visibility of the various degrees can also be used for identifying the modes in the second step. For instance the $l = 3$ modes are significantly damped by the integration over the stellar disk (Christensen-Dalsgaard and Gough, 1982 for velocity and Toutain and Goutebroze, 1993 for intensity); it renders the ($l = 0 - 2$) ridges significantly different from those of the ($l = 1 - 3$) ridges in the solar case. The rotational splitting is even more challenging: it does require a few months of observation in the case of the Sun, as its rotation period is of the order of a month.

The order identification requires more technological developments. In the not too distant future space based interferometers may provide the first low-resolution images of the stars. If it were possible to make higher-resolution images of the stars, the f-mode identification would certainly be achieved for stars; this is likely to a be dream for the XXIst century and beyond.

At the time of writing, only two stars have been observed showing unambiguously solar-like p modes: the Sun and α Cen. These two stars are the test bed of the identification process described above. Figure 1 shows what can be obtained by observing the Sun (Grec and Fossat, 1977) and α Cen (Bouchy and Carrier, 2002) over a few days. The large and small frequency separation are easily identified for both stars. As for the splitting, the time series are still too short for α Cen for revealing the splitting. Careful attempts have been made by Bouchy and Carrier (2002) for deriving the rotational splitting of the modes of α Cen but more observing time is required. The measurements of the rotational splitting for low-degree modes is rather difficult. That is only in the mid-90's, that reliable splittings could be obtained for the Sun using the theory of power spectrum fitting. It is likely that such measurements for the stellar case will require extreme care before getting some confidence in the results, but as we show hereafter the theory of power spectrum fitting is now well mastered.

4. Theory of Peak Bagging

After tagging each ridge in the echelle diagramme with a degree l, we proceed with fitting the power spectrum or performing peak bagging. The theory of power spectrum fitting for the Sun as a star is now well developed. It has been described in numerous articles and is now well understood; it is based on the use of Maximum Likelihood Estimators (MLE) (See Duvall and Harvey, 1986 and Appourchaux et al., 1998 for a review). The power spectrum is fitted using MLE that is by assuming:
- a statistics for the spectrum
- a model for the modes

The statistics of the spectrum for uninterrupted time series is a χ^2 with 2 degree of freedom (Gabriel, 1994). The model of the modes includes the mode frequency, the mode linewidth, the mode amplitude, the mode profile and the background noise. In some case, modes of different degrees overlaps ($l = 0 - 2, l = 1 - 3$), they are usually fitted together. The simultaneous fitting of these modes requires to know the visibility of the modes. For intensity this is given by Toutain and Gouttebroze (1993) while for velocity this is given by Christensen-Dalsgaard and Gough (1982). The rotational mode splitting and the inclination of the star have to be taken into account. There are up to $2l + 1$ components depending on the inclination of the star. The star inclination is taken into account depending on the way the modes are observed (velocity or intensity). The spherical harmonics eigenfunctions has to be weighted by the projection onto the line of sight or by the limb darkening. For intensity, the visibilities of the modes can simply be approximated using the decomposition given by quantum mechanics rotation matrices (Toutain and Gouttebroze, 1993).

Error bars on the parameters are derived using the inverse of the Hessian (or curvature matrix) (Appourchaux et al., 1998; Toutain and Appourchaux, 1994). The error bars derived give a good estimate of the true error bars (Appourchaux et al., 1998). The validity of such error bars can be tested using the z test as described by Chaplin et al. (1998); this is a test comparing internal and external error bars.

Finally, the significance of the fitted parameters (under the H_0 hypothesis) can be checked using the likelihood ratio test as described by Appourchaux et al. (1998). This test is rather useful for assessing the adequacy of the model fitted to the data.

5. Mode Identification and Peak Bagging in Practice: Hare-and-Hound Exercise

As coined by an anonymous scientist: 'In theory there is no difference between theory and practice; in practice there is'. The theoretical approach described above needs to be tested with *real* data. In the history of helioseismology, the *real* data have been sometimes fabricated in a such a way that it looked like a road runner chase. The term hare and hound* appeared in the GONG** Newsletter #9 of 1988 when the GONG Inversion team performed simulated inversion on data fabricated by a hare (Douglas Gough).

Within the COROT team, we found that it would be useful to have such a hare-and-hound exercise that would simulate (as well as stimulate) the mode identification and the peak bagging. The steps for this exercise are the following:
- a Team A generates theoretical mode frequencies and synthetic time series
- a Team B analyzes the time series, performs mode identification, peak fitting and structure inversion

The two teams have no access to any other information but the time series and the known characteristics of the star. Nothing else is allowed.

5.1. TEAM A: THE MAKING OF SYNTHETIC TIME SERIES

The steps for making the artificial time series are very similar to those needed for using MLE. You need to assume:
- a statistics for the (Fourier) spectrum
- a model for the modes

The statistics of each component of the Fourier spectrum is assumed to be normally distributed with a zero mean and an rms value described by the model for the modes. For the model of the modes, the following data are needed:
- the characteristics of the solar-like star and the theoretical mode frequencies (or asymptotic mode frequencies - à la Tassoul)
- the mode visibility (Christensen-Dalsgaard and Gough, 1982; Toutain and Gouttebroze, 1993)
- the excitation profile and linewidth (Houdek et al., 1999; Samadi and Goupil, 2001; Samadi et al., 2001)
- the rotational splitting and stellar inclination
- the background stellar noise derived from that of Trampedach et al. (1998) or of the Sun as in Harvey (1993)
- any other parameter (asymmetry, trick, mind twister or anything else to make it more *real*)

As soon as each Fourier component is properly modeled, an inverse Fourier transform provides the necessary time series that will be passed to the next team.

5.2. TEAM B: POWER SPECTRUM ANALYSIS

The power spectrum analysis can be done in various ways. Here I recommend to perform the steps as described in Section 3 and 4, i.e. to make an echelle diagramme

---
* In French we would rather call it: *le jeu du chat et de la souris*
** Global Oscillation Network Group

of the power spectrum, perform mode identification and then to fit the modes according to a model. When the p-mode parameters are obtained the last task of this team is to invert the mode frequencies in order to derive the stellar structure. This is beyond the scope of this talk but this is discussed by Berthomieu et al in these proceedings in the frame work of the COROT HH exercise described hereafter.

5.3. THE COROT HH EXERCISES

The Asteroseismology and Exoplanet-search mission of the French space agency (CNES) is going to be launched at the end of 2005. There are several scientific groups being involved in the preparation of the mission. The Seismology Working Group (SWG) prepares the data analysis and scientific interpretation of the seismology data. The Data reduction group of the SWG is more precisely in charge of the data analysis aspects. This group has set up three different hare-and-hound exercises[*]:
– HH#1 Validation of power spectrum fitting technique (no inversion, asymptotic frequencies)
– HH#2 Recovery of the initial stellar model for synthetic stars (full cycle as described above)
– HH#3 Choice of targets

The first exercise is over and lead to the validation of power spectrum fitting performed by different groups (Institut d'Astrophysique Spatiale, Orsay: Boumier; Observatoire de Nice: Toutain; European Space Agency: Appourchaux). The last exercise is on going. The second HH exercise is over and I report on some of the results obtained.

5.3.1. *The teams*
There were three teams involved in the process:
– *Meudon*: Observatoire de Meudon (stellar model and inversion) + Appourchaux (time series and power spectrum fitting)
– *Nice*: Observatoire de Nice (stellar model and inversion) + Toutain (time series and power spectrum fitting)
– *Queen Mary*: Queen Mary (stellar model and inversion) + Barban (time series and power spectrum fitting)

Each team produced synthetic time series and passed it on to the 2 other groups for power spectrum fitting and structure inversion.

5.3.2. *A piece-of-cake case:* Nice *synthetic data*
Figure 2 shows the echelle diagramme obtained for the *Nice* synthetic data. The time series are 150-day long and sampled at 60 sec. The frequencies fitted are obtained within about 0.1 to 0.2 μHz of the theoretical frequencies; the splittings

---

[*] The activity related to these HH exercise can be found in
http://virgo.so.estec.esa.nl/html/corot/datagroup/hh.html

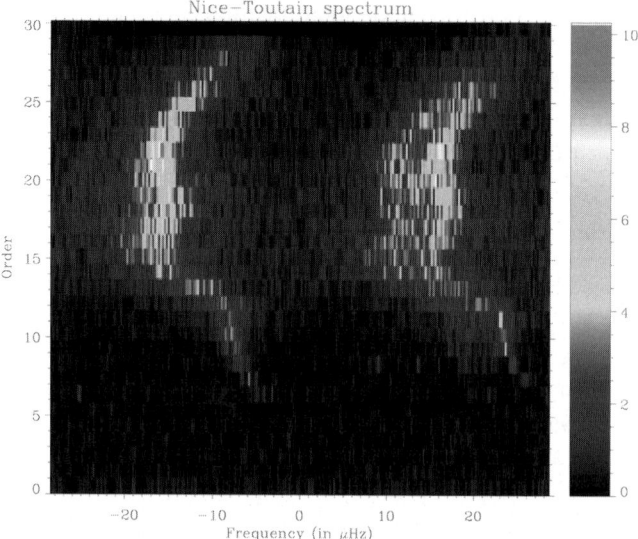

Figure 2. Echelle diagramme of the *Nice* time series. The large separation is about 58 μHz. The ridge at the left hand side is due to the $l = 1$ modes. The identification is rather easy because the ridge at the right hand side show a more complex structure related to the splitting of the $l = 2$ modes interfering with the $l = 0$ mode ridge. The order labeling the y axis have no absolute meaning.

and the star inclination are also recovered. Similar results were obtained with an other piece-of-cake time series; that of the *Meudon* team. The frequency comparisons and inversion associated with this time series are presented elsewhere in these proceedings by Berthomieu et al.

5.3.3. *A difficult case: Queen Mary synthetic data*

Figure 3 shows the echelle diagramme for the *Queen Mary* synthetic data. The time series are 150-day long and sampled at 32 sec. The identification as explained in the caption is somewhat more difficult, to say the least. The main problem is that it seemed that the mode visibility was not according to what is usually expected, in addition the large 'apparent' rotational splitting confused even more the data fitters. The $l = 0$ and $l = 1$ modes were identified with some confidence. De facto we did not assume a common amplitude for the $l = 2$ modes constrained by the visibility of the multiplet as given by Toutain and Gouttebroze (1993). Instead we assumed that the $l = 2$ multiplet had 5 different amplitudes not constrained by geometrical visibility. This was as matter of fact a correct assumption verified *a posteriori*, but not sufficiently correct as it turned out that the $l = 3$ mode ridge was improperly identified as being $l = 2$.

Figures 4 and 5 shows the results obtained when comparing the output fitted parameters with those of the input theoretical parameters. The results obtained for Figure 4 are typical of what we obtained for other time series (See Berthomieu et

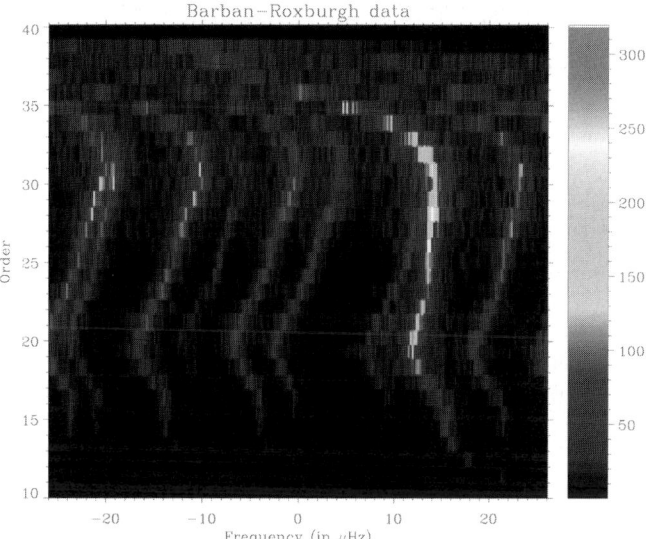

Figure 3. Echelle diagramme of the *Queen-Mary* time series. The large separation is about 52 μHz. The bright ridge at +12 μHz is identified as due to the $l = 0$ modes; it is the only one ridge with this shape. The 3 parallel ridges at -23μHz, -13μHz and -3μHz are attributed to the $m = -1$, $m = 0$ and $m = +1$ modes of $l = 1$; the splitting is about 10μHz. The ridges left over (5 of them) were attributed to the $l = 2$ modes; the $m = 0$ mode ridge crosses the $l = 0$ mode ridge. The $m = -2$ and $m = -1$ mode ridges seemed to be on either side of the $l = 1$, $m = +1$ mode ridge, while the $m = +1$ and $m = +2$ mode ridge are on either side of the $l = 1$, $m = -1$ ridge.

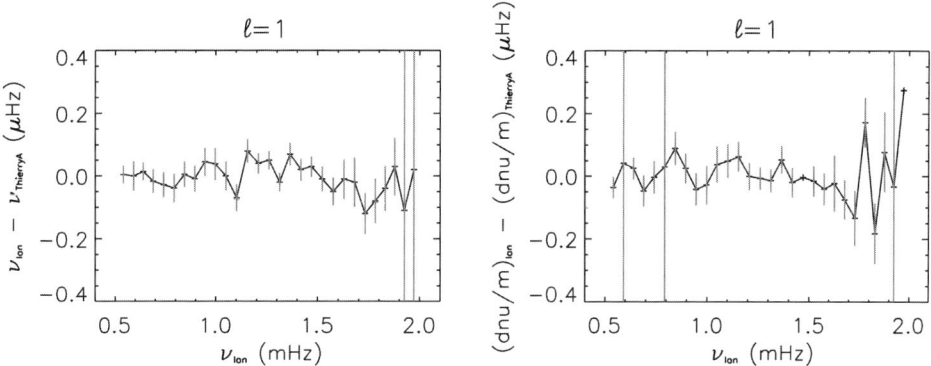

Figure 4. Comparison between the parameters fitted by the *Meudon* team with the theoretical parameters of the *Queen Mary* team; (left) for the $l = 1$ mode frequencies, (right) for the $l = 1$ mode splitting. The agreement between the input and the fit is excellent for $l = 1$; a similar agreement was obtained for $l = 0$. The error bars on the the freqeuncies are typically ranging from 0.05 μHz to 0.1 μHz.

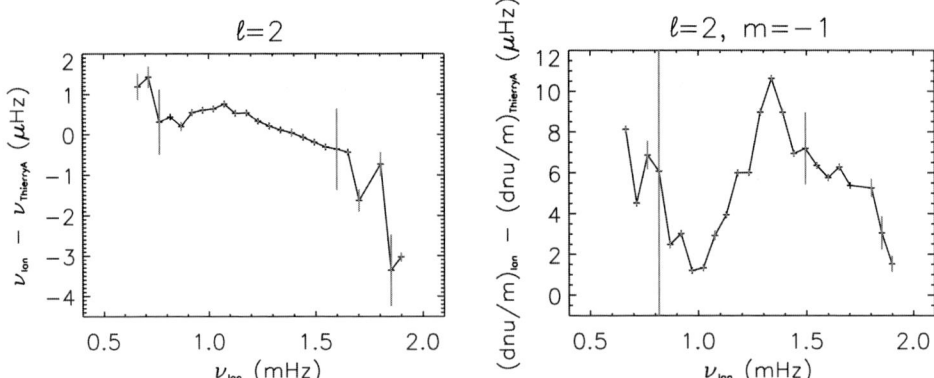

Figure 5. Comparison between the parameters fitted by the *Meudon* team with the theoretical parameters of the *Queen Mary* team; (left) for the $l = 2$ mode frequencies, (right) for the $l = 2, m = 1$ mode splitting. The agreement between the input and the fit is extremely poor. The misidentification of the $l = 2$ is obvious; the ridge identified as an $l = 2, m = 0$ was as a matter of fact a $l = 3, m = -2$.

al in these proceedings), or for the $l = 0$ of the *Queen-Mary* time series. Figure 5 is an example of what happens when modes are misidentified.

6. Conclusion

I have shown that mode identification and power spectrum fitting for solar-like stars benefit from 20 years of helioseismic experience. The methodology explained here can easily be applied when the signal-to-noise ratio is rather high. Hands-on experience for solar-like stars awaits the availability of space-based measurements (COROT, MOST). Within the COROT team, the waiting has been replaced by the use of hare-and-hound exercises. Examples of the results obtained within the COROT team show what can happen when one encounters the expected and the unexpected.

Acknowledgements

I would like to thank the SOC for the kind invitation for giving this review, and also the LOC for a perfect organization (with special thanks to Margarida and Mario).

This review benefited from the fruitful collaboration with the Data Analysis Team of the Seismology Working Group of COROT; many thanks to Gabrielle Berthomieu, Janine Provost, Thierry Toutain, Marie Jo Goupil, Frederic Baudin, Patrick Boumier, Ian Roxburgh and Caroline Barban for many interesting discussions. Last but not least thanks to Eric Michel for managing us all and for coping with my humor and humeur.

Thanks also to Reza Samadi and Fabio Favata for teaching a solar physicist the arcane of stellar structure. My presentation would have not been made possible without the contribution of Olga Moreira who made the pretty movies for the presentation; thank you so much Olga.

References

Appourchaux, T., Gizon, L. and Rabello-Soares, M.C.: 1998, *A&AS* **132**, 107.
Audard, N. and Provost, J.: 1994, *A&A* **282**, 73.
Baglin, A. and The COROT Team: 1998, in: F.-L. Deubner, J.Christensen-Dalsgaard and D.Kurtz (eds.), IAU Symp. 185: New Eyes to See Inside the Sun and Stars, Kluwer Academic Publishers, Dordrecht, The Netherlands, 301.
Bouchy, F. and Carrier, F.: 2002, *A&A* **390**, 205.
Chaplin, W.J., Elsworth, Y., Isaak, G.R., Lines, R., McLeod, C.P., Miller, B.A. and New, R.: 1998, *MNRAS* **300**, 1077.
Christensen-Dalsgaard, J. and Gough, D.O.: 1982, *MNRAS* **198**, 141.
Claverie, A., Isaak, G., McLeod, C., van der Raay, H. and Roca Cortés, T.: 1979, *Nat* **282**, 591.
Cox, J.P. and Giuli, R.T.: 1968, in: *Principles of stellar structure*, New York, Gordon and Breach.
Deubner, F.-L.: 1975, *A&A* **44**, 371.
Duvall, T.L. and Harvey, J.W.: 1986, in: *Seismology of the Sun and the Distant Stars*, 105.
Favata, F., Roxburgh, I. and Christensen-Dalsgaard, J.: 2000, in: T.C.Teixeira and T.R.Bedding (eds.), *The Third MONS Workshop: Science Preparation and Target Selection*, Aarhus Universitet, 49.
Fossat, E., Grec, G. and Pomerantz, M.: 1981, *Solar Phys.* **74**, 59.
Gabriel, M.: 1994, *A&A* **287**, 685.
Grec, G.: 1981, 'Thèse de doctorat', Ph.D. thesis, Université de Nice.
Grec, G. and Fossat, E.: 1977, *A&A* **55**, 411.
Grec, G., Fossat, E. and Pomerantz, M.: 1980, *Nat* **288**, 541.
Grec, G., Fossat, E. and Pomerantz, M.A.: 1983, *Solar Phys.* **82**, 55.
Harvey, J.: 1993, *ASP Conference Series* **42**, 111.
Houdek, G., Balmforth, N.J., Christensen-Dalsgaard, J. and Gough, D.O.: 1999, *A&A* **351**, 582.
Iben, I. and Mahaffy, J.: 1976, *ApJ* **209**, L39.
Iben, I.J. and Ehrman, J.R.: 1962, *ApJ* **135**, 770.
Kjeldsen, H., Bedding, T.R. and Christensen-Dalsgaard, J.: 2000, *A.S.P. Conf. Ser.* **203**, 73.
Matthews, J.M., Kuschnig, R., Walker, G.A.H., Pazder, J., Johnson, R., Skaret, K., Shkolnik, E., Lanting, T., Morgan, J.P. and Sidhu, S.: 2000, *A.S.P. Conf. Ser.* **203**, 74.
Samadi, R. and Goupil, M.-J.: 2001, *A&A* **370**, 136.
Samadi, R., Goupil, M.-J. and Lebreton, Y.: 2001, *A&A* **370**, 147.
Tassoul, M.: 1980, *ApJS* **43**, 469.
Toutain, T. and Appourchaux, T.: 1994, *A&A* **289**, 649.
Toutain, T. and Gouttebroze, P.: 1993, *A&A* **268**, 309.
Trampedach, R., Christensen-Dalsgaard, J., Nordlund, A. and Stein, R.F.: 1998, in: H. Kjeldsen and T. Bedding (eds.), *The First MONS Workshop: Science with a Small Space Telescope*, Aarhus Universitet, 59.

MODE IDENTIFICATION FROM LINE PROFILES USING THE DIRECT FITTING TECHNIQUE

LUIS A. BALONA
South African Astronomical Observatory, PO Box 9, Observatory 7935, Cape Town, South Africa

Abstract. The method of moments and the direct fitting method are the only spectroscopic methods of mode identification which allow a determination of all pulsational parameters. The pulsation parameters are required to predict the light amplitude and phase which can be important discriminants in mode identification. The direct fitting method has several advantages over the method of moments. It is not restricted to low spherical harmonic degree or form of the eigenfunction and is not sensitive to the placement of the continuum. In the last few years the method has been applied to several different types of stars. We briefly describe the method and give some examples of its application.

Keywords: line: profiles; stars: oscillations

1. Introduction

One way of obtaining confidence that a particular spectroscopic mode identification is correct is to calculate the amplitude and phase of the light variation using the parameters obtained from line profile fitting. The parameters are as follows: the vertical velocity amplitude, V_p and phase, ϕ_p; the ratio of relative temperature to relative radius variation, f, and its phase, ψ, and the angle of inclination, i. For computational purposes, it is more convenient to use the pseudo velocity and phase, V_f, ϕ_f, rather than the theoretical quantities f and ψ. The relationship between these quantities can be found in Balona (2000). We assume that the projected rotational velocity, $v \sin i$, the intrinsic line profile and the limb darkening coefficients are known. We also need to know how the line strength depends on temperature and gravity.

There are only two methods which allow determination of all pulsational parameters – the method of moments (Balona, 1987; Aerts et al., 1992) and the direct fitting technique (Balona and Kambe, 1999). The direct fitting technique has several advantages over the method of moments. Firstly, it is not limited to low values of ℓ or the particular form of the eigenfunction. Secondly, it is less susceptible to how the continuum is positioned, since the fitting may be restricted to a region near the line core. The drawback is that it is far more computationally intensive, a problem which is no longer as restrictive as it used to be. In applying the method to multiperiodic stars, the assumption is made that the line profile variations from other modes cancel out within a phase bin. This restriction is closely related to the sampling and aliasing problems associated with time series analysis and is not unique to the direct fitting method.

In this article, we report on the application of the method to four very different types of stars. I do not believe that it is necessary to use as many different methods as possible in mode identification. Rather, I believe that it is important to use the *best possible method* suited to the data. If the data is limited, then rough estimates of ℓ and m are all that can be deduced. Given enough data, however, a more rigorous method such as the moment or direct fitting methods is to be preferred so that the results may be checked via the photometry.

2. An Outline of the Method

To apply the direct fitting method, one needs to compute line profile variations for a particular (ℓ, m) given the pulsational parameters. Because there is a large number of parameters, it is important to fix as many of them as possible. If the star is slowly rotating, then temperature perturbations have a negligible effect on the line profile variations, so that V_f and ϕ_f can be neglected. These parameters are important in rapidly rotating stars and may, in fact, have a much larger effect than the pulsational velocity, V_p. In this case it may be possible to fix the ratio V_f/V_p by making use of theoretical values of f and ψ (e.g. Balona et al., 2002).

The observed line profiles are binned and averaged in a number of phase bins (5 to 10 bins are normally used) for the period under consideration. The line profile for a given (ℓ, m) is calculated for these phases and a goodness of fit criterion, σ, is determined. One or more of the pulsational parameters is then adjusted and the process repeated. One needs to be careful not to locate a local minimum of σ. To avoid this, a coarse grid of σ values is calculated over all physically possible values of the pulsational parameters. Having obtained the approximate global minimum, accurate values of the pulsational parameters are then calculated using one of many techniques, such as the method of steepest descent.

Once the best parameters for suitable ranges of ℓ and m have been found, together with their associated goodness of fit, it is important to calculate the predicted light amplitude and phase in a given passband. This cannot be done unless one knows the values of f and ψ from models. In principle, f and ψ could be determined empirically. Except in rapidly rotating stars, the values obtained have such large error margins as to render them valueless. Assuming, however, that f and ψ are known (even approximately), comparison of the predicted to observed light amplitudes and phases can be crucial in selecting which of several solutions with nearly the same goodness of fit is likely to be correct.

3. Examples

3.1. 19 MON

The simplest example is a star with a single period and a mode of low degree. For this purpose, we consider the application to the rapidly rotating star B1V star 19 Mon (Balona et al., 2002). Although three periods have been detected in this star, one of them has by far the highest amplitude, so the star can be considered as monoperiodic. In this β Cep star values of f and ψ were available from models using Dziembowski's NADROT code, so that V_f/V_p could be treated as a fixed parameter. Furthermore, the large value of $v \sin i = 274$ km s^{-1} could be used to constrain the inclination angle $i \approx 90°$. The free parameters are V_p, ϕ_p and ϕ_f.

The results showed that the most likely mode is (2, –2), but (3, –3) and (1, –1) gave values of σ which were almost as small. However, calculations of the light amplitude and phase are more consistent with (2, –2), giving a good level of confidence to this identification. One of the difficulties here is that the angular velocity of rotation, Ω, is not well known. This means that there is considerable uncertainty in determining the period for zero rotation, which is required for direct comparison with the models. Moreover, the eigenfunction is likely to depart significantly from a spherical harmonic because the ratio of rotation to pulsation frequency is not very small.

3.2. 1 MON

The δ Scuti star 1 Mon is of particular interest because it has a frequency triplet with practically equal frequency spacing. It is very tempting to regard this as an example of a rotationally split $\ell = 1$ mode. However, many years ago, Balona and Stobie (1980) identified the central component as a *radial* mode on the basis of photometric mode identification. Balona et al. (2001) obtained simultaneous spectroscopy and photometry of the star. Application of the direct fitting technique together with photometric mode identification confirmed that the central component is indeed a radial mode. One of the other components is identified as (1, –1), but the third member of the triplet was not seen, presumably because the amplitude had decreased in the 20 years since it was last observed.

In this case, rotation is very small so that V_f and ϕ_f can be neglected. The free parameters are V_p, ϕ_p and i. Results did not produce a unique solution for the central component, ν_1. The identifications (0, 0), (1, 0) and (1, –1) give almost identical values of σ. However, the photometry clearly supports the (0,0) mode, illustrating the value of combining both spectroscopy and photometry in mode identification.

3.3. ζ Oph

ζ Oph is a rapidly rotating O9V star in which features moving from blue to red cross the line profile, indicating prograde modes of high degree. Balona and Kambe (1999) identified two frequencies and applied the direct fitting method to obtain $\ell = 8$ and $\ell = 4$. An interesting aspect of this analysis is that the line profile variations are caused mainly by the temperature perturbation and not by the pulsational velocity (i.e. $V_f \gg V_p$). This may be true for all high degree modes in rapidly rotating stars. The light variations are not detected, presumably because ℓ is too large.

Another interesting aspect is that, whereas the observations clearly rule out retrograde modes, there is little discrimination between (8, –8), (8, –7), (8, –6) in the one case or (4, -4), (4, –3) (4, –2) in the other case. The true identification need not necessarily be the pure sectorial modes (8, –8) and (4, –4). It seems that for modes of high order it becomes increasingly difficult to discriminate between neighbouring values of m. This is very unfortunate because the exact value of (ℓ, m) as well as the rotational frequency, Ω, is necessary for asteroseismology.

References

Aerts, C., De Pauw, M. and Waelkens, C.: 1992, *A&A* **266**, 294.
Balona, L.A. and Stobie, R.S.: 1980, *MNRAS* **190**, 931.
Balona, L.A.: 1987, *MNRAS* **224**, 41.
Balona, L.A. and Kambe, E.: 1999, *MNRAS* **308**, 1117.
Balona, L.A.: 2000, in: M. Breger and M.H. Montgomery (eds.), *Delta Scuti and Related Stars*, A.S.P. Conf. Ser. **210**, 170.
Balona, L.A. et al.: 2001, *MNRAS* **321**, 239.
Balona, L.A., James, D.J., Motsoasele, P., Nombexeza, B., Ramnath, A. and van Dyk, J.: 2002, *MNRAS* **333**, 952.

STATISTICAL REVISION OF THE MOMENT METHOD

JORIS DE RIDDER, GEERT MOLENBERGHS and CONNY AERTS

Institute of Astronomy, K.U. Leuven, Celestijnenlaan 200 B, B-3001 Leuven, Belgium
Center for Statistics, L.U.C., Universitaire Campus, B-3590 Diepenbeek, Belgium

Abstract. The moment method is a well known technique, which uses a time series of the first 3 moments of a spectral line, to estimate the (discrete) mode parameters ℓ and m. The method, contrary to Doppler imaging, also yields other interesting (real-valued) parameters such as the inclination angle i, or $v \sin i$, during its identification procedure. In this paper, we are not only interested in the estimation of these real-valued parameters themselves but also in reliable estimates for their uncertainty. We designed a statistical formalism for the moment method based on the so-called *generalized estimating equations (GEE)*. This formalism aims to estimate the uncertainty of the real-valued parameters taking into account that the different moments of a line profile are correlated and – more importantly – that the uncertainty of the observed moments depends on the pulsation parameters. The latter property of the moment method makes the least-squares technique a poor choice to estimate the uncertainty of the real-valued parameters. We implemented the GEE method and present an application to a high-resolution spectroscopic dataset of the slowly pulsating B star HD181558.

Keywords: Mode identification, moment method, slowly pulsating B star

1. Introduction

Mode identification is an important step in the proces of asteroseismology. For some κ-driven pulsators, like the β Cephei stars, no helpful asymptotic relation for the frequencies is available, so that for these stars we have to determine the mode numbers with other methods. One such method is the *moment method* which compares the first three observational moments of a spectral line with theoretical moments.

The moment method has several advantages. First, it is a spectroscopic mode identification technique, and one can expect that line profile variations contain more detailed information than photometric variations. Secondly, the moment method is computationally faster than direct line profile fitting, and is still useful for low degree ℓ modes. Briquet and Aerts (these proceedings) show that the moment method is also useful for multiperiodic stars. Thirdly, contrary to, e.g., Doppler imaging, the moment method is not only able to estimate the discrete mode numbers ℓ and m, but also some real-valued parameters such as the amplitude v_p of the oscillation, the inclination angle i, the equatorial rotational velocity v_e and the width σ of the local gaussian line profile.

In this research project, we address the estimation of the uncertainties of the model parameters mentioned above. As this is the first attempt ever to do so, we concentrate on the real-valued parameters $\mathbf{b} = (v_p, \sigma, v_e, i)$ which are easier to handle statistically, and we restrict ourselves to the monoperiodic case.

2. Statistical Challenges

It is natural to start exploring whether we can use the method of least squares. The relevant statistical point to note is that the three moments are dependent, and that their covariance matrix is unknown. We do know, however, that the covariance matrix depends on the unknown parameters, i.e., not only the moments but also the uncertainties of the moments depend on the parameters $\mathbf{b}$. For example, the larger $v_e \sin i$, the broader the line, and the less precise the first moment is known. This property of the covariance matrix of the observations, makes the method of least-squares inappropriate because this method would yield an inconsistent estimate of $\mathbf{b}$ and its uncertainty (Seber and Wild, 1989).

For this reason, we turned to the method of the *generalized estimating equations*, or the *GEE* formalism, as developed by Liang and Zeger (1986).

3. The GEE Formalism

The most important assumptions of the GEE formalism are the following. First, it assumes that observations made at different times are independent. Secondly, although the formalism does not require the exact covariance matrix of the observations, it does assume that a working approximation is available. Thirdly, the GEE formalism assumes that expected values of the observations (i.e., the model) are well specified.

In return, the GEE formalism gives a consistent and asymptotically normal estimate of the real-valued parameters $\mathbf{b}$, and a robust estimate of the uncertainty of $\mathbf{b}$.

For a detailed outline of the GEE formalism applied to the moment method we refer to De Ridder et al. (submitted to *JASA*). What follows is a brief summary. First, for each couple (ℓ, m), $\mathbf{b}$ is estimated by locating the root of a 4-dimensional so-called quasi-score function $\mathbf{U}$ which contains the observational and theoretical moments and the model derivatives. This is not an easy task and it requires a good initial guess. For this reason we scanned for each (ℓ, m) the 4-dimensional parameters space for plausible $\mathbf{b}$ sets with a simple goodness-of-fit function $g(\mathbf{b})$.

As a result, we obtain for each (ℓ, m) couple a GEE estimate of the real-valued parameters $\hat{\mathbf{b}}$. Next, we compute for each mode a goodness-of-fit G^2 to see how well the solutions fit the observations. We now use a sum of squares, weighted with the higher moments, because we use the higher moments to measure the

TABLE I
The goodness-of-fit values g of the best fitting parameter sets **b** after scanning the 4D parameter space for each mode (ℓ, m). The lower the g value, the better the fit

g_{min}	$\ell = 1$	$\ell = 2$	$\ell = 3$	$\ell = 4$
$m = +4$				11.9
$m = +3$			7.52	11.0
$m = +2$		6.57	6.71	11.3
$m = +1$	4.72	4.74	5.85	10.8
$m = 0$	6.37	6.37	6.79	11.6
$m = -1$	4.79	6.57	7.05	10.7
$m = -2$		4.68	6.86	10.5
$m = -3$			6.92	11.3
$m = -4$				11.3

uncertainty of the lower moments. Consequently, we can attach a weight to each mode.

Finally, to compute the final estimate $\tilde{\mathbf{b}}$ of the real-valued parameters, we compute a weighted mean over all modes. For the final estimate for the uncertainty on $\tilde{\mathbf{b}}$, we not only compute the intra-mode uncertainty, which is a weighted mean of the uncertainties of $\hat{\mathbf{b}}$ of each mode, but we also add the inter-mode uncertainty which takes into account that the $\hat{\mathbf{b}}$ values themselves can vary quite a lot from mode to mode.

4. An Application: HD181558

HD181558 is a slowly pulsating B star. Although the star is multi-periodic, it has a very dominant first mode, which allows us to use a mono-periodic model in a good approximation. We applied the GEE formalism to an observational time series of 30 high-quality spectra of the SiII (412.805 nm) line, gathered by De Cat and Aerts (submitted to *A&A*).

For each mode (ℓ, m) with $0 < \ell \leq 4$, we scanned the 4D parameter space for plausible **b** sets with a simple goodness-of-fit function $g(\mathbf{b})$. We also list the $g(\mathbf{b})$ value of the best fitting set **b** for each of the modes (ℓ, m) in Table I. The modes $(\ell, m) = (1, \pm 1), (2, 1)$ and $(2, -2)$ seem to stand out as possible candidates. It is therefore clear that in this case, it is not meaningful to only mention the best fitting mode. Rather, one should try to find other independent mode identifications to shorten the list of candidate modes. For the star HD 181558 we mention that a

TABLE II

The final estimate $\tilde{\mathbf{b}}$ for the real-valued parameters and their uncertainties (between brackets). We also list the separate contributions of the intra- and the inter-mode variance to the total variance. $\tilde{v}_p, \tilde{\sigma}, \tilde{v}_e$ are expressed in km/s, and the inclination angle $\tilde{\imath}$ is expressed in degrees

$\tilde{\mathbf{b}}_i$	Weighted Mean	Intra-mode Variance	Inter-mode Variance
$\tilde{v}_p$	2.0 (1.0)	0.19	0.71
$\tilde{\sigma}$	5.3 (2.0)	0.82	3.1
$\tilde{v}_e$	16 (12)	100	37
$\tilde{\imath}$	170 (137)	8342	10450

photometric mode identification of De Cat et al. (in preparation) points towards an $\ell = 1$ mode, which is consistent with our results.

For each of the modes listed in Table I, we tried to compute a GEE estimate $\hat{\mathbf{b}}$ for the real-valued parameters together with their uncertainties. From these estimates we then computed our final estimate $\tilde{\mathbf{b}}$ and its uncertainty, taking into account that we do not know the true mode (ℓ, m). The results are listed in Table II. Clearly, the relative uncertainties can be rather large. From simulations with artificial datasets we find that these uncertainties are quite typical, except for the width σ which is usually more precisely estimated. We also point to the special case of $\tilde{v}_e$ for which the intra-mode variance is larger than the inter-mode variance. This means that the estimates $\hat{v}_e$ for the different modes (ℓ, m) do not differ that much, but that each of these estimates has a large uncertainty.

References

Liang, K.Y. and Zeger, S.: 1986, *Biometrika* **73**, 13.
Seber, G.A.F. and Wild, C.J.: 1989, in: *Nonlinear Regression*, Wiley series in Probability and Mathematical Statistics, John Wiley and Sons, Inc.

AN IMPROVED METHOD OF PHOTOMETRIC MODE IDENTIFICATION: APPLICATIONS TO SLOWLY PULSATING B, β CEPHEI, δ SCUTI AND γ DORADUS STARS

M.-A. DUPRET, R. SCUFLAIRE, A. NOELS and A. THOUL
Institut d'Astrophysique et de Géophysique de l'Université de Liège, Belgium

R. GARRIDO and A. MOYA
Instituto de Astrofísica de Andalucía-CSIC, Granada, Spain

J. DE RIDDER, P. DE CAT and C. AERTS
Instituut voor Sterrenkunde, Katholieke Universiteit Leuven, Belgium

Abstract. We present an improved version of the method of photometric mode identification based upon the inclusion of non-adiabatic eigenfunctions determined in the stellar atmosphere, according to the formalism recently proposed by Dupret et al. (2002). We apply our method to β Cephei, Slowly Pulsating B, δ Scuti and γ Doradus stars. Besides identifying the degree ℓ of the pulsating stars, our method is also a tool for improving the knowledge of stellar interiors and atmospheres, by imposing constraints on the metallicity for β Cephei and SPBs, the characteristics of the superficial convection zone for δ Scuti and γ Doradus stars and the limb-darkening law.

1. Magnitude Variation of a Non-Radial Pulsator

Methods of mode identification based on multi-colour photometry have been derived by different authors: Dziembowski (1977), Stamford and Watson (1981), Watson (1988), Garrido et al. (1990), Garrido (2000), Heynderickx et al. (1994), Cugier et al. (1994) and Balona and Evers (1999). In our study, we assume that the geometrical distortion of the photosphere is given by the Lagrangian displacement at $T = T_{\text{eff}}$ (one-layer approximation) and that the atmosphere remains in radiative equilibrium during the pulsation. Under these hypotheses, the monochromatic magnitude variation for the mode (ℓ, m) and wavelength λ is given by :

$$\delta m_\lambda = -\frac{2.5}{\ln 10} \epsilon \, P_\ell^m(\cos i) \, b_{\ell\lambda} \left[(1-\ell)(\ell+2) \cos(\sigma t) \right.$$
$$\left. + f_T \cos(\sigma t + \psi_T)(\alpha_{T\lambda} + \beta_{T\lambda}) - f_g \cos(\sigma t)(\alpha_{g\lambda} + \beta_{g\lambda}) \right], \quad (1)$$

where ϵ is the amplitude of the radial displacement, i is the inclination angle of the star, $b_{\ell\lambda}$, $\alpha_{T\lambda}$, $\alpha_{g\lambda}$, $\beta_{T\lambda}$ and $\beta_{g\lambda}$ are defined in Balona and Evers (1999), f_T is the amplitude of local effective temperature variation $(\delta T_{\text{eff}}/T_{\text{eff}})$ and f_g is the amplitude of local effective gravity variation $(\delta g_e/g_e)$ for a normalized radial displacement at the photosphere and ψ_T is the phase difference between the effective temperature variation and the radial displacement.

In our method, the coefficients f_T, ψ_T and f_g are computed by a non-adiabatic code including the interaction with the atmosphere, as derived by Dupret et al. (2002). Our determination of the effective gravity variation is similar to the one proposed by Cugier and Daszynska (2001), but not assuming the Cowling approximation. We do not assume that the effective temperature variation is equal to the Lagrangian temperature variation, as Dupret et al. (2002) showed that these two quantities can be very different.

We improved the mode identification code of Garrido et al. (1990) which uses the phase-lag information, as well as the code of Heynderickx et al. (1994) based on amplitude ratios, so that they can now take our non-adiabatic predictions into account, following Eq. (1).

The stellar models used in this study were computed with the new Code Liégeois d'Evolution Stellaire (CLES) written by R. Scuflaire.

2. β Cephei Stars

We applied our method to the β Cephei star EN (16) Lac, using the photometric amplitudes with Johnson filters derived by Jerzykiewicz (1993). We identified the degrees ℓ of the 3 frequencies $f_1 = 5.9112 \, \text{c} \, \text{d}^{-1}$, $f_2 = 5.8551 \, \text{c} \, \text{d}^{-1}$ and $f_3 = 5.5033 \, \text{c} \, \text{d}^{-1}$ as $\ell = 0$, 2 and 1, respectively. Our photometric mode identification is fully compatible with the spectroscopic mode identification (see Aerts et al. and Briquet et al., these proceedings). Moreover, we could constrain the metallicity of EN Lac, because the non-adiabatic results are dependent on it. The higher the metallicity, the more efficient the κ mechanism, which implies a more important decrease of the luminosity variation in the driving region and thus a smaller effective temperature variation (for a normalized displacement). The best agreement between the theoretical and observed amplitude ratios was obtained for a model with $Z = 0.015$. Values below $Z = 0.015$ do not lead to excitation of the modes.

3. Slowly Pulsating B Stars

We applied our method to 13 SPBs observed with Geneva photometry by De Cat and Aerts (2002). We derived the effective temperatures and gravities of these stars, using the recent calibrations of Künzli et al. (1997). We computed theoretical models closest to these values. We performed non-adiabatic computations and we did the photometric mode identification of the dominant modes. In Figure 1, we illustrate the results obtained for the star HD 74560. We see that the dominant frequency of this star ($f = 0.64472 \, \text{c} \, \text{d}^{-1}$) is identified as an $\ell = 1$ mode. This result is in full agreement with the spectroscopic mode identification achieved with the moment method (De Cat et al., 2003).

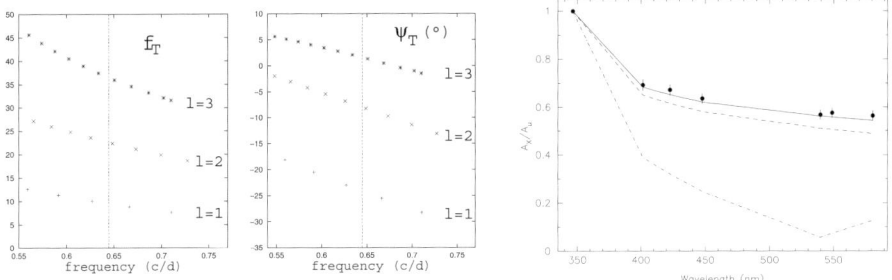

Figure 1. Model calculations for the SPB star HD 74560. On the left (resp. middle), non-adiabatic values of f_T (resp. ψ_T) for different modes. On the right, observed amplitude ratios (bullets) and theoretical predictions (solid, dashed and dot-dashed lines for $\ell = 1$, 2 and 3).

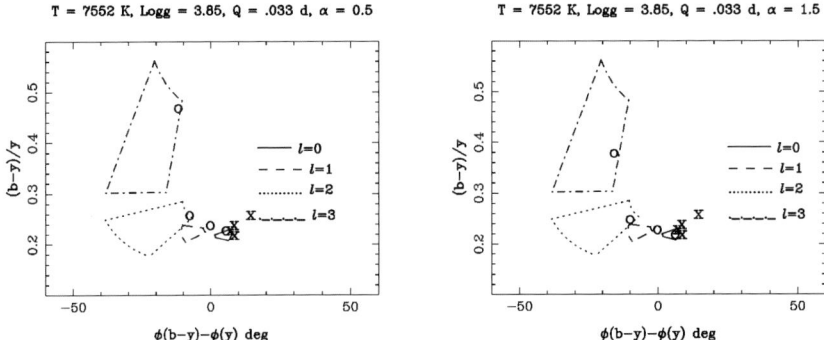

Figure 2. Strömgren filters phase-amplitude diagrams for two δ Scuti models with $\alpha = 0.5$ (left) and $\alpha = 1.5$ (right). The different regions are for different degrees ℓ, with R and ψ_T as free parameters (Garrido et al., 1990). The circles are our non-adiabatic predictions and the crosses are for observations of radial pulsators.

4. δ Scuti Stars

We performed non-adiabatic computations for δ Scuti models with different values of the mixing length parameter α. Our non-adiabatic results (mainly the phase-lag ψ_T) are very sensitive to the size of the thin superficial convection zone (linked to α), confirming the results of Balona and Evers (1999). In Figure 2, we illustrate typical results obtained for two δ scuti models with $\alpha = 0.5$ (left) and $\alpha = 1.5$ (right). We refer to Moya et al. (these proceedings) for more details and a comparison between the non-adiabatic results obtained with and without inclusion of the atmosphere in the pulsation equations.

5. γ Doradus Stars

We also performed non-adiabatic computations for γ Doradus models with different α and these results are extremely dependent on the size of the superficial convection zone. At present, no theory is able to explain at the same time the excitation mechanism and the photometric observations. We refer to Moya et al. (these proceedings) for more theoretical results and to Aerts et al. (these proceedings) for the observations of two multi-periodic γ Doradus stars, with Geneva photometry.

6. Conclusions

Multi-colour photometry can be used efficiently for the identification of the degree ℓ of pulsation modes. These observables are very sensitive to the degree of non-adiabaticity of the superficial layers. The theoretical non-adiabatic predictions depend much on the metallicity for β Cephei and SPBs and on the mixing-length parameter α for δ Scuti and γ Doradus stars. Therefore, a precise confrontation between theory and observations can constrain these parameters.

References

Balona, L.A. and Evers, E.A.: 1999, *MNRAS* **302**, 349.
Cugier, H. and Daszynska, J.: 2001, *A&A* **377**, 113.
Cugier, H., Dziembowski, W. and Pamyatnykh A.: 1994, *A&A* **291**, 143.
De Cat, P. and Aerts, C.: 2002, *A&A* (in press).
De Cat, P., Aerts, C. et al.: 2003, *A&A* (in press).
Dupret, M.-A., De Ridder, J., Neuforge, C., Aerts, C. and Scuflaire, R.: 2002, *A&A* **385**, 563.
Dziembowski, W.: 1977, *Acta Astron.* **27**, 203.
Garrido, R.: 2000, in: M. Montgommery, M. Breger (eds.), *The 6th Vienna Workshop on δ Scuti and related stars*, A.S.P. Conf. Ser. **210**, 67.
Garrido, R., Garcia-Lobo, E. and Rodriguez, E.: 1990, *A&A* **234**, 262.
Heynderickx, D., Waelkens, C. and Smeyers, P.: 1994, *A&AS* **105**, 447.
Jerzykiewicz, M.: 1993, *Acta Astron.* **43**, 13.
Künzli, M., North, P., Kurucz, R.L. and Nicolet, B.: 1997, *A&AS* **122**, 51.
Stamford, P.A. and Watson, R.D.: 1981, *Ap&SS* **77**, 131.
Watson, R.D.: 1988, *Ap&SS* **140**, 255.

PHOTOMETRIC NONADIABATIC OBSERVABLES IN ROTATING β CEPHEI MODELS

J. DASZYŃSKA-DASZKIEWICZ[1,2], W.A. DZIEMBOWSKI[2,3] and
A.A. PAMYATNYKH[2,4]

[1] *Astronomical Institute of the Wrocław University, ul. Kopernika 11, 51-622 Wrocław, Poland*
[2] *Copernicus Astronomical Center, Bartycka 18, 00-716 Warsaw, Poland*
[3] *Warsaw University Observatory, Al. Ujazdowskie 4, 00-478 Warsaw, Poland*
[4] *Institute of Astronomy, Russian Academy of Sciences, Pyatnitskaya Str. 48, 109017 Moscow, Russia*

Abstract. We study how moderate rotation affects photometric observables in β Cephei stars. The most important effect is the rotational mode coupling of modes with harmonic degree, ℓ, differing by 2 and the same azimuthal order, m, if the frequencies are close. This is not an uncommon situation among unstable modes in β Cep stars. Positions of the coupled modes in the amplitude ratio *vs.* phase difference diagrams are aspect, i, – and m-dependent. Inference from the diagrams becomes more complicated.

Keywords: β Cep stars, oscillation, rotation

1. Introduction

Photometric nonadiabatic observables, that is, calculated amplitude ratios and phase differences for individual mode seen in two passbands are popular tools for assessing spherical harmonic degree, ℓ (see e.g. Cugier et al., 1994 or Balona and Evers, 1999, where references to earlier related works may be found). If effects of rotation are ignored, as it has been done in all these works, then the nonadiabatic observables are both aspect and m-independent. Furthermore, modes of different ℓ degrees are located in well separated regions of the amplitude ratio *vs.* phase difference diagnostic diagrams. So that inferring the ℓ-value is then straightforward.

Even at moderate rate of, say, 100 km/s, which may be regarded typical for β Cep stars, rotation may significantly complicate this simple picture. Complications arise if frequency separation between modes of ℓ's differing by 2 and of the same m is small. Consequences of such a situation have been studied in our recent paper (Daszyńska-Daszkiewicz et al., 2002). Here we summarize its main results.

2. Rotational Mode Coupling

Figure 1 shows that frequencies of $\ell = 0$ and $\ell = 2$ modes are indeed very close over extended ranges of the β Cephei instability strip. Occasionally also $\ell = 4$

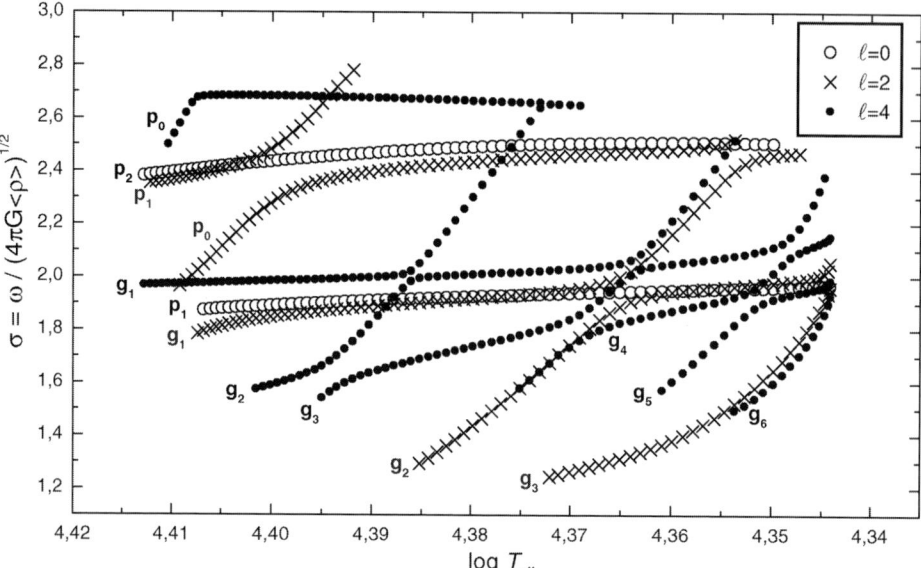

Figure 1. Frequency evolution of unstable modes with $\ell = 0, 2, 4$ in a sequence of β Cephei star models with mass of 12 $M_\odot$.

comes close to the two. A very similar situation occurs for the first three even-ℓ modes. This is precisely the reason why, even at modest rotation rates, mode coupling is important in these stars.

If the frequency separations are of the same order as the angular frequency of rotation, then the displacement eigenvectors for coupled modes, $\vec{\xi}_j^c$, must be searched in the form of the linear superposition,

$$\vec{\xi}_j^c = \sum_k a_{jk} \vec{\xi}_{0k}, \tag{1}$$

of the eigenvectors, $\vec{\xi}_{0k}$, calculated in the nonrotating model. The a_{jk} coefficients, which are determined together with the rotational frequency perturbation, are of the same order of magnitude for each of the coupled modes. The nominal mode degree ℓ' could be assigned to these modes according to the avoided crossing principle, which ensures continuity of the eigenfrequencies with varying stellar parameters. Usually, but not always, absolute values of the diagonal a_{jk} coefficients are the largest.

An interesting situation arises at near coincidence of three modes. Then the $\ell' = 4$ mode acquires a substantial $\ell = 0$ and $\ell = 2$ components becoming more easily detectable by means of photometry. Similarly, the $\ell' = 5$ mode may become detectable through its low degree components.

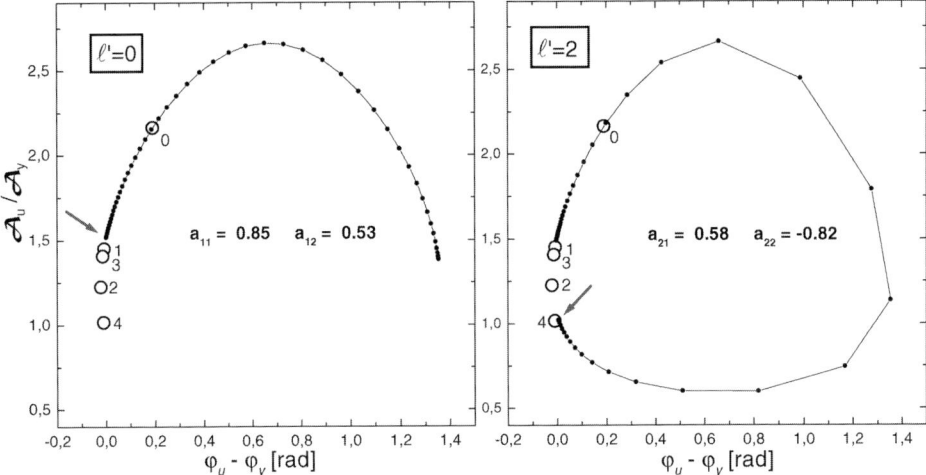

Figure 2. The diagnostic diagrams with u and y Strömgren passbands for coupled $\ell = 0$, p_1 and $\ell = 2$, g_1 modes in the model at $\log T_{\text{eff}} = 4.374$ (see Figure 1). Circles mark positions of single ℓ modes and numbers give the ℓ-values. Arrows correspond to observations from the polar direction. Spacing between consecutive dots is 0.02 in $\cos i$.

3. Diagnostic Diagrams for Coupled Modes

For a mode (denoted k) described by a single spherical harmonic, Y_ℓ^m, the amplitude of the flux variation in the passband centered at wavelength λ, may be expressed in the following complex form

$$A_{k,\lambda}(i) = \varepsilon Y_\ell^m b_{\ell,\lambda}(D_{s,\ell} + D_{g,\lambda} + D_{T,\lambda}), \qquad (2)$$

where ε is a small parameter determining intrinsic mode amplitude, i is the inclination angle of the stellar pulsation axis to the observer, $b_{\ell,\lambda}$ is the disc-averaging coefficient, the $D_{s,\ell}$ term describes the surface distortion effects, the $D_{g,\lambda}$ term the effects of g_{eff} perturbation, and the $D_{T,\lambda}$ term those of T_{eff} perturbation.

First two terms may be arbitrarily set real. The third term, which requires nonadiabatic calculations, is then complex. In calculations of the last two terms we rely on use of static atmosphere models and we include the perturbation of the limb-darkening law. If it is nonlinear, then the last two terms become ℓ-dependent, but the dependence is very weak. We obtain the nonadiabatic observables from the ratios of complex amplitudes at two values of λ. The $\varepsilon Y_\ell^m(i, 0)$ coefficient is eliminated and there is no i- and m-dependence left. For a coupled mode (denoted j) the complex amplitude is given by

$$A_{\lambda,j}^c(i) = \sum_k a_{jk} A_{\lambda,k}(i). \qquad (3)$$

It is easy to see that the amplitude ratios become both m and i dependent. The latter dependence is quite interesting. An example is presented in Figure 2. We show there the case of coupling between a close $\ell = 0$ and 2 pair at equatorial velocity of about 100 km/s.

Note that in a certain range of the inclination angles, an unaware observer could report an $\ell = 0$ doublet. In other ranges both modes could be identified as the $\ell = 1$ modes. We have found that in many cases coupled $\ell = 0$ and $\ell = 2$ modes appear in the $\ell = 1$ domain of the diagram. Note, however, that such modes may also appear in the parts of the diagram not accessible to any single ℓ mode, and not between the $\ell = 0$ and $\ell = 2$ domains, as one might expect.

4. Conclusion

Rotational mode coupling complicates usage of the diagnostic diagrams. No longer the diagrams may safely be used to infer the ℓ values of excited modes. In fact, while the ℓ value may be formally assigned to a coupled mode, its surface amplitude is not described by a single spherical harmonic. Furthermore, the mode positions in the diagram become aspect dependent.

Nonadiabatic observables always provide valuable constrains on stars and their oscillation modes. It is not our aim to discourage their usage for rotating stars. Our message is only that interpretation of the diagnostic diagrams for stars rotating with rate $\sim 10^2$ km/s must be done carefully, having in mind the effect of the rotational mode coupling.

Acknowledgements

The work was supported by KBN grant No. 5 P03D 012 20.

References

Balona, L.A. and Evers, E.A.: 1999, *MNRAS* **302**, 349.
Cugier, H., Dziembowski, W.A. and Pamyatnykh, A.A.: 1994, *A&A* **291**, 143.
Daszyńska-Daszkiewicz, J., Dziembowski, W.A., Pamyatnykh, A.A. and Goupil, M-J.: 2002, *A&A* **392**, 151.

GRAVITY MODES IN DELTA SCUTI STARS

MICHEL BREGER

Institut für Astronomie, Universität Wien, Türkenschanzstr. 17, A-1180 Wien, Austria

Abstract. This paper examines the observational evidence for the detection of gravity modes in δ Scuti stars, which are p-mode pulsators. Low-order gravity modes have also been found in at least one star (FG Vir). Some reports of gravity modes may be due to systematic errors in the absolute magnitude calibrations for slowly rotating stars. Furthermore, many detected low frequencies are not high-order gravity modes, but linear combinations, f_i-f_j, of the main pulsation modes. Other low frequencies are caused by a close binary companion leading to tidal deformation as well as tidally excited gravity modes.

Keywords: δ Scuti stars, nonradial pulsation

1. Introduction

δ Scuti stars are radial and nonradial pulsators situated inside the classical instability strip with low-order pressure (p) modes (e.g., Breger, 2000). The γ Doradus stars, on the other hand, are gravity (g) mode pulsators, partially overlapping the δ Scuti stars in the HRD (e.g., Handler and Shobbrook, 2002). The question arises whether g modes are also found in δ Scuti stars, especially in stars with temperatures and luminosities similar to those of γ Doradus stars.

2. Pulsation in Different Frequency Regions

We can divide the frequency region in which variability is found for a δ Scuti star into several regions, in which the observed variability has different properties. The δ Scuti stars cover different evolutionary states from the zero-age-main-sequence to evolved stars. Consequently, it is useful to work with a unit of period normalized for different stellar densities, viz., the pulsational constant, Q. For δ Scuti stars, the radial fundamental mode has a Q value of 0.033d.

(i) The main pulsation region is found between the Q values of 0.012 and 0.033d. Here we find the pressure (p) modes, which may have a mixed character: the modes are p modes in the envelope and g modes in the center of the star.

(ii) At higher frequencies, we find the frequency combinations from the main pulsation region, viz., f_i+f_j. Furthermore, 2f, 3f, etc. terms from the Fourier decomposition are also found.

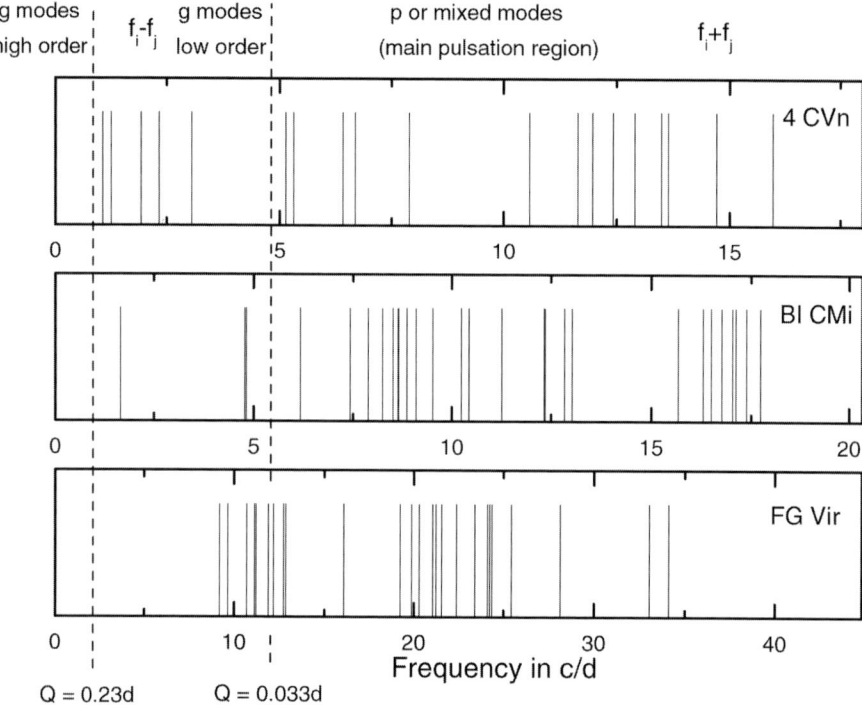

Figure 1. Frequency distribution of the excited modes in three well-studied δ Scuti stars.

(iii) At slightly lower frequencies (corresponding to Q values larger than 0.033d), low-order g modes are also predicted to be excited by pulsation models (e.g., see Breger et al., 1999a), although the most precise terminology to be used for these modes can be discussed.

(iv) At even lower frequencies, we also find the other half of the combination frequencies, viz., f_i-f_j. Again, they can be recognized by the numerical agreement of their frequencies with those found in the main pulsation region. This allows us to distinguish them from gravity modes.

(v) Finally, in the very low-frequency region around 1 c/d we can look for high-order g modes, which are responsible for the variability of the γ Doradus stars. A limit of $Q = 0.23$d has been proposed (Handler and Shobbrook, 2002) in the sense that γ Doradus-type pulsation is found at lower frequencies (higher period and Q values).

Figure 1 illustrates the different frequency regions for three well-studied stars: 4 CVn, BI CMi and FG Vir (Breger et al., 1998, 1999b, 2002).

3. Low-Order Gravity Modes

Since the radial fundamental mode has a Q value of 0.033d for δ Scuti stars, observed Q values higher than this limit suggest that g modes might also be excited. In particular, a number of stars with Q values up to 0.05d (such as δ Scuti itself) have led to the suspicion that low-order g modes modes might be involved.

However, many of these high Q values in the literature are erroneous because of a systematic error in the calibrations of $uvby\beta$ photometry for slowly rotating A/F stars. This was demonstrated in Figure 7 of Rodríguez and Breger (2001), where the difference between the Hipparcos and $uvby\beta$ luminosities were shown as a function of $v \sin i$. In fact, for a 'typical' shift of ΔM_v of 0.5 mag, the radial fundamental Q value of 0.033d would become 0.047d and might be falsely interpreted as a low-order g mode!

To examine the existence of modes with Q values larger than 0.033 d, it is necessary to choose stars with well-determined absolute magnitudes. This suggests cluster members or field stars with high-precision Hipparcos parallaxes. For the field star FG Vir an accurate Hipparcos parallax is available, allowing the determination of $M_v = 1.95 \pm 0.13$ mag. This confirms the identification of the 12.15 c/d mode as the radial fundamental as well as the 9.20 and 9.66 c/d as low-order g modes. More details can be found in Figure 1 and in Breger (2000). FG Vir may at this stage present the best evidence for the successful detection of low-order gravity modes.

4. High-Order Gravity Modes – The γ Doradus-Type Pulsation

A number of δ Scuti stars show peaks in the low-frequency (γ Doradus) part of the power spectrum. However, these peaks do not permit us to conclude a priori that in a normal star both high-order gravity modes as well as pressure modes are excited. The peaks detected so far tend to have more mundane origins or are found in unusual stars, such as close binaries. Some examples are:

(i) HD 209295: This F0V single-lined 3.1d binary was announced by Handler et al. (2002) to be the first member of both the δ Scuti and γ Doradus groups. This means that both p modes and low-order g modes are excited in this star. In addition to the δ Scuti p mode near 26.0 c/d, the star also shows of frequencies in the γ Doradus domain a number of low pulsation frequencies, of which 5 were found to be exact multiples of the orbital frequency. The remarkable study suggests that these gravity modes are tidally excited by the companion.

(ii) θ Tuc: In this star the 0.14 c/d peak is matched by the orbital period (Paparó et al., 1996; De Mey, Daems and Sterken, 1998) and corresponds to tidal distortions in a close binary system, rather than high-order g modes.

(iii) 4 CVn: All seven low-frequency peaks in the power spectrum are combination modes. These frequencies are easily recognized because of the numerical agreement with the two modes from the main pulsation region (Breger et al., 1999b).

(iv) BI CMi: Over 1000 hours of photometry of BI CMi led to the detection of a rich pulsation spectrum in the main pulsation region, but also revealed the existence of a low frequency peak at 1.66 cd^{-1} (Breger et al., 2002). It can be shown from velocity arguments that rotational modulation is very improbable as an origin. Since only one such peak was found, its origin remains unclear.

It is too early to form conclusions concerning the excitation of high-order gravity modes in δ Scuti stars. This question remains a fascinating field of research.

Acknowledgements

This investigation has been supported by the Fonds zur Förderung der wissenschaftlichen Forschung under project number P14546-PHY.

References

Breger, M.: 2000, *A.S.P. Conf. Ser.* **210**, 3.
Breger, M., Zima, W., Handler, G. et al.: 1998, *A&A* **331**, 271.
Breger, M., Pamyatnykh, A.A., Pikall, H. and Garrido, R.: 1999a, *A&A* **341**, 151.
Breger, M., Handler, G., Garrido, R. et al.: 1999b, *A&A* **349**, 225.
Breger, M., Garrido, R., Handler, G. et al.: 2002, *MNRAS* **329**, 531.
De Mey, K., Daems, K. and Sterken, C.: 1998, *A&A* **336**, 527.
Handler, G. and Shobbrook, R.R.: 2002, *MNRAS* **333**, 521.
Handler, G., Balona, L.A., Shobbrook, R.R. et al.: 2002, *MNRAS* **333**, 262.
Paparó, M., Sterken, C., Spoon, H.W.W. and Birch, P.W.: 1996, *A&A* **315**, 500.
Rodríguez, E. and Breger, M.: 2001, *A&A* **366**, 178.

SEISMIC INFERENCE USING GENETIC ALGORITHMS

TRAVIS S. METCALFE
Theoretical Astrophysics Center, Aarhus University, Denmark

Abstract. A flood of reliable seismic data will soon arrive. The migration to larger telescopes on the ground may free up 4-m class instruments for multi-site campaigns, and several forthcoming satellite missions promise to yield nearly uninterrupted long-term coverage of many pulsating stars. We will then face the challenge of determining the fundamental properties of these stars from the data, by trying to match them with the output of our computer models. The traditional approach to this task is to make informed guesses for each of the model parameters, and then adjust them iteratively until an adequate match is found. The trouble is: how do we know that our solution is unique, or that some other combination of parameters will not do even better? Computers are now sufficiently powerful and inexpensive that we can produce large grids of models and simply compare *all* of them to the observations. The question then becomes: what range of parameters do we want to consider, and how many models do we want to calculate? This can minimize the subjective nature of the process, but it may not be the most efficient approach and it may give us a false sense of security that the final result is *correct*, when it is really just *optimal*. I discuss these issues in the context of recent advances in the asteroseismological analysis of white dwarf stars.

Keywords: numerical methods, stellar interiors, stellar oscillations, white dwarfs

1. Wampler's Screwdriver

Most scientists are familiar with the concept of Occam's razor – the idea that if you have to choose between competing explanations for some physical phenomenon, the simplest explanation is most likely to be correct. My thesis supervisor, Ed Nather, told me a story about another less widely known scientific tool that may be just as important as Occam's razor. He calls it 'Wampler's screwdriver' (Nather, 1995).

In the early 1970's, Ed was attending a conference of the Astronomical Society of the Pacific in California, and Joe Wampler was giving a presentation about the first discovery of a double quasar (Wampler et al., 1973). The standard procedure at the time was to identify blue objects inside the relatively large positional error box of a newly discovered radio point source, and then take spectra of them, one by one, until you found one with a big redshift. What Joe decided to do was go back to the fields where quasars had been discovered in this way, and take spectra of *all* of the blue objects, even after he found one of them to be a quasar. Joe's double quasar turned out to be an accidental alignment of two quasars at different distances, but later on others repeated what he had done and found a double quasar that was the result of gravitational lensing – so his method was an important contribution to the field. At the end of Joe's presentation, he posed a simple question to the audience: 'Why do you always find a lost screwdriver in the last place you look?' The answer, of course, is because you stop looking.

In this paper, I will review a method of fitting models to seismological data that *keeps looking*, even after it has found a pretty good fit to the observations. This is drawn from work I have been doing over the past few years to develop a model-fitting method based on a genetic algorithm, helping us to learn more about pulsating white dwarf stars (Metcalfe et al., 2000, 2001). In section 4, I will discuss what this method has allowed us to learn about white dwarfs, but for the bulk of this review I will focus on the method and related issues.

1.1. Reality, Models and Physics

Before I get to that, I would like to step back and take a philosophical look at the general process that we use to learn anything about the objects we study (see Figure 1). Out there somewhere there is a 'real world' that we pass through various filters and selection effects to get 'observations'. As observers, we try our best to compensate for every effect between the real world and the data point, but it's important to realize that we use models in this process too.

With the observations in hand, we devise computer models to try to explain them, doing our best to include all of the relevant physical processes that are in principle detectable. Generally these models have a number of tunable parameters, and we do our best to adjust them until the predictions of the model agree as closely as possible with the observations – in our case, generally the pulsation periods of a star. When we have found a model that adequately reproduces the observations, we assume that the values of the parameters tell us something about the properties of the actual star. But we should never forget that what we are actually dealing with are *models* of reality, and not reality itself. When we derive values of $\log g$ and $T_{\rm eff}$ from spectral lines, for example, we are not measuring the mass and temperature of the star – we are (at best) deriving the optimal match between our models of stellar atmospheres and the extracted spectrum over a finite wavelength interval. We call this approach the 'forward method', and it can only tell us what is best *within the context of the models we use*. It cannot tell us that we are using the wrong models, unless or until we actually try different models.

2. Optimization and Objectivity

With this caveat in mind, the process of trying to adjust our model parameters to fit the observations is really just an optimization problem. Mathematicians have devised various methods, each with their strengths and weaknesses, to approach such problems. Imagine that the two axes of the plots shown in Figure 2 represent the two parameters of a model, and that the height of the surface for each combination

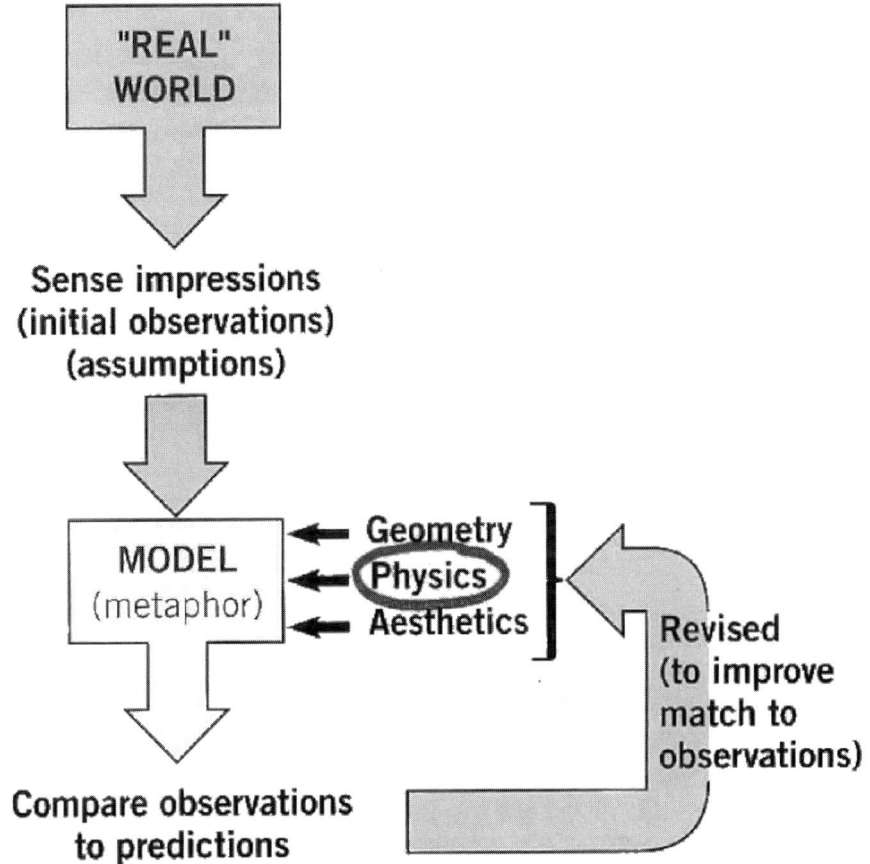

Figure 1. A representation of the logical distinction between reality, observations, and models of reality. Note that the ultimate goal of this process is an improved understanding of the constitutive physics.

of parameters is some measure of how well the model matches the observations. If the surface looks like the plot on the left, then just about any optimization method will work. But if the surface looks more like the plot on the right, then most traditional optimization methods will fail miserably, most likely ending up on top of one of the smaller peaks – effectively finding a locally optimal fit, rather than the globally optimal solution. Of course, the trouble is that in general we do not *know* what the shape of this surface will be for a given model unless we actually evaluate it at each of these points. And that's exactly what we want to avoid by using an optimization method in the first place: we want the solution as quickly as possible, but we also want it to be the global solution. In section 3.1, I will demonstrate how a genetic algorithm offers a nice tradeoff between these two competing demands.

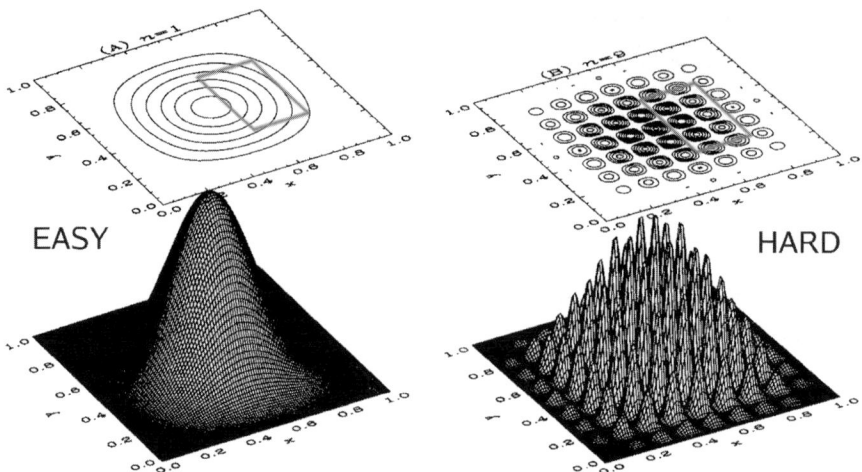

Figure 2. Example model-spaces that would be easy (left) and hard (right) for traditional optimization methods. Note how the likely outcome in the two cases would differ if the search were confined to the region specified by the grey rectangle in the contour plots (Adapted from Charbonneau, 1995).

Before even choosing an optimization method, there is a more fundamental way that we can bias our final result: by defining the range of our search too narrowly. Once again, if the model space is simple, as in the left side of Figure 2, we will probably get a final result at the edge of our search range if we have defined it too narrowly. But if the model space is more complicated, we can find a 'global' solution well inside our search range that is not really globally optimal. For this reason, it is important that we define the limits of our search as broadly as possible – constrained only by the physics of the model, and by observations.

For the models of pulsating white dwarfs that I have been using, for example, we adjust five different parameters: the stellar mass, the effective temperature, the mass of the surface helium layer, and two parameters to describe the internal carbon/oxygen profile. Our search range includes masses between 0.45 and 0.95 $M_\odot$: white dwarfs with lower masses are expected to have a helium core, and almost all white dwarfs with known masses are below our upper limit (Napiwotzki et al., 1999). For now we have been concentrating on the simplest, helium-atmosphere (DB) white dwarfs, so we allow temperatures between 20,000 and 30,000 K, which easily encompasses the spectroscopic temperatures of all known pulsators whether or not trace amounts of hydrogen are included in the atmospheres (Beauchamp et al., 1999). We allow the surface helium layer mass to be anywhere from 10^{-2} M_* (a larger mass would theoretically lead to nuclear burning at the base of the layer) down to a few times 10^{-8}, close to the limit where our models no longer pulsate (Bradley and Winget, 1994). We allow the internal composition to range from pure carbon to pure oxygen, and we use the fifth parameter to specify the fractional mass location where the composition begins to change from a uniform C/O mixture in the center, which is what we expect from evolutionary models (Salaris et al., 1997). The location of this composition transition is expected to be reasonably close to the half-mass point of the model, but we allow it to be anywhere from 0.1 to 0.85 in fractional mass.

3. Model Fitting

Once we have defined the search range, the only 100% reliable method of finding the globally optimal set of model parameters for a given set of observations is to calculate an entire grid. The main thing to decide at this point is the appropriate resolution of the grid in each parameter. It is common to do this by deciding how long we want to wait for the computer to finish, and which parameters are the most interesting. But even if we had infinite computing power, it makes no sense to calculate a grid so fine that adjacent points give fits that differ by less than the observational noise. A good way to quantify the uncertainties on observed pulsation periods is to identify all of the combination frequencies in the power spectrum, and see how much they differ from what is expected based on the measured parent frequencies.

When we did this for a pulsating DB white dwarf, we found deviations of a few hundredths of a second on pulsation periods that were between 400-800 seconds (Metcalfe et al., 2000). This implied that it made sense to calculate about 100 points along each of the five parameters within the ranges specified above. The full grid at this resolution would require 10^{10} model evaluations, which would take more than a year to finish even if we had 1000 of today's fastest processors. If we wanted this kind of resolution for just 1, 2, or possibly 3 of the five model parameters, then maybe a full grid would make sense, especially if we could reuse it for observations of many objects. But certainly for problems with a higher number of free parameters, and to avoid recalculating the grid every few years when the physics gets updated significantly, genetic algorithms can provide almost everything that a grid search can, at a small fraction of the computational cost.

3.1. GENETIC ALGORITHMS

The basic idea behind a genetic algorithm is fairly simple: it is just an iterative Monte Carlo method that samples the model space randomly, but keeps a sort of memory of what worked well in the past. It accomplishes this through a computational analogy with the idea of biological evolution through natural selection. The model parameters serve as the genetic building blocks, and the observations provide the selection pressure. It starts just like a simple Monte Carlo, where we generate N random sets of parameters, evaluate the model for each set, and then compare them to the observations. The genetic algorithm treats each set of parameters as an individual in a population, and assigns each a 'fitness' based on how well it matches the observations. Next, it selects from this population at random, with the fittest individuals more likely to survive. Here comes the weird part: it encodes the parameters into simple strings of numbers, sort of like chromosomes; it pairs them up and performs operations that are analogous to breeding and mutation, and then decodes the strings back into numerical values for the parameters. Now we have a new population, so we evaluate the model for each case again, and continue the whole process until the population converges to one region of the model space.

This is a bit abstract, so let me illustrate how it works in practice, using a 3-parameter example from the white dwarf problem. The top panel of Figure 3 shows the initial random sample of the model space: each point in the left side of the plot corresponds one-to-one with a point in the right side – this is a front and side view of the model-space. There are 128 points in the sample, and the best one is shown as a black open square. Initially the best root-mean-square (rms) difference between the observed and calculated pulsation periods is 4 seconds, but after 80 iterations (or 'generations' of the genetic algorithm) the sample is starting to narrow in on one region of the model-space, and the best solution is substantially better than anything in the original sample. At this point, if we were to compare the observed and calculated periods, we would judge the fit to be pretty good. But as I mentioned at the beginning, in the context of Wampler's Screwdriver: the genetic algorithm *keeps looking*. After 200 generations, it has found the globally optimal solution within this modeling framework.

It is worth noting that along the way to the final solution, the genetic algorithm has evaluated quite a few models, and most of them fall around better-than-average areas of model-space. So, in the end we also get a fairly detailed map giving some sense of the uniqueness of the final solution (see Figure 4). In this case we see that our optimal solution has a relatively thick surface helium layer, but there is also a family of models with thin helium layers that do not match the observations as well, but still match them better than average. This may be telling us something about unmodeled structure in the real star (Dehner and Kawaler, 1995; Brassard and Fontaine, 2003), or possibly about the limitations of the models we are using (Montgomery, 2003).

3.2. HARE & HOUND

One question that I have not answered yet is: how do we decide when to stop the genetic algorithm? The answer is what is usually called a 'Hare & Hound' exercise. If you search the Internet for 'Hare & Hound', you will probably find some rather gruesome photographs of a pack of hunting dogs chasing after a little rabbit and collectively ripping it apart when they catch it. I do not like the implication of such an image, but I will use the term anyway. What is usually meant by 'Hare & Hound' is a blind test of the analysis method, where you pass simulated data through a process to see how well (and how often) you can recover the input data.

SEISMIC INFERENCE USING GENETIC ALGORITHMS 147

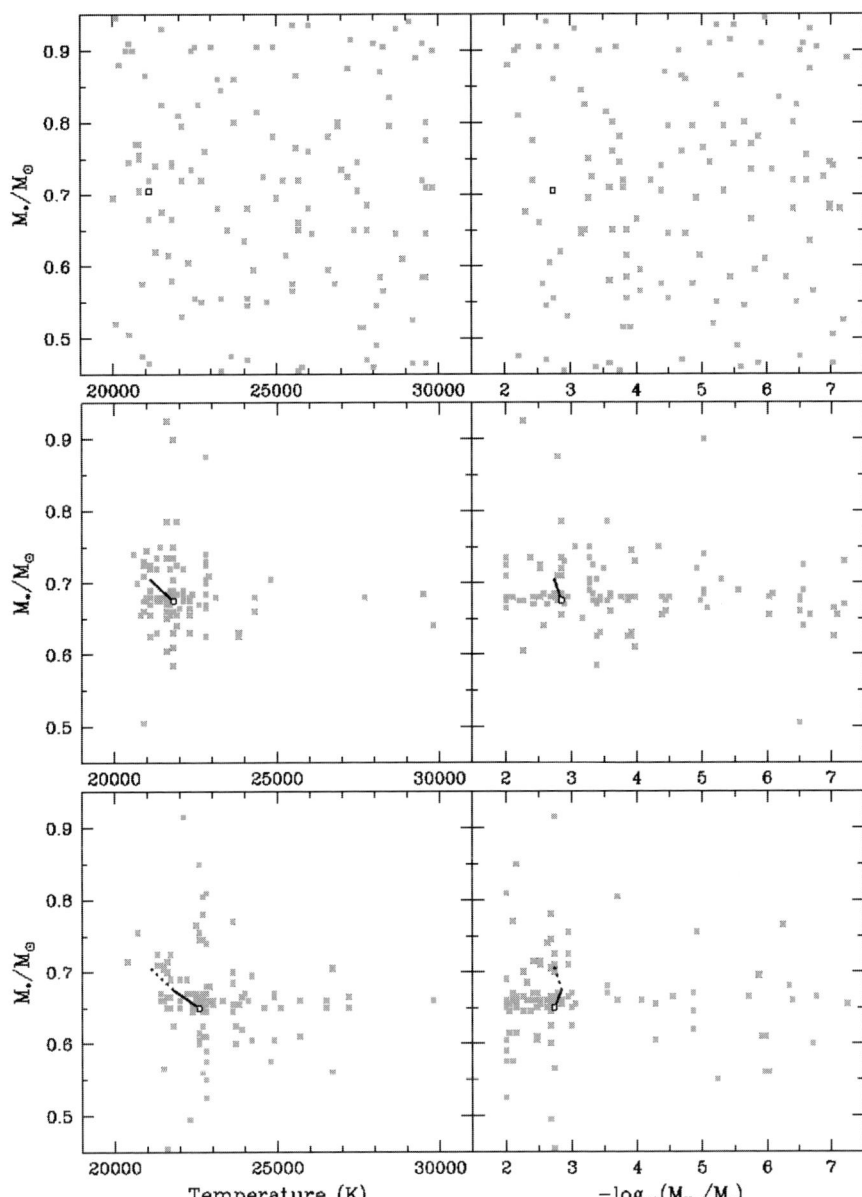

Figure 3. Front and side views of the white dwarf model-space, with snapshots for generation 0 (top), 80 (middle), and 200 (bottom). The best parameter-set in each case is indicated by a black open square, and yields root-mean-square residuals of 4.0, 1.8, and 1.5 seconds in the three panels respectively. The initial convergence is reasonably fast, but the genetic algorithm *keeps looking* for a better solution, and eventually finds the globally optimal set of model parameters.

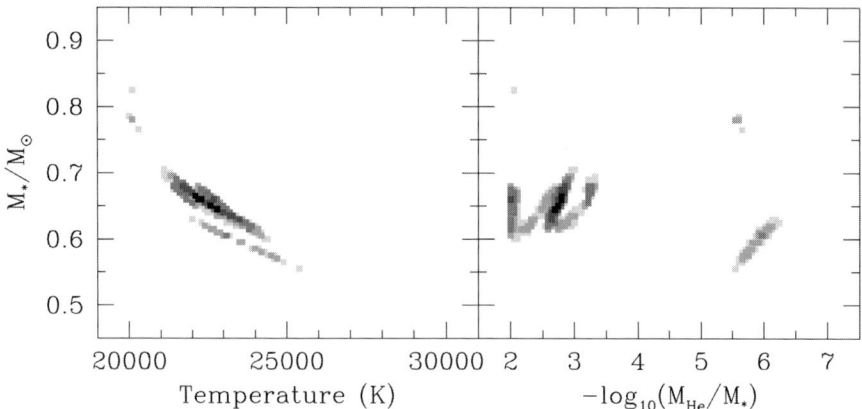

Figure 4. Front and side views of the 3-parameter white dwarf model-space, showing every model evaluated by the genetic algorithm with rms residuals smaller than 3 seconds. The darkness of each point indicates the relative quality of the fit. Note the existence of a second family of solutions with thin helium layer masses.

Continuing with our example from the 3-parameter application to white dwarf models, we used the model to calculate the pulsation periods of a theoretical white dwarf with reasonable values for the mass, temperature, and surface helium layer mass. Then we picked out the pulsation periods corresponding roughly to those we had observed in a real white dwarf and used the genetic algorithm to try to find the model parameters that most closely reproduced those periods. Since the genetic algorithm relies on many random processes, it is inevitable that in a finite number of generations it will sometimes fail to find the input model parameters from this test. To quantify the success rate, we repeated this experiment 20 times with different random number sequences (Metcalfe et al., 2000). In each case, we saw a rapid improvement after the initial sample, followed by a series of incremental improvements brought about by random processes. Out of the 20 tests, 9 found the exact input parameters within 200 generations, and an additional 4 finished with parameters that were close enough to the input model that a small grid, 11 points on a side, would reveal the exact values. The other 7 runs got stuck in local minima. So, for any given run we have a success probability of the method – genetic algorithm plus small grid – of about 65%. If we only run it once, there is a 35% chance that we will not find the globally optimal solution. But by running it several times, we gradually increase the probability that we will find the globally optimal solution, though we pay for this with more model evaluations. If we do five independent runs, there is a 99.5% chance that we will succeed at least once.

3.3. PROBLEM SCALE

The five independent runs, each with 200 generations of 128 trial parameter sets, required that we compute the same number of models as a grid with half the resolution in each dimension. The lesson here is that for small enough problems, the genetic algorithm is not much more efficient than a grid. However, when we scale up to 4 or more parameters, it becomes *much* more efficient (Metcalfe et al., 2001). So our hare and hound test established two things: (1) it told us how many generations we should let the genetic algorithm run, and (2) it implied that there is some minimum problem size that should be attempted with the genetic algorithm – for smaller problems, a model grid might be more efficient.

As an interesting aside, there is also a maximum problem size that should ever be attempted, and this is set by the rate at which computing power increases. There is a well known empirical relation called Moore's Law, which notes that computing power roughly doubles every 18 months – this has been true since the 1960's. Because of this trend, if a problem is big enough you can actually get it done faster by waiting for the computing power to increase before you start it. There was a very humorous paper published on astro-ph a couple of years ago on this topic (Gottbrath et al., astro-ph/9912202), but the basic conclusions were sound. The authors found that any problem requiring more than 26 months on the fastest computer presently available should not be attempted until the future. As a rule of thumb for the genetic algorithm, a model requiring about 5 minutes to run on today's fastest available processor will not finish within this 26 month limit on a single machine.

3.4. PARALLEL COMPUTING

Even if the problem were able to finish in say 24 months, few of us would be willing to wait for two years to get the final answer anyway. Fortunately, we do not have to confine ourselves to one of today's fastest processors: we can use many of them. This is one of the great things about genetic algorithms – or even model grids for that matter: they are inherently parallelizable. We need to calculate many models, and each of them is independent of the others. So the number of available processors sets the number of models that we can evaluate in parallel. Also, there is very little communication overhead in this process: we send parameter values out to each processor, and we get back either a list of pulsation periods, or just a goodness of fit measure if the periods have already been compared to the observations.

As part of my PhD thesis, I built a 64-processor Linux cluster at the University of Texas (Metcalfe and Nather, 2000). At the time, dual processor machines were much more expensive than single-processor systems, so it was actually less expensive to have a separate box for each processor. Now, because dual-processor systems are much cheaper, and because computing power has more than quadrupled, we have built a new system that exceeds the speed of the old system in a much smaller space. We can scale this new system up to many more units if we need the computing power – but our problems just don't demand it yet.

In the future, I would guess that stability and space considerations will be the most important factors for machines like this, and the so-called 'bladed Beowulf' concept has already reached the point where it is now possible to put 24 1-GHz processors – none of which require a fan – into a small cabinet the size of a single desktop computer case (Feng et al., 2002). The speed of these processors is upgradeable with improved *software*, written as part of his day job by Linus Torvalds, the creator of the Linux operating system. The big market for these machines, of course, is for massive web servers – but as a nice side-effect they will also provide cheap parallel processing for scientists.

4. Application to White Dwarfs

To give you a sense of the potential of this method, let me summarize what this has done for white dwarf model-fitting. Allowing five adjustable parameters (as outlined in section 2), the genetic algorithm can find the optimal core composition with a precision of a few percent. This is a very significant result in itself, because the core composition can affect the derived ages of white dwarfs by up to 3 Gyr (Fontaine et al., 2001), which has implications for using white dwarfs as chronometers to date stellar populations. But the central C/O ratio in a white dwarf formed through single-star evolution can also lead to a measurement of the important $^{12}C(\alpha, \gamma)^{16}O$ nuclear reaction rate.

When a white dwarf is being formed in the core of a red giant star during helium burning, the $^{12}C(\alpha, \gamma)^{16}O$ and 3α reactions compete for the available helium nuclei. As a result, the final central C/O ratio is primarily determined by the relative rates of these two reactions. The 3α rate is relatively well determined, but the $^{12}C(\alpha, \gamma)^{16}O$ reaction is much harder to measure, and so its rate is quite uncertain – leading to a broad range of expected central C/O ratios. By applying the genetic algorithm method, we can now *measure* the central C/O ratio. By combining this result with evolutionary models that produce internal chemical profiles, we can tune the $^{12}C(\alpha, \gamma)^{16}O$ rate for a model with a given mass until we match the derived central C/O ratio (Metcalfe et al., 2002).

We have now applied this method to observations of two pulsating white dwarfs, and in the latest results they both yield a reaction rate consistent with the value derived from laboratory measurements, though they are at best still marginally consistent with each other (Metcalfe, 2003). We still have more work to do, and crucial tests will come as we apply the method to additional white dwarfs as observations become available. But we have come a long way from the subjective, local model-fitting methods of the past.

5. Conclusions and Discussion

I hope that I have convinced you that genetic algorithms are potentially a very powerful tool for asteroseismology. They can provide objective global optimization for problems with more than a few parameters, and in the process they yield fairly detailed maps of the model-space, which we can use to judge the uniqueness of the final result. To convince yourself that a genetic algorithm is more efficient than a large grid, and to convince others that the final result can be trusted, you should definitely perform a 'Hare & Hound' exercise – trying to match your models to the observables from another model, and demonstrating that it can be done. Remember that the speed of your computer effectively determines the size of the problems you can attack. If you want to solve a specific problem, you can and should determine how much computing power you will need to solve it in a reasonable time. Finally, the payoff can be quite high, as it was when we applied this method to our white dwarf models.

I have seen this method begin to transform white dwarf asteroseismology from a field where the theory was being driven by the observations, to one where new observations are being driven by the theory. I hope that it can help launch revolutions in the seismological analysis of other types of stars too.

References

Beauchamp, A. et al.: 1999, *ApJ* **516**, 887.
Bradley, P.A. and Winget, D.E.: 1994, *ApJ* **421**, 236.
Brassard, P. and Fontaine, G.: 2003, in: R. Silvotti and D. de Martino (eds.), *Proc. of the 13th European Workshop on White Dwarfs*, (in press).
Charbonneau, P.: 1995, *ApJS* **101**, 309.
Dehner, B.T. and Kawaler, S.D.: 1995, *ApJ* **445**, L141.
Feng, W. et al.: 2002, *LANL Technical Report LA-UR 02-1210*.
Fontaine, G., Brassard, P. and Bergeron, P.: 2001, *PASP* **113**, 409.
Metcalfe, T.S.: 2003, in: R. Silvotti and D. de Martino (eds.), *Proc. of the 13th European Workshop on White Dwarfs*, (in press).
Metcalfe, T.S. and Nather, R.E.: 2000, *Baltic Astronomy* **9**, 479.
Metcalfe, T.S., Nather, R.E. and Winget, D.E.: 2000, *ApJ* **545**, 974.
Metcalfe, T.S., Winget, D.E. and Charbonneau, P.: 2001, *ApJ* **557**, 1021.
Metcalfe, T.S., Salaris, M. and Winget, D.E.: 2002, *ApJ* **573**, 803.
Montgomery, M.H.: 2003, in: R. Silvotti and D. de Martino (eds.), *Proc. of the 13th European Workshop on White Dwarfs*, (in press).
Napiwotzki, R. et al.: 1999, *ApJ* **517**, 399.
Nather, R.E.: 1995, *Baltic Astronomy* **4**, 117.
Salaris, M. et al.: 1997, *ApJ* **486**, 413.
Wampler, E.J. et al.: 1973, *Nat* **246**, 203.

STELLAR INVERSIONS

SARBANI BASU

Astronomy Department, Yale University, PO Box 208101, New Haven, CT 06520-8101, USA

Abstract. Inversions of oscillation frequencies have proved to be extremely powerful in studying the structure of the Sun. We examine the conditions under which we can invert oscillation frequencies of other stars to study their structure. We show that despite the very limited number of modes that are expected to be observed in other stars, we can perform inversions that will give localized information of the stellar core. These results can limit the space of admissible models of a given star.

1. Introduction

Helioseismology has proved to be a powerful tool in studying the structure and dynamics of the Sun. Not much attention has been paid so far to astero-seismology of Sun-like stars, mainly because of the lack of data. Seismology of solar-type stars is expected in the not-too-distant future to provide information of relevance for understanding stellar structure. Already there are ground based observations of oscillations of stars such as α Centauri (Bouchy and Carrier, 2001, 2002), even though individual frequencies have not been determined. The situation is going to change soon with data from space-based experiments. Observations using the guide telescope of the WIRE satellite have already succeeded in determining the mode frequencies of a giant (α UMa; Buzasi et al., 2000). Thus it is now time to develop tools that will allow us to use the full potential of the information provided by the seismic observations of other stars. In the case of the Sun, inversion of oscillation frequencies have given us detailed information about the structure and dynamics of the Sun (see e.g., Gough et al., 1996; Basu et al., 1997, 2000; Schou et al., 1998; etc.). It is not yet clear whether inversions will prove as valuable in asteroseismology as it has with helioseismology. The advantage of doing an inversion is that the results provide *localized* information about the structure of the star under study.

In this review we discuss the conditions required to invert meaningfully the frequencies of stars other than the Sun. We first discuss how frequencies are inverted and some early attempts using artificial data sets. We then discuss how the number of modes available for the inversions, the data-errors etc., affect the inversion results. I shall restrict the discussion to linear inversion methods. These methods have the advantage that the statistical properties of the solution, such as errors, are well defined. Non-linear inversion techniques are discussed elsewhere (see Roxburgh and Vorontsov, this volume).

2. The Inverse Problem

Stellar oscillation frequencies are characterized by three numbers, the radial order n, the degree l and the azimuthal order m. The number n is basically the number of nodes in the radial direction, l the numbers of nodes along the circumference of the star, and m the number of nodes along the equator. The value of m ranges from $-l$ to l. In case of a spherically symmetric star, all modes with the same value of n and l have the same frequency, and this frequency depends on the structure of the star. In the presence of rotation or magnetic fields, each mode of a given n and l splits into $2l+1$ modes, one for each value of m. If rotation is slow, or the magnetic field is weak, then the frequencies of the $2l+1$ modes in an (n, l) multiplet can be averaged to get a single frequency for the multiplet characterized by a given n and l and this frequency can be used to determine the structure of the star. For stars other than the Sun, we expect to observe low degree modes ($l \leq 3$) only.

Inversions for stellar structure are based on linearizing the equations of stellar oscillations around a known reference model. If we consider two functions f_1 and f_2, then the relative differences of f_1 (i.e., $\delta f_1/f_1$) and the relative difference in f_2 (i.e., $\delta f_2/f_2$) between a star and a reference stellar model can be related to the differences in the frequencies of the star and the model ($\delta\omega_i/\omega_i$),

$$\frac{\delta\omega_i}{\omega_i} = \int K^i_{f_1,f_2}(r)\frac{\delta f_1}{f_1}dr + \int K^i_{f_2,f_1}(r)\frac{\delta f_2}{f_2}dr + \frac{F_{\rm surf}(\omega_i)}{Q_i}, \tag{1}$$

where r is the normalized distance to the centre. The index i numbers the multiplets (n, l). The kernels $K^i_{f_1,f_2}$ and $K^i_{f_2,f_1}$ are known functions of the reference model. The term in $F_{\rm surf}(\omega_i)$ is the contribution from the uncertainties in the near-surface region (e.g. Christensen-Dalsgaard and Berthomieu 1991); here Q_i is the mode inertia, normalized by the inertia of a radial mode of the same frequency.

In the case of the Sun, the squared sound-speed c^2 and density ρ are used as functions f_1 and f_2. However, other variables can be used too. If it is assumed that the equation of state of the star is known, then the equation can be re-written in terms of $u \equiv P/\rho$ (where P is the pressure), and the helium abundance Y. In the case of the Sun, this combination is often used to determine the helium abundance.

For linear inversion methods, the solution at a given point r_0 is determined by a set of inversion coefficients $c_i(r_0)$, such that the inferred value of, say, $\delta f_1/f_1$ is

$$\left\langle\frac{\delta f_1}{f_1}(r_0)\right\rangle = \sum_i c_i(r_0)\frac{\delta\omega_i}{\omega_i}. \tag{2}$$

From the corresponding linear combination of equations (1) it follows that the solution is characterized by *the averaging kernel*, obtained as

$$\mathcal{K}(r_0, r) = \sum_i c_i(r_0) K^i_{f_1,f_2}(r), \tag{3}$$

and also by the cross-term kernel:

$$\mathcal{C}(r_0, r) = \sum_i c_i(r_0) K^i_{f_2, f_1}(r) , \qquad (4)$$

which measures the influence of the contribution from $\delta f_2/f_2$ on the inferred $\delta f_1/f_1$.

The surface term in equation (1) may be suppressed by assuming that F_{surf} can be expanded in terms of polynomials ψ_λ, and constraining the inversion coefficients to satisfy $\sum_i c_i(r_0) Q_i^{-1} \psi_\lambda(\omega_i) = 0$, $\lambda = 0, 1, ..., \Lambda$ (Däppen et al. 1991), where Λ is the degree of the highest degree polynomial used.

The goal of the inversions is to obtain a localized averaging kernel, while suppressing the contributions from the cross term and the surface term in the linear combination in equation (2), and limiting the error in the solution. Also, $\mathcal{K}(r_0, r)$ must have unit integral with respect to r. If this can be achieved, then

$$\left\langle \frac{\delta f_1}{f_1}(r_0) \right\rangle \simeq \int \mathcal{K}(r_0, r) \frac{\delta f_1}{f_1} dr \qquad (5)$$

defines a proper, localized, average of $\delta f_1/f_1$.

3. Earlier Work

There have been several early attempts at inverting artificial sets of low-degree modes. Gough and Kosovichev (1993, henceforth GK93) inverted a set of 64 modes of degrees 0-2. They were attempting an inversion to obtain the u profile of a 1.1M$_\odot$ mass and $q \equiv M/R^3 = 0.468$ with a solar model ($M = 1$M$_\odot$, $q = 1$) as the reference model. It could be argued that their assumed errors (0.1μHz) in the mode frequencies was too low to be realistic. However, their results were encouraging.

Roxburgh et al. (1998, henceforth R98) inverted a set of 120 modes of degrees 0-3. The error on each mode frequency was assumed to be 0.3μ Hz. They inverted to find the u profile of a 0.8M$_\odot$ star of age 1.9×10^{10} years with a reference model of the same mass but a different age (and hence different radius and q), 1.2×10^{10} years. The u profile thus obtained was used to calculate the density profile under the assumption of hydrostatic equilibrium. The results are shown in Figure 1. We can see that the profile obtained by inversions is very close to the actual profile. There is some mismatch and that shall be discussed in Section 7.

Basu et al. (2001, henceforth BA01) tried to invert for the squared sound-speed (c^2) profile of two sets. Set 1 had 47 modes of with l =0-2, and Set 2 has 65 modes with l =0-3. They tried to invert for the sound-speed profile and were unsuccessful in their attempts. Basu et al. (2002, henceforth BA02) were however successful in inverting the same two mode-sets to obtain u. The reason for this is discussed in Section 4.

Berthomieu et al. (2001) did not adopt an *ad hoc* set of modes, but tried to determine the mode set and errors from a simulated power-spectrum. They carried

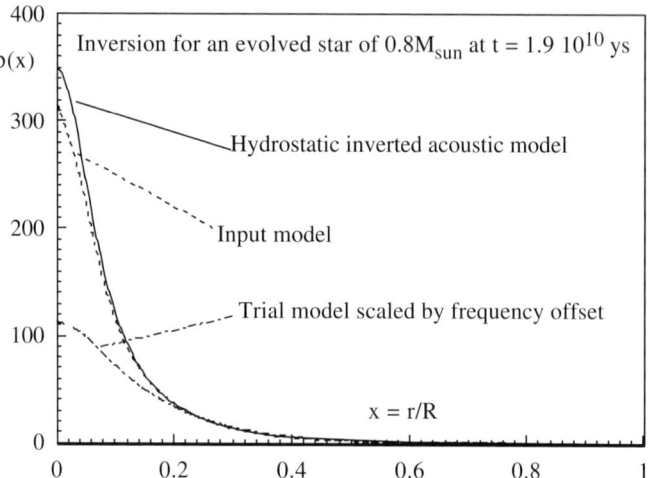

Figure 1. The inverted and actual density profiles of a 0.8 $M_\odot$ star.

out a careful analysis of the results to be expected in inversions of solar-like oscillations, taking into account the expected errors in the COROT observations and mode amplitudes in a 1.45 $M_\odot$ star. They obtained two sets of modes. With a conservative selection criterion, they obtained a set of 89 modes, and with a less conservative one they obtained 105 modes. The errors ranged from a few μHz to a minimum of 0.1μHz. Although they did not show any inversion results, they did show that they can form well-localized averaging kernels for c^2 in the core. However, despite their larger mode-set (they used the set with 105 modes) their averaging kernels were not as good as those of GK93. Thus, the early results have been very mixed.

4. Effect of Variables Chosen

As seen from from the preceding section, different groups have had varying amounts of success. What is clear though is that inversions for u have been successful, while those for c^2 have either not been successful, or the averaging kernels have not been as localized as one would wish for.

The kernels for Y are almost zero everywhere except at the helium ionization zones, this makes it easy to reduce the influence of the helium abundance on the u inversion results. It is known that density kernels (at fixed c^2) have much larger amplitudes than Y kernels (at fixed u), and thus it is difficult to minimize the effect of density in sound-speed inversion results. This makes c^2 inversions more difficult than u inversions. The smaller Y kernels leads to better averaging kernels and smaller cross-term kernels for u than for c^2 for the same mode-set. This can be seen in Figure 2. The better averaging kernels lead to better inversion results.

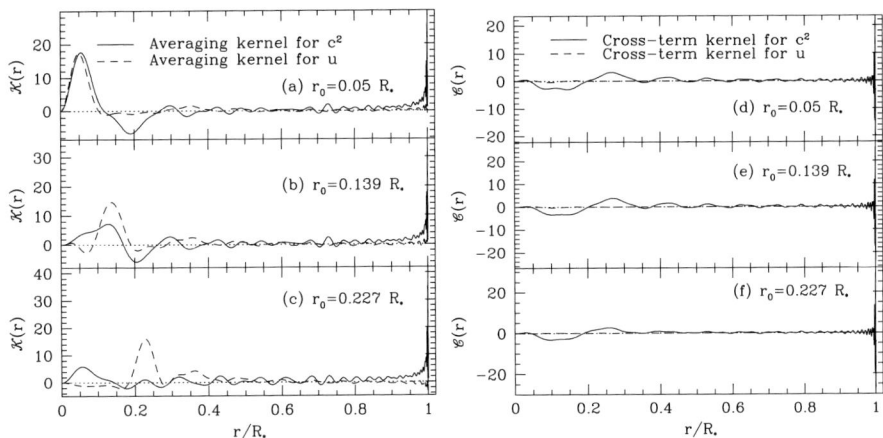

Figure 2. The averaging kernels (left column, panels a–c) and cross term kernels (right column, panels d–f) obtained for c^2 and u inversions with Set 2 of BA02.

The question then is why not use u always? The construction of kernels for the c^2, ρ combination does not involve any assumption about the physical inputs in the stellar model (except for hydrostatic equilibrium). Calculation of kernels for the u, Y combination on the other hand requires the assumption that the equation of state of stellar material is known exactly, since the calculation involves derivatives of the equation of state. It has been shown by Basu and Christensen-Dalsgaard (1997) that this introduces systematic errors in the case of inversions to determine the solar u profile. The reason for this is that the equations of state used for constructing solar models deviate from the equation of state of solar plasma in some regions (mainly in the ionization zones). In the case of the Sun, the systematic errors are larger than the random errors, and hence need to be avoided. For other stars, the errors in the inversion results due to data errors are expected to be larger than the systematic erors introduced by assuming an equation of state and hence the systematic errors can be ignored. Thus, our chances of successfully inverting asteroseismic data will increase if we use the u, Y combination for inversions.

5. Effect of Mode Set

The effect of the mode set used has been discussed in detail by BA02. We review those results here. In order to compare the effect of the mode-sets used, the mode-sets of R98, GK93, and Sets 1 and 2 of BA02 were used. The modes sets were all used to invert for u (to eliminate differences due to differences in the kernels used for the inversions). All sets were assumed to have an error of 0.3μHz. To avoid uncertainty in results due to uncertainty in the radius and mass of the star under study, we invert one solar model with respect to another. The reference model used in this work is the solar Model S of Christensen-Dalsgaard et al. (1996). This is

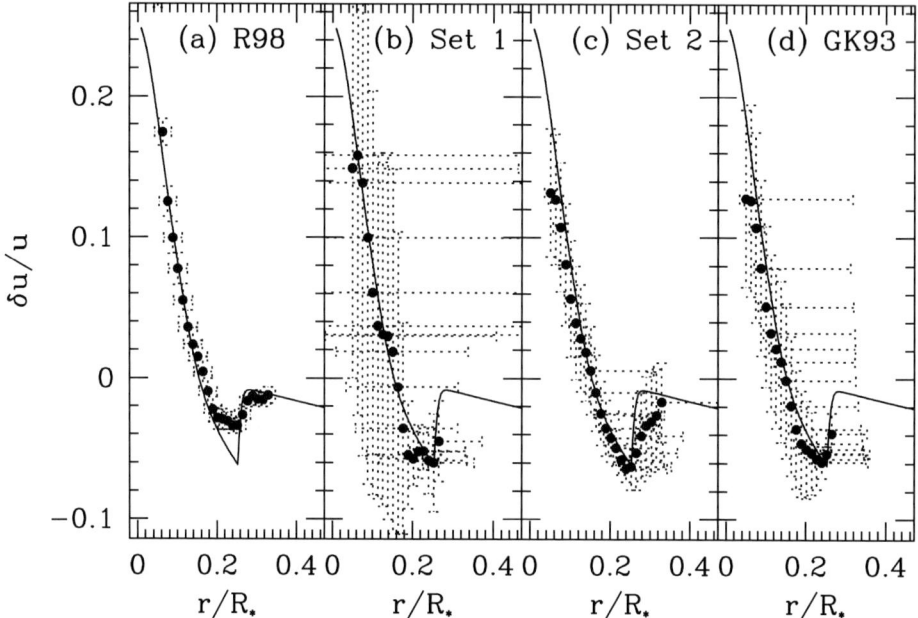

Figure 3. Results of u inversion for the mode-sets of R98 (panel a) Set 1 (panel b) and Set 2 (panel c) of BA02 and the mode set of GK93 (panel d). The continuous line is the exact difference between the models. Inversion results are plotted at the corresponding target radius. The vertical error bars show the 1-σ uncertainty in the inversion result, obtained by propagating the mode uncertainties through the inversion process. The horizontal error bars provide an indication of the resolution of the inversion; they extend from the first to the third quartile point of the averaging kernels.

a standard solar model. As the test model (the proxy star) we use model MIX of Basu, Pinsonneault and Bahcall (2000). This is a non-standard solar model, with an artificially mixed core. This model was selected because the differences in u with respect to the reference model are large, somewhat along the lines of what we expect for stars other than the Sun.

The results of the inversions can be seen in Figure 3. It can be seen that the inversion results of the R98 set has the best resolution as well as the smallest errors. The results of Set 1 are the worst, while GK93 is acceptable, though the errors are large and the resolution degrades with increasing radius very quickly. The results from Set 2 are very reasonable, though not as good as those obtained from the R98 set. The R98 set is perhaps a bit too optimistic and we do not expect to get such a mode-set from observations. The other sets are more realistic. If we can get sets like Set 2 and GK93, we should be able to invert the sets successfully. Although Set 2 and GK93 have almost the same number of modes (65 and 64 respectively), GK93 does not have any $l = 3$ modes. The $l = 3$ modes have a larger lower turning point than $l = 0, 1$ or 2 modes and these help in making the averaging kernels mode localized towards the high r end, thereby improving the resolution

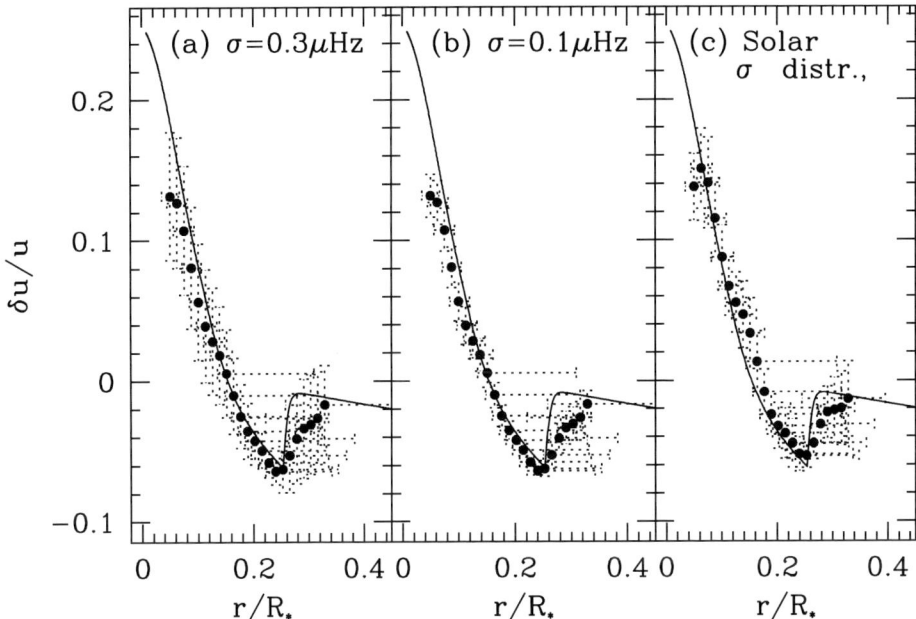

Figure 4. Results of u inversion with Set 2. Different error distributions were used for the inversions whose results are shown in the different panels.

obtained for Set 2. Our chances of success are lower if the observed set turns out to be more like Set 1.

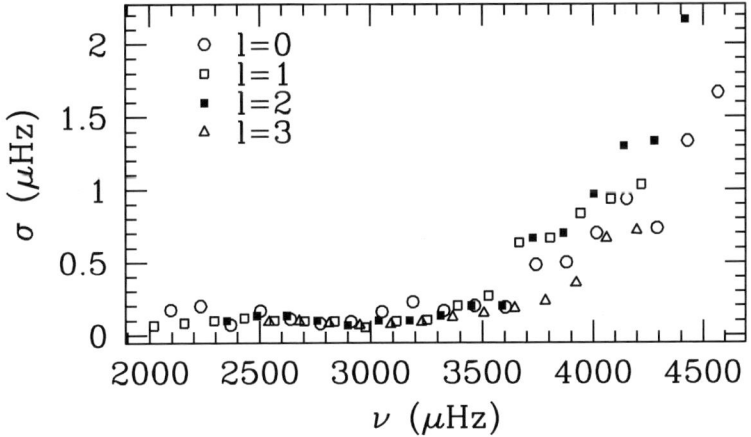

Figure 5. The distribution of errors used for the inversions shown in Figure 4(c). The errors were scaled from the error distribution of solar oscillation frequencies so that the low-frequency part of the distribution is around $0.1\,\mu$Hz.

6. Effect of Errors

Lowering the errors on the mode frequencies will enable us to get better results. The effects can be felt in two ways: one could improve the resolution, while keeping the error on the solution the same; or, one could keep the averaging kernel the same and decrease the error on the solution. Figure 4 shows the inversion results for Set 2 for three different error-distributions: (1) uniform errors of $0.3\,\mu$Hz, (2) uniform errors of $0.1\,\mu$Hz, and (3) an error-distribution which mimics the distribution of errors in solar oscillations frequencies. The distribution was scaled to be around $0.1\,\mu$Hz at low frequency (see Figure 5 for the error distribution).

As can be seen from the above figure, decreasing the errors, improves the inversion results. The reduction of errors gives a more dramatic improvement for Set 1 results. If the errors are reduced by a factor of 5, Set 1 can be inverted to give results that are acceptable.

7. Effect of Differences in q

One of the serious difficulties facing stellar inversions when compared to solar inversions (apart from the much smaller number of modes) is that we usually do not know the mass and radius of the star under study to a satisfactory degree of accuracy. As a result, we cannot construct reference models which have the same mass and radius as the star. As we shall see in this section, this affects what we can do. Independent measures of mass and radius thus are important.

One of the most noticeable effects of the difference in M and R is the frequency differences between the reference model and the star in question. Frequencies of stars scale as $\sqrt{(M/R^3)}$, and hence a difference in $q \equiv M/R^3$ results a constant relative frequency difference between two stars. This is over and above any frequency difference due to other differences between the stars. Thus in general, the right-hand side of equation (1) may also contain a term $\frac{1}{2}\delta \ln(M/R^3)$ to absorb scaling of the frequencies with stellar mass M and radius R. GK93 and Berthomieu et al. (2001) treated the term as an addition to equation (1) and obtained the inversion result by minimizing the influence of this term on the solution in a manner similar to that of minimizing the effect of the second function (f_2 in Eq. 1). R98 on the other hand effectively subtracted out a constant from the relative frequency difference and inverted the residual.

In Figure 6 we show the relative frequency differences between Model S and the model of a $0.9 M_\odot$ star which has a radius of $0.9 R_\odot$ (therefore, with $q = 1.235$). The frequency differences for the R98 set are shown. As can be seen, the predominant trend is a constant frequency difference. Also shown in the figure, the residual relative frequency difference left after subtracting out a constant. The constant was determined by a least-squares fit to modes with frequency greater than 1500μHz. The result of inverting the frequencies is shown is Figure 7. Results are shown for

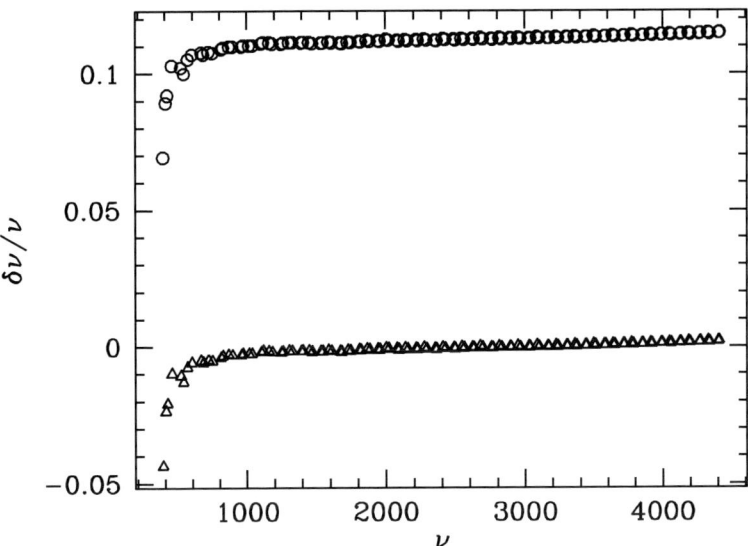

Figure 6. The relative frequency differences between a $0.9M_\odot$, $0.9R_\odot$ model and a standard solar model (circles). The residuals obtained after subtracting a constant are shown as triangles. The results obtained by inverting the residuals are shown in Figure 7.

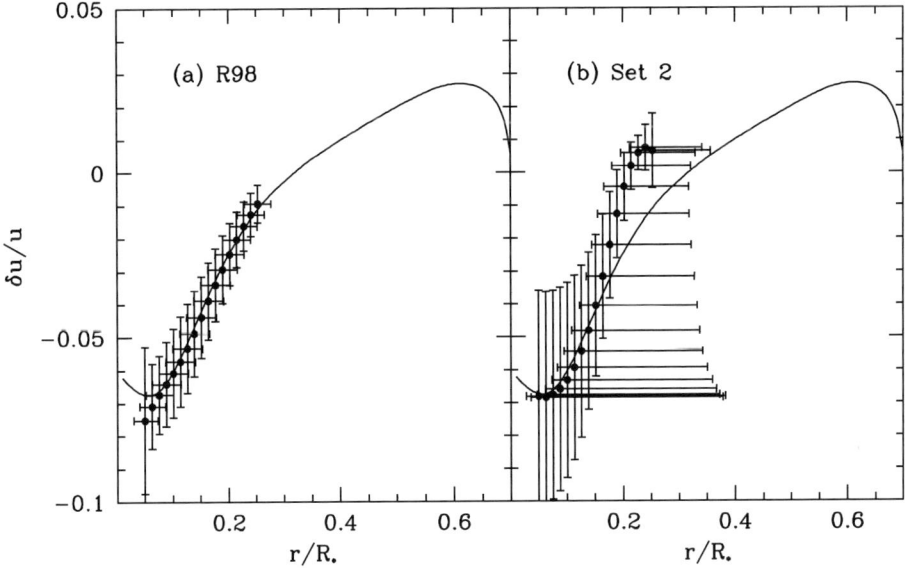

Figure 7. The inverted (points) and actual (continuous line) relative u difference between a $0.9M_\odot$, $0.9R_\odot$ model and a standard solar model. Results obtained with the R98 mode-set (panel a) and Set 2 of BA02 (panel b) are shown.

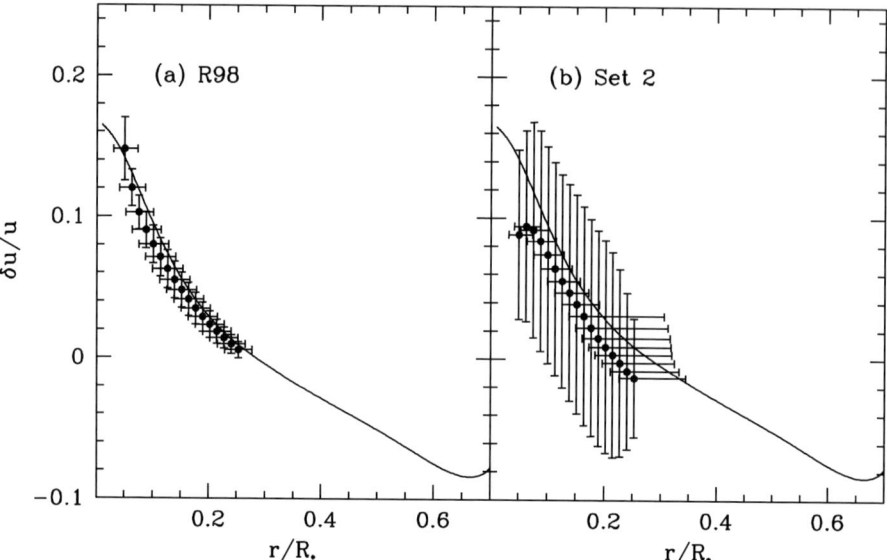

Figure 8. The inverted (points) and actual (continuous line) relative u difference between a $1.1 M_\odot$, $1.1 R_\odot$ model and a standard solar model. Results obtained with the R98 mode-set (panel a) and Set 2 of BA02 (panel b) are shown.

both R98 and Set 2 mode-sets. As we can see, despite the large difference in q, we get decent inversion results.

A set of inversion results for a $1.1\,M_\odot$, $1.1\,R_\odot$ ($q = 0.826$) stellar model is shown in Figure 8. Again we see that we get decent inversion results.

The situation is very different though for the inversion of a $1.1\,M_\odot$, $1.16 R_\odot$ ($q = 0.7047$) model (see Figure 9). As can be seen in the figure, there is a systematic offset between the actual difference and the inverted difference. The difference is clearer for the R98 set than for Set 2 because of the lower errors and better resolution of the former inversions. A similar effect can be seen in Figure 1, where the inverted profile is higher than the input profile. The reason behind this turns out to be very simple: inversion kernels are usually calculated using the dimensionless form of variables, the inverted result obtained in the process is the relative difference between the dimensionless form of the variables. This effect is not seen in Figure 8 and Figure 7 because the dimensionless form of u is given by $u' = uR/M$, and R/M for the reference and test models is the same in these cases. In Figure 10 we compare the inversion results shown in Figure 9 with $\delta u'/u'$, the relative difference of the dimensionless u and see that now the results match. Thus we can conclude that unless we have independent measures of the mass and radius of a star, inversions will only give us localized information about the dimensionless form of the variables. This has never been an issue in helioseismic inversions since there the reference model is always constructed with $1 M_\odot$ and $1 R_\odot$.

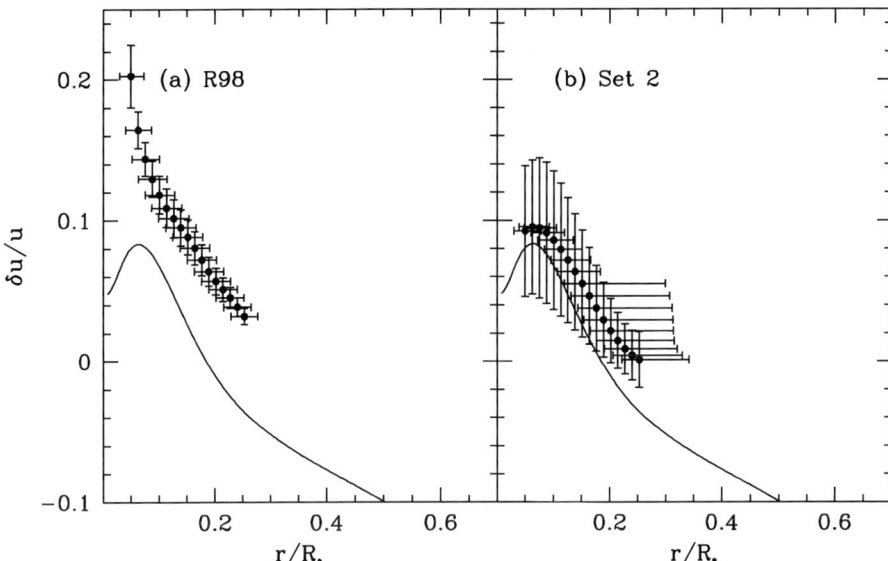

Figure 9. The inverted (points) and actual (continuous line) relative u difference between a 1.1M$_\odot$, 1.16R$_\odot$ model and a standard solar model. Results obtained with the R98 mode-set (panel a) and Set 2 of BA02 (panel b) are shown.

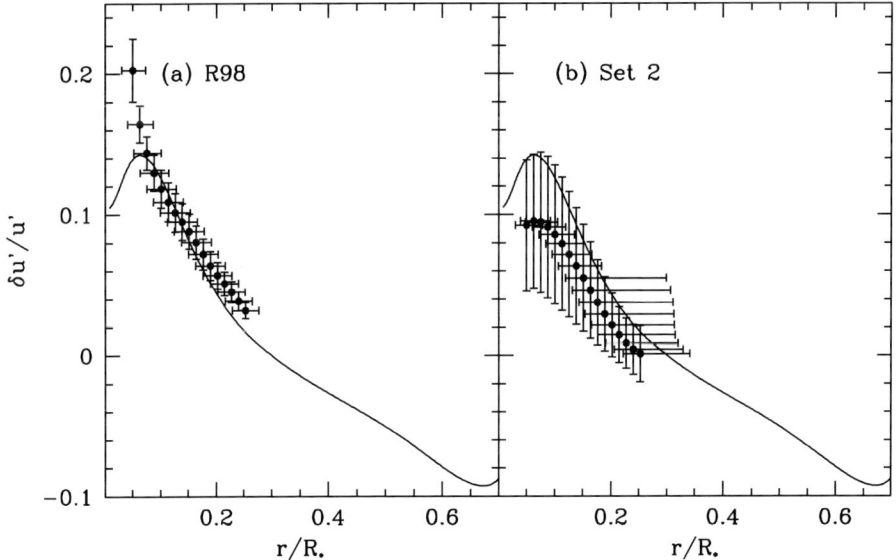

Figure 10. The inverted (points) and actual (continuous line) relative differences of u', the dimensionless u between a 1.1M$_\odot$, 1.16R$_\odot$ model and a standard solar model. Results obtained with the R98 mode-set (panel a) and Set 2 of BA02 (panel b) are shown.

8. Conclusions

From the results presented above, we conclude that it is possible to invert oscillation frequencies of solar-types stars. The larger the number of modes that we have frequencies for, the better is the result and we find that the presence of $l = 3$ modes in the data-sets help in improving the solution. Lowering data-errors is also important. We also find that one can obtain better inversion results when one inverts for $u \equiv P/\rho$ with the helium abundance Y as the second variable, instead of inverting for the squared sound-speed c^2 with the density ρ as the second variable. While inverting stars other than the Sun, one must keep in mind that the inversions give us information only about the dimensionless form of the variables, and we need an independent estimate of the mass and radius of the star in order to recover the actual measures of the variables.

References

Basu, S. and Christensen-Dalsgaard, J.: 1997, *A&A* **322**, L5.
Basu, S., Pinsonneault, M.H. and Bahcall, J.N.: 2000, *ApJ* **529**, 1084.
Basu, S., Christensen-Dalsgaard, J., Monteiro, M.J.P.F.G. and Thompson, M.J.: 2001, in: A. Wilson (ed.), *Proc. SOHO 10/GONG 2000 Workshop: Helio- and Asteroseismology at the Dawn of the Millennium*, ESA SP-**464**, 407 [**BA01**].
Basu, S., Christensen-Dalsgaard, J. and Thompson, M.J.: 2002, in: F. Favata, I.W. Roxburgh, D. Galadi-Enriquez (eds.), *Proc. 1st Eddington Meeting 'Stellar Structure and habitable Planet Finding'*, ESA SP-**485**, 249 [**BA02**].
Basu, S., Chaplin, W.J., Christensen-Dalsgaard, J., Elsworth, Y., Isaak, G.R., New, R., Schou, J., Thompson, M.J. and Tomczyk, S.: 1997, *MNRAS* **291**, 243.
Basu, S., Turck-Chièze, S., Berthomieu, G., Brun, A.S., Corbard, T., Gonczi, G., Christensen-Dalsgaard, J., Provost, J., Thiery, S., Gabriel, A.H. and Boumier, P.: 2000, *ApJ* **535**, 1078.
Berthomieu, G., Toutain, T., Gonczi, G., Corbard, T., Provost, J. and Morel, P.: 2001, in: A. Wilson (ed.), *Proc. SOHO 10/GONG 2000 Workshop: Helio- and Asteroseismology at the Dawn of the Millennium*, ESA SP-**464**, 411.
Bouchy, F. and Carrier, F.: 2001, *A&A* **374**, 5.
Bouchy, F. and Carrier, F.: *A&A* **390**, 205.
Buzasi, D.L., Catanzarite, J., Laher, R., Conrow, T., Shupe, D., Gautier, T.N., III, Kreidl, T. and Everett, D.: 2000, *ApJ* **532**, 133.
Christensen-Dalsgaard, J. and Berthomieu, G.: 1991, in: A.N. Cox, W.C. Livingston and M. Matthews (eds.), *Solar Interior and Atmosphere*, University of Arizona Press, Tucson, 401.
Christensen-Dalsgaard, J., Däppen, W., Ajukov, S.V. et al.: 1996, *Science* **272**, 1286.
Däppen, W., Gough, D.O., Kosovichev, A.G. and Thompson, M.J.: 1991, in: D.O. Gough and J. Toomre (eds.), *Lecture Notes in Physics vol. 388*, Springer, Heidelberg, 111.
Gough, D.O. and Kosovichev, A.G.: 1993, in: W.W. Weiss and A. Baglin (eds.), *Proc. IAU Colloq. 137: Inside the Stars*, A.S.P. Conf. Ser. **40**, 541 [**GK93**].
Gough, D.O., Kosovichev, A.G., Toomre, J. et al.: 1996, *Science* **272**, 1296.
Schou, J., Antia, H.M., Basu, S. et al.: 1998, *ApJ* **505**, 390.
Roxburgh, I.W., Audard, N., Basu, S., Christensen-Dalsgaard, J. and Vorontsov, S.V.: 1998, in: J. Provost and F.-X. Schmider (eds.), *Proc. IAU Symp. 181: Sounding Solar and Stellar Interiors (poster vol.)*, Nice Observatory, 245 [**R98**].

ON THE PRINCIPAL ASTEROSEISMIC DIAGNOSTIC SIGNATURES

DOUGLAS GOUGH

[1] *Institute of Astronomy, Madingley Road, Cambridge, CB3 0HA, UK*
[2] *Department of Applied Mathematics and Theoretical Physics, Silver Street, Cambridge, CB3 9EW, UK*
[3] *Jila, University of Colorado and National Institute of Science and Technology, Boulder, Colorado, 80309, USA*

Abstract. Many stellar model calibrations against seismic data will utilize just the so-called large and small frequency separations. In this report I explain in physical terms, based on resonance conditions for constructive interference between simple waves, why those frequency separations depend on the variation of sound speed through the star in the manner in which they do. By so doing I hope to increase appreciation of the nature of the stellar calibrations, and thereby provide a means for assessing the credibility of the results.

Keywords: Asteroseismology, wave theory

1. Introduction

The basic structure of the seismic spectrum of any star exhibiting solar-like oscillations (i.e. a range of high-order intrinsically stable acoustic modes – necessarily of low degree because the high-degree modes cannot be detected – excited to low amplitude by turbulence in the convective envelope immediately beneath the photosphere where the convective timescales are comparable with the oscillation periods) is generic: it is composed of a row of groups of peaks, the central frequencies of which are approximately uniformly spaced, with spacing Δ, say, the individual peaks within a group being produced by subgroups of modes of alternately even and odd degree – $l = 0$ and 2, and $l = 1$ and possibly 3 – which are separated in frequency by $d_l \simeq (1 + 2l/3)d$, in which $l - 0$ for the even-degree group and $l = 1$ for the odd-degree group, and d is almost constant. The solar spectrum (e.g. Chaplin et al., 1996) is a typical example. The different mode groups of a given parity arise from modes of different orders n. Moreover, there is, in addition, fine structure in the subgroups produced by the splitting of modes of like degree l and order n but different azimuthal orders m, which is caused by agents such as rotation or a magnetic field which break the symmetry of an otherwise spherically symmetrical star. Under a broadish envelope, which has a single maximum and a characteristic width comparable with but typically somewhat less than the mean frequency, the total power in each of the groups of peaks is expected to be about the same, save for statistical fluctuations with standard deviation approximately equal to the mean. In this brief report I shall address the diagnostic value of only the frequencies of the modes.

In the case of the Sun, both the angular velocity and the magnetic field intensity are dynamically small on a global scale, and the frequency splitting is on the whole less than the line widths in the acoustical spectrum. Consequently the fine structure is difficult to measure. This is bound to be the case too of many of the stars whose seismic spectra will be measured in imminent asteroseismic missions. Accordingly, I shall restrict attention here to only the dependence of the mode frequencies on n and l. Moreover, I shall not even discuss subtle inferences that might be drawn from the detailed distribution of the frequencies of the spectral peaks, but instead I shall concentrate on just the two principal quantities that describe it: Δ and d. These are commonly called, respectively, the large and the small frequency separations. They are the quantities that are the most readily measured, and they will be (and have been) the first quantities used to compare with, and calibrate, theoretical models. My intention is to explain in physical terms the properties of the star that determine the values of those parameters, to elucidate the meaning of the calibration procedure, and thereby assist in the appreciation of the outcome. Specifically, I represent a mode of oscillation as a standing interference pattern of (oppositely directed) propagating sound waves (or, as in my first example, of portions of a single wave) that are reflected in the surface layers of the star, showing that Δ is determined principally by the sound travel time of such a wave from one point of reflection to the next, and that d is determined principally by refraction induced by the sound-speed gradient.

The dependence of Δ and d on stellar mass M_* and age t_* (for given chemical composition) has been discussed by Christensen-Dalsgaard (e.g. 1988) for post ZAMS stars near the main sequence: both Δ and d decrease as M_* increases at fixed t_*, at least when t_* is small; and both decrease as t_* increases at fixed M_*. The partial derivatives with respect to M_* and t_* are not equal, so in principle a simultaneous calibration for both M_* and t_* can be carried out, provided that all other pertinent properties of the star (such as initial chemical composition X_0) can be determined. I shall not discuss in detail how Δ and d depend on M_*, t_* and X_0; instead I shall address simply how they depend on the variation of sound speed c through the star. How that is related to the global parameters M_*, t_* and X_0 of the star has been discussed by Christensen-Dalsgaard (1993), and some rough relations, based on homology scaling, have been presented by Gough (1990).

The dependence of Δ and d on sound speed can be deduced from the formula

$$\omega_{n,l} \sim \left(n + \tfrac{1}{2}l + \varepsilon\right)\omega_0 - [Al(l+1) - \delta]\omega_0^2/\omega \quad \text{as } n \to \infty, \tag{1}$$

which can be obtained from the asymptotic analysis by Tassoul (1980) of high-order modes of low degree, where

$$\omega_0 = \left(\frac{1}{\pi}\int_0^R c^{-1} dr\right)^{-1} \tag{2}$$

and

$$A = \frac{1}{2\pi\omega_0}\left[\frac{c(R)}{R} - \int_0^R \frac{1}{r}\frac{dc}{dr}dr\right],\qquad(3)$$

in which r is a radial coordinate and R is some fiducial radius of the star, and ε and δ are integrals of the structure of the star (depending predominantly on conditions near the reflecting surface) that do not depend on the properties of the mode of oscillation.

From Eq. (1) can be deduced approximate formulae for the large and the small frequency separations:

$$\Delta \simeq \omega_0 \qquad(4)$$

and

$$d \simeq 6A\omega_0^2/\omega \simeq 6A\omega_0/n',\qquad(5)$$

where $n' = n + \frac{1}{2}l + \varepsilon$.

Equations (1)–(3) depend principally on the sound speed; that is to be expected of acoustic waves, of course. Strictly speaking, acoustic propagation in a star depends also on density variation and buoyancy (and the perturbation to the gravitational potential induced by the waves, which was ignored in deriving Eqs (1)–(3) but was included by Tassoul (1990) in a subsequent analysis). But when ω is large these produce only small perturbations to the dynamics, except near the surface where rapid density variation induces reflection: consequently the integrals ε and δ depend on density ρ, but I shall pay no attention here to how they so depend. I can represent the asymptotic dynamics of the waves by the standard wave equation which, when the waves have constant frequency ω, reduces to the Helmholtz equation:

$$\nabla^2\Psi + \frac{\omega^2}{c^2}\Psi = 0.\qquad(6)$$

The characteristics of the reflection at the surface will be incorporated into the boundary condition.

The principles that I explain in this discussion can be generalized to the more realistic conditions in which buoyancy, density stratification and the perturbation to the gravitational potential are included (e.g. Gough, 1993), but I refrain from such generality here in order not to obfuscate my arguments with superfluous algebraic complication.

I shall develop the theory by considering a sequence of four idealized examples of increasing complexity, each of which requires the introduction of just one or two new concepts. Moreover, to keep matters simple, I assume the background state to be steady, and in all but the last example I apply the condition $\Psi = 0$ on the boundary. In the final example I reproduce essentially Eqs (1)–(3), save that the formula for A is changed somewhat, apparently providing a somewhat more faithful representation of the values of the true eigenfrequencies and of how those values depend on the sound-speed gradient.

2. Oscillations of a Two-Dimensional Waveguide

Consider longitudinal linearized standing waves in a two-dimensional waveguide occupying with respect to a Cartesian coordinate system (x, y) the narrow domain $-R \leq x \leq R$ and $-d \leq y \leq d$ with $d \ll R$. (The width d is not to be confused with the small frequency separation of the previous section.) In some respects this may be regarded as a primitive one-dimensional model of a star, but I shall not pursue that analogy here. Instead I shall later use the model to represent a great-circle strip around a spherical star. The waveguide is uniform and steady, aside from the oscillations, supporting a constant wave speed c, so that the perturbation Ψ to some scalar physical quantity satisfies equation (6) precisely to linear order. I regard Ψ to be the Lagrangian pressure perturbation δp, and I impose the condition that Ψ vanishes on the boundary of the domain:

$$\Psi = 0 \text{ on } x = \pm R \text{ and on } y = \pm d. \tag{7}$$

Of course, the Cartesian components of the associated displacement or velocity **u** also satisfy Eq. (6), so long as c is constant, but later I shall consider waves in circular or spherical geometry, in which the components of a vector with respect to the natural polar coordinate systems do not behave like scalars, so for didactic purposes it is preferable to restrict attention from the outset to variables Ψ that can survive subsequent generalization. The reason for having δp in mind is that it generalizes perhaps the most easily to stellar conditions, although of course when the background state is strictly uniform, as it is in this example, there is no difference between Lagrangian and Eulerian perturbations, and perturbations of all thermodynamical variables are proportional to each other in linearized theory. However, there is no harm here in drawing attention to the conceptual distinction, because even in the recent literature spurious conclusions have been drawn by adopting Lagrangian constraints on Eulerian variables.

As is well known, the solutions of Eq. (6) are wave-like with wavenumber $\mathbf{k} = (k_x, k_y)$ satisfying $k^2 := |\mathbf{k}|^2 = \omega^2/c^2$. They can be either standing modes or propagating waves, each being obtained as an appropriate linear superposition of examples of the other. Thus, we may have

$$\Psi = \Psi_0 \cos k_y y \, [\cos(-k_x x - \omega t) + \cos(k_x x - \omega t + \hat{\delta})]$$
$$= 2\Psi_0 \cos k_y y \, \cos(k_x x + \tfrac{1}{2}\hat{\delta}) \cos(\omega t - \tfrac{1}{2}\hat{\delta}), \tag{8}$$

where the amplitude Ψ_0 is constant. The first term in the first line is a wave propagating in the negative x direction which is incident on the boundary at $x = -R$;

the second term is the reflected wave, which has been presumed to have suffered a phase shift $\hat{\delta}$. There are similar solutions, antisymmetric in y, with dependence $\sin k_y y$. The imposition of the boundary conditions (7) requires of solutions (8) that $k_x = n'\pi/2R$, $\hat{\delta} = -\pi$ or 0 according as n' is even or odd, and that $k_y = l'\pi/2d$, where n' and l' are integers (which, without loss of generality, I may take to be positive) and in which, having longitudinal propagation (motion almost in the x direction) in mind, I have selected only the set of modes that are fundamental in y (i.e., with the gravest nontrivial y variation: $l' = 1$), which is why I do not consider explicitly the y-wise antisymmetric solutions; and I am particularly interested in the limit as $n' \to \infty$. The dispersion relation $\omega = kc$ (for the propagating waves) becomes the eigenfrequency equation for the standing waves, namely

$$\omega = \left(n + \tfrac{1}{2}l - \tfrac{1}{2}\right)\frac{\pi c}{R}\sqrt{1 + \frac{l'^2 R^2}{n'^2 d^2}} \simeq \left(n + \tfrac{1}{2}l - \tfrac{1}{2}\right)\omega_0 + \hat{A}l'^2\omega_0^2/\omega, \qquad (9)$$

in which I have set $n' = 2n + l - 1$, $n = 1, 2, \ldots$, with $l = 0$ pertaining to x-wise symmetric modes and $l = 1$ to antisymmetric modes; also $\omega_0 = \pi c/R$ and $\hat{A} = R^2/2d^2$. I hasten to add that, strictly speaking, l here is simply a parameter to denote symmetry, and does not represent the cross-wise (i.e., y-wise) variation; that is given by the second term in the square root in Eq. (9): it is quadratic in the ratio l'/n', a dependence characterizing rectangular geometry. It will be evident later that in spherical geometry the frequencies of the high-n modes first sense modest horizontal variation linearly, rather than quadratically; moreover, the quadratic dependence has the opposite sign to that in Eq. (9).

The two components in the first of Eqs (8) may be regarded as guided waves with y-varying amplitude trapped in $|y| < d$, propagating (oppositely) in the x direction and subject to the dispersion relation (9). The phase speed (when $l' = 1$) is given by

$$c_\Phi \equiv \omega/k_x = c\sqrt{1 + R^2/n'^2 d^2} = c\sqrt{1 + k_y^2/k_x^2} > c; \qquad (10)$$

it is the speed at which the wave pattern travels. The group velocity $\mathbf{v}$ is in the x direction and has magnitude

$$v = \frac{\partial \omega}{\partial k_x} = \frac{c^2 k_x}{\omega} = \frac{c^2}{c_\Phi} < c; \qquad (11)$$

it is the velocity at which conserved quantities, such as ω, would be carried by a packet of waves generated with a (narrow) range of frequencies (e.g. Gough, 1993), such as by forcing the boundary at $x = -R$, say, to oscillate with amplitude $\tilde{\Psi}_0(t)$ and frequency $\omega(t)$ that vary (slowly) with time.

In preparation for the analysis of the subsequent examples, and, I hope, to shed a little more light on what is happening in this, I now present a second discussion of the eigenfrequency Eq. (9), this time in terms of rays of resonant waves. The principle is to compute how phase is propagated about the region, and to insist that wherever a wave that has been reflected from the boundary returns to the neighbourhood of any point to interfere with itself, it does so constructively. Since the background state is independent of time, and because the boundary condition permits no loss of energy from the system, ω is real (and constant), and I need not be concerned that the wave might have changed its character during the time it takes to return; only its phase changes, and that change is all that I need to compute in order to determine the eigenfrequencies. From a mathematical point of view, it is perhaps most prudent to compute the phase variation over the region directly from the asymptotic wave equation. This is what was done originally by Brillouin (1926, corrected by Keller, 1958) on the basis of earlier work by Einstein (1917). Its attraction is that with the help of some general results about phase one can transform phase integrals to simplify the evaluation of the resonance conditions (quantization conditions) that result. However, here I take a more physically motivated but less rigorous approach, which is probably conceptually simpler, at least if one doesn't try to think too deeply about it (cf. Gough, 1984). It has the advantage of being directly related to telechronoseismology (time-distance seismology) which is used on the Sun.

I imagine phase to be propagated along a ray, which is considered to be an infinitesimally thin tube whose axis is a line of the group velocity $\mathbf{v}$ – that is to say, it is everywhere parallel to $\mathbf{v}$ – and whose surface is composed of lines of $\mathbf{v}$. When it is necessary I shall refer to the 'central' line of $\mathbf{v}$ (actually any line of $\mathbf{v}$ in the limit of an infinitesimally thin tube) as the ray path. Moreover, I use the word tube in preparation for my future discussion in three dimensions; in two dimensions one might more appropriately call it a strip. Throughout most of my discussion it will be adequate to refer to the rays (ray tubes) and the ray paths both simply as rays.

I now have in mind that propagating along the ray is a simple (locally plane) wave, with sinusoidal variation $\Psi_0 \cos \Phi$, where $\Phi = \int k \mathrm{d}s - \omega t + \hat{\delta}$ is the phase at a distance s along the ray ($\hat{\delta}$ being a constant) and in which ω and k are related by the dispersion relation $\omega = kc$. In principle the amplitude Ψ_0 could be spatially varying, although in this (and only this) example it is not. Since the dispersion relation (which has been chosen for presentational simplicity) is actually nondispersive, the group velocity and the phase velocity of the basic wave are the same, namely c. The arguments that follow can easily be generalized to truly dispersive propagation (Gough, 1993).

Before proceeding with the discussion of the resonance conditions it is worth noting that because the surface of a ray tube is an envelope of lines of $\mathbf{v}$, no line of $\mathbf{v}$ can cross it. Geometrically, the ray tube is like a thin magnetic flux tube: magnetic flux through the tube is invariant along the length of the tube, and is equal to the product of the magnetic intensity and the cross-sectional area of the tube. In wave theory the ray flux is the product of the wavenumber and the square of the amplitude Ψ_0 of Ψ, which after multiplication by the cross-section a of a ray tube is constant along the tube (Keller, 1958): $\Psi_0^2 k a = $ constant.

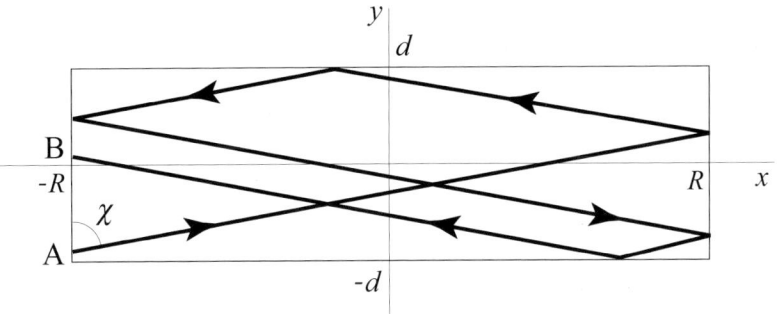

Figure 1. Portion of a ray (thick lines with arrows indicating the direction of propagation) in a rectangular waveguide. The interference pattern formed by a simple wave travelling along the ray is the standing wave given by equation (8).

A portion of a typical ray is illustrated in Figure 1. It starts from point A on the boundary at $x = -R$ and returns to that segment of the boundary at point B after reflecting, in the example illustrated, twice at the far end $x = R$ of the waveguide (and once at $x = -R$, of course), suffering single reflections on each long portion of the boundary at $y = \pm d$. Thus, in total, there are n_{rx} (=4) reflections in the x direction (counting $\frac{1}{2}$ at each of A and B) and n_{ry} (=2) reflections in the y direction. All reflections of the ray are specular, so all rays are inclined, either positively or negatively, from the y direction by the same angle χ. Since the ray is parallel to $\mathbf{k}$, $\tan \chi = k_x/k_y$. The wave function Ψ is the interference pattern formed by the infinite continuation of the ray illustrated in Figure 1, which in general covers the entire domain (except when the ratio R^2/d^2 is rational, which amongst all rectangular domains has probability zero). The time to propagate phase from A to B is $T = n_{rx}(2R/c)\operatorname{cosec}\chi = 2n_{rx}(R/c)\sqrt{1 + k_y^2/k_x^2}$.

In the neighbourhood of any point in the domain there are four types of ray segment, one directed positively in x and positively in y, one directed positively in x and negatively in y, and two antiparallel to these, but I think it is probably simpler not to spell out the distinction every time, hoping that it will be obvious from the context whether I actually mean one type or all four types of ray segment. I should point out, however, that it is the interference between only parallel ray segments that can be constructive everywhere. It should therefore be clear now why the length of the portion of the ray depicted in Figure 1 was so chosen: it follows one of the shortest ray paths, starting on the left short segment of the boundary, that ends immediately before being reflected into a ray segment that is parallel to the first.

It must be recognized that because Ψ is constrained to vanish on the boundary, the amplitude of the reflected wave should have the opposite sign to that of the incident wave. It is probably more convenient, however, to regard the amplitude always to be positive, and instead to consider the phase of the wave to be retarded by

π on reflection. (One can easily convince oneself that it is appropriate to consider the phase to be retarded and not advanced by considering there to be a medium outside the domain in which the density decreases with distance from the boundary with a scale length H that is sufficiently small ($H < 2\omega/c$) for the mode to be evanescent – see Figure 3 – and then letting $H \to 0$ to recover the condition that Ψ vanishes on the boundary; although in this case whether the phase be regarded as advanced or retarded makes no difference to the final eigenfrequency equation.) Thus the phase $\Phi(B, t)$ at B is $\Phi(A, t - T) - n_r \pi = \Phi(A, t) + \omega T - n_r \pi$, there having been a reduction of π from each of the n_r (here $n_{rx} + n_{ry} = 6$) reflections. But, given the sinusoidal variation of the interference pattern, particularly on the boundary, the phase at B can be related to that at A at time t according to $\Phi(B, t) = \Phi(A, t) + k_y \Delta_y$, where Δ_y is the distance (counted positively in the y direction) from A to B, namely $2n_{rx} R \cot \chi - 2n_{ry} d = 2n_{rx} R k_y/k_x - 2n_{ry} d$. For constructive interference these two computations of the phase at B must contribute identically to the wave function Ψ, which implies that the phases can differ by only an integral multiple $\hat{n}$ of 2π. Thus

$$2n_{rx}\omega(R/c)\sqrt{1 + k_y^2/k_x^2} - n_r\pi - k_y\Delta_y = 2\hat{n}\pi . \tag{12}$$

Similarly one may consider a different portion of the ray, beginning and ending on one of the long segments $y = \pm d$ of the boundary, to obtain

$$2n_{ry}\omega(d/c)\sqrt{1 + k_x^2/k_y^2} - n_r\pi - k_x\Delta_x = 2\hat{n}\pi , \tag{13}$$

where $\Delta_x = 2n_{ry}d \tan \chi - 2n_{rx} R = 2n_{ry}dk_x/k_y - 2n_{rx} R$. After setting $2\hat{n} = (n' - 1)n_{rx} + (l' - 1)n_{ry}$, Eqs (12) and (13), together with the dispersion relation $\omega^2/c^2 = k^2 = k_x^2 + k_y^2$, may be solved for k_x, k_y and ω to yield $k_x = n'\pi/2R$, $k_y = l'\pi/2d$, and the eigenfrequency Eq. (9).

Because neighbouring ray paths are parallel, the cross-sections of all rays are constant along the ray paths, as is k, and consequently the amplitude Ψ_0 of the waves is constant too.

This ray calculation is superficially more elementary than my first calculation because it does not require the solution of a partial differential equation. The justification for its validity rests on the asymptotic properties of the solutions of Eq. (6) as $\omega \to \infty$, which, in this simple example (because ω^2/c^2 is constant), are actually exact – but to establish that justification in general requires considerable further analysis. However, if one is prepared simply to accept the basic wave-interference idea, one is (almost) ready to proceed to more complicated, and pertinent, examples. But before I describe them, I summarize, at the risk of boring the reader by repetition, what has been learned so far from the ray picture: The phase carried by a simple wave along a ray travels with (constant) speed c, as does any other conserved quantity, and hence moves in the x direction at speed $v = c \sin \chi = c/\sqrt{1 + k_y^2/k_x^2}$. This is the group velocity of the interference pattern produced by the wave along the entire (infinitely long) ray, which constitutes what I have called the guided wave. The pattern itself propagates in the x direction at the phase speed, which is simply the speed of the intersection of a wave front (which is perpendicular to the ray) with a line parallel to the x direction, namely $c \operatorname{cosec} \chi$. The frequency of the waves (both the guided waves propagating in the x direction and the basic constituent simple waves propagating along the ray) is given by $\omega = ck$, where k is the total wavenumber.

3. Oscillations of a Circular Drum

As a small step towards a realistic representation of a star, I now consider the two-dimensional acoustic oscillations of an isothermal circular cylinder that do not vary under translation parallel to the axis (i.e., the component of the wavenumber parallel to the axis is zero). This problem is equivalent to determining the oscillations of a circular drum with a uniform isotropic skin that is isotropically stretched; in that case the wave function Ψ is the displacement of the skin (presumed to be perpendicular to the equilibrium plane of the skin), which behaves as a scalar in the two-dimensional space of the equilibrium state. The solution to Eq. (6) satisfying $\Psi = 0$ on $r = R$ is

$$\Psi = J_l \left(j_{ln} \frac{r}{R} \right) \begin{matrix} \cos \\ \sin \end{matrix} l\phi \begin{matrix} \cos \\ \sin \end{matrix} \omega t \tag{14}$$

and

$$\omega = j_{ln} \frac{c}{R}, \tag{15}$$

where J_l is the Bessel function of the first kind of order l, and j_{ln} is the nth zero of J_l. If $n \gg 1$ with moderate l, the case of principal interest here, $j_{ln} \simeq \pi n' - (l^2 - \frac{1}{4})/2\pi n'$ where $n' = n + \frac{1}{2}l - \frac{1}{4}$, whence

$$\omega = \omega_{n,l} \simeq n'\omega_0 - (l^2 - \tfrac{1}{4})\omega_0^2/2\pi^2\omega, \quad n = 1, 2, \ldots, \tag{16}$$

where $\omega_0 = \pi c/R$. The structure of this equation is similar to that of Eq. (1). The large frequency separation $\omega_{n,l} - \omega_{n-1,l}$ is approximately ω_0, and the small separation is given by

$$d_{n,l} = \omega_{n,l} - \omega_{n-1,l+2} \simeq 2(l+1)\omega_0^2/\pi^2\omega_{n,l}. \tag{17}$$

The eigenvalue problem is also solved quite easily by ray theory. A typical ray with fairly large n/l is illustrated in Figure 2(a). The rays are all inclined from the radial direction by a small angle χ, whose tangent is essentailly the ratio of the azimuthal to the radial component of the wavenumber vector. There is consequently a small zone of avoidance around the centre of the drum which the waves do not sense, bounded by a caustic circle which is the envelope of the rays and on which

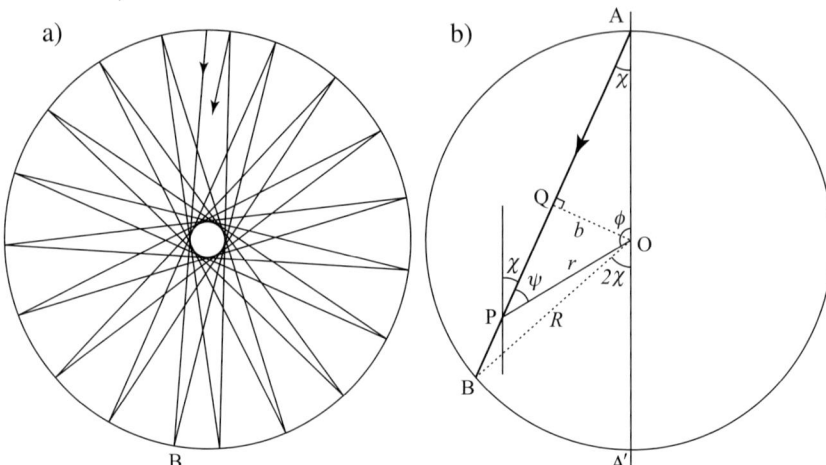

Figure 2. (a) Portion of a ray propagating in a circular drum. It also represents a ray in a plane through the centre of an isothermal sphere. A simple wave propagating in the direction indicated rotates about the centre of the drum always in the anticlockwise sense; unlike the waveguide, the drum has no corners that can cause the sense to change. The interference pattern that is caused by the wave appears as a standing wave in a frame of reference rotating anticlockwise about the centre of the drum with frequency ω/l. The standing wave (14) in an inertial frame is produced by the interference of two similar waves propagating in opposite directions. (b) Segment of the ray illustrated by (a), but with χ increased, for clarity; the thin fiducial line through P is parallel to the axis A'OA.

the amplitude of the waves is asymptotically infinite (although for any real wave, having finite ω, Ψ_0 is finite too – see Figure 3).

The segment of the ray connecting the two points A and B on the boundary is shown in Figure 2(b). It is similar to the corresponding ray segment in Figure 2(a), except that χ has been increased, for clarity. The resonance condition can be computed in a manner similar to that for the waveguide. The time taken to propagate phase from A to B is $T = (2R/c)\cos\chi$, over which passage I presume there to have been a phase retardation of $\alpha\pi$. Thus $\Phi(B, t) = \Phi(A, t) + \omega T - \alpha\pi$; I shall address the value of α later. The interference pattern on $r = R$ has the sinusoidal form $\genfrac{}{}{0pt}{}{\cos}{\sin}l\phi$, and, to ensure that the wave function is unchanged after a rotation of 2π, l is an integer (which, without loss of generality, I may take to be non-negative). Thus the phase difference $\Phi(B, t) - \Phi(A, t)$ is equal, modulo 2π, to $l\phi(B) = (\pi - 2\chi)l$. Therefore

$$\frac{2\omega R}{c}\cos\chi - \alpha\pi - l(\pi - 2\chi) = 2\hat{n}\pi, \quad \hat{n} = 0, 1, 2, \dots. \tag{18}$$

The sine of the angle χ between the ray path and the axis (which at A is perpendicular to the bounding circle) is essentially the ratio of the lateral wavenumber l/R at the surface to the total wavenumber $k = \omega/c$: $\sin\chi = lc/\omega R$. Using this to eliminate χ from Eq. (18) would then close the equation for ω once α is known. It is interesting to note, in passing, that the ϕ component of the wavenumber at any

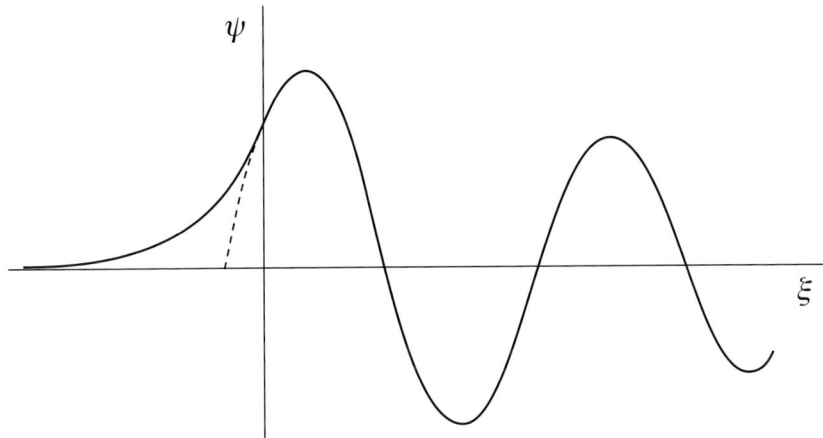

Figure 3. Illustration of the wave function Ψ (continuous curve) in the vicinity of a caustic (at $\xi = 0$). The coordinate ξ represents distance away from the caustic. The dashed curve represents roughly what Ψ might have been had it continued through the caustic in a sinusoidal fashion.

point P on the ray is $k \sin \psi = \omega b/cr \propto r^{-1}$, where b is the perpendicular distance OQ of the ray from the centre O of the circle, a result which generalizes to the isothermal sphere, and also to a sphere in which c varies with r (see §5).

To evaluate α one observes first that there are two half contributions from reflections at A and B, which sum to $2 \times \frac{1}{2}\pi = \pi$. But there is a further term arising from the caustic circle, which has radius $b = R \sin \chi = lc/\omega$. Here the rays cross, and so, algebraically, the cross-section a of the ray changes sign. Recalling that $\Psi_0^2 ka$ is constant along a ray, and that $v = c = $ constant, it follows that Ψ_0^2 changes sign, which formally renders it imaginary after passing the caustic. However, I wish to preserve the reality of Ψ_0, and that can be achieved, regarding $\Psi_0 \cos \Phi$ as $\Re(\Psi_0 e^{i\Phi})$, by introducing a further phase retardation of $\pi/2$. That it be a retardation, rather than an advance, in phase can be established by considering the variation of the interference pattern perpendicular to the caustic and requiring that its amplitude decays (exponentially) inward from the caustic circle. To be strict, that requires an analysis of the appropriate solution of the Helmholtz equation (6) in the vicinity of the caustic (Keller, 1958), but one can see at least the sign of the phase shift, if one accepts the evanescence, simply by matching smoothly the oscillatory behaviour of Ψ on one side of the caustic to a decaying function on the other (see Figure 3). The caustic is at a point of inflexion of Ψ where Ψ and $\partial\Psi/\partial\xi$ have the same sign. The ξ dependence of Ψ is similar to the x dependence in Eq. (8), whence $\Psi \partial\Psi/\partial\xi \simeq -k\hat{\Psi}_0^2 \sin\frac{1}{2}\hat{\delta} \cos\frac{1}{2}\hat{\delta} > 0$ at $\xi = 0$ (where $\hat{\Psi}_0$ is the product of Ψ_0 and the function describing the variation of Ψ with the coordinate s along the ray), whence, assuming $\hat{\delta}$ to lie between $-\pi$ and π, $\hat{\delta} < 0$: phase is indeed retarded (by $\pi/2$), and hence $\alpha = 3/2$.

One can now substitute for χ and α in Eq. (18) to give

$$\frac{\omega}{\omega_0}\left(1 - \frac{l^2\omega_0^2}{\pi^2\omega^2}\right)^{1/2} = n + \tfrac{1}{2}l - \tfrac{1}{4} - \tfrac{1}{\pi}\sin^{-1}\left(\frac{l\omega_0}{\pi\omega}\right), \quad n = 1, 2, \ldots, \quad (19)$$

in which I have replaced $\hat{n}$ by $n - 1$. Our interest is in the asymptotic regime $n \gg l$, in which Eq. (19) may be expanded in powers of l/n to yield

$$\frac{\omega}{\omega_0} \simeq n' - l^2/2\pi^2 n', \quad (20)$$

where $n' = n + \tfrac{1}{2}l - \tfrac{1}{4}$, which almost reproduces Eq. (16). Granted that we are representing the system by a superposition of locally plane waves, both n and l should formally be large, and so there is no significant difference between the numerator l^2 in the second term in Eq. (20) and the factor $l^2 - \tfrac{1}{4}$ in Eq. (16). Note that this difference does not influence the formula (17) for the small frequency separation.

Note that because there is no preferred orientation, and because, in general, the ray path is not closed, the points of reflection are distributed with uniform density on the boundary, and consequently the amplitude Ψ_0 of the interference pattern on the boundary is constant. However, because two neighbouring inwardly directed ray paths converge (and therefore the width of a ray strip decreases inwards), the amplitude Ψ_0 of the wave increases inwards.

4. Oscillations of an Isothermal Sphere

Solutions of Eq. (6) in a sphere of radius R satisfying $\Psi = 0$ on $r = R$ are of the form

$$\Psi = \Psi_0 \, r^{\tfrac{1}{2}} J_L\left(j_{Ln}\frac{r}{R}\right) P_l^m(\cos\theta) \, \frac{\cos}{\sin} m\phi \, \frac{\cos}{\sin} \omega t \quad (21)$$

with respect to spherical polar coordinates (r, θ, ϕ), where P_l^m is the associated Legendre function of the first kind, of degree l and (azimuthal) order m, and $L = l + \tfrac{1}{2}$. The eigenfrequencies are given by

$$\omega_{n,l} = j_{Ln}\frac{c}{R}; \quad (22)$$

they are independent of m because m determines the azimuthal component of the wavenumber, and in a spherically symmetric system the azimuthal direction is arbitrary. When $n \gg l$, the eigenfrequencies are approximated by Eq. (16) with l replaced by L, namely

$$\omega = \omega_{n,l} \simeq (n + \tfrac{1}{2}L - \tfrac{1}{4})\omega_0 - (L^2 - \tfrac{1}{4})\omega_0^2/2\pi^2\omega$$

$$= (n + \tfrac{1}{2}l)\omega_0 - l(l+1)\omega_0^2/2\pi^2\omega. \quad (23)$$

The large frequency separation is again approximately ω_0, and the small separation is

$$d_{n,l} \simeq (2l+3)\omega_0^2/\pi^2\omega_{n,l}. \tag{24}$$

The ray theory is very similar to that for the circular drum. Because reflection is specular, any ray lies in a plane through the centre of the sphere; the ray path is simply that illustrated in Figure 2. Accordingly, the modes that I shall find it convenient to describe are those high-order oscillations whose wave functions are confined to a neighbourhood of such a plane. These are the sectoral modes (having $m = l$), in the language of spherical harmonics. Any other spherical harmonic of degree l can be constructed as a linear combination of sectoral modes of degree l about different axes. For example, asymptotically, at least, the spherical harmonic of Eq. (21) is an integral with respect to azimuth ϕ of sectoral modes in planes inclined by an angle $\cos^{-1}(m/L)$ from the (fixed) equatorial plane, each with phase chosen to match with $\cos m\phi$ at the points where they intersect any line of latitude (Keller and Rubinow, 1960; see also Gough, 1993).

One can develop a crude ray theory for sectoral modes by regarding the oscillation to be confined in a thin circular disc of thickness $2d$, bounded by the surface of the sphere itself and by two parallel planes at distance $\pm d$ from the equatorial plane of the sphere. By considering this disc to be a waveguide, the modes of oscillation can be estimated by combining the results of §2 and §3. The analysis of the projection of the waves onto the equatorial plane (the interference pattern of the waveguide) is identical to that of the circular drum, except that now (as we learned from the waveguide) the effective ϕ component of the wavenumber vector must be the magnitude k_h of the total wavenumber vector perpendicular to the radial direction (a direction which, from here on, I shall refer to as horizontal), rather than the actual ϕ component $k_\phi = l/R$ of the basic constituent simple waves. Hence we expect $k_h = \hat{L}/r$, where, because the waveguide has uniform thickness, $\hat{L}$ is a constant (to be determined). Consequently, the eigenfrequency equation is simply Eq. (23) with L replaced by $\hat{L}$.

It behoves us finally to estimate $\hat{L}$. Figure 4 shows a cross-section of the disc, cut by a plane through the centre of the sphere. The bounding planes of the disc are at a constant colatitudes, θ_0 and $\pi - \theta_0$, and the edge is a curved surface which, if cut along a line of longitude and flattened, would resemble the waveguide of §2. The sectoral mode of the sphere is the mode of the waveguide having the gravest lateral variation, as I considered in §2. The wave fronts of the interference pattern are along lines of longitude, and I must fit an integral number $m \, (= l)$ of wavelengths around the equator. But the circumferences of other circles of constant latitude are shorter, and therefore the waves do not fit exactly; interference is no longer perfectly constructive and the amplitude of the interference pattern is reduced progressively as distance from the equatorial plane increases. When the latitude is reached at which the number of half-wavelengths around the circumference of the disc is diminished by exactly unity, interference is totally destructive, and Ψ

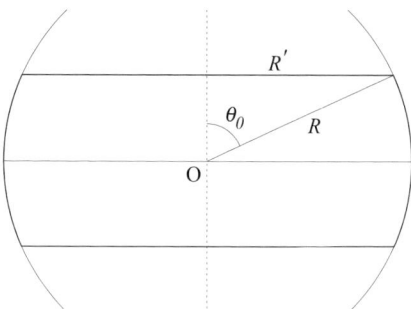

Figure 4. Partial meridional cross section through the sphere. The thin horizontal (with respect to the page) line through O lies in the equatorial plane; the vertical dotted line is the axis of the sectoral mode. The two other horizontal lines denote the characteristic latitudinal extent of the mode. Their locations are determined by Eq. (25). This figure is also considered in the text to be part of a (flattened) equatorial strip on the surface of the sphere bounded by lines of latitude and longitude.

vanishes asymptotically. This condition therefore determines the thickness $2d$ of the waveguide. It may be written

$$2\pi R' / \left(l - \tfrac{1}{2}\right) = 2\pi R \sin\theta_0 / \left(l - \tfrac{1}{2}\right) = 2\pi R / l, \qquad (25)$$

from which $\theta_0 = \sin^{-1}(1 - 1/2l)$. The effective lateral wavenumber k_θ is therefore approximately $1/R\cos\theta_0 = \sqrt{l}/R\sqrt{1 + 1/4l}, \simeq \sqrt{l}/R$, whence $Rk_h = \hat{L} = R\sqrt{k_\phi^2 + k_\theta^2} \simeq \sqrt{l(l+1)} \simeq l + \tfrac{1}{2} = L$, as required.

As in the case of the drum, the amplitude Ψ_0 is constant on any spherical surface $r = $ constant, and increases inwards, becoming formally infinite at the caustic.

5. Oscillations of a Stratified Sphere

In the three previous sections I developed techniques that are required to study the asymptotic (wave-like) oscillations of a star by ray theory. I applied the techniques to three simple problems, all of them simple enough to solve exactly by other means, in order to compare the predictions of the ray theory with the exact solutions, and thereby provide some degree of confidence in the theory. With such confidence, I hope, now established, I progress to the oscillations of a sphere in which c varies with radius r, and which cannot, in general, be solved exactly. Equation (6) still holds asymptotically, with $\Psi = r\rho^{-1/2}\delta p$. I shall assume that c varies smoothly, with a characteristic scale much greater than k^{-1}. Thus formally I exclude from consideration convection zones, at the edges of which the chemical composition, and hence also the sound speed, are typically almost discontinuous, for discontinuities require special attention (transmission and reflection, and subsequent interference between transmitted and reflected waves), which space here does not permit me to pursue. The only new phenomenon that I shall need to

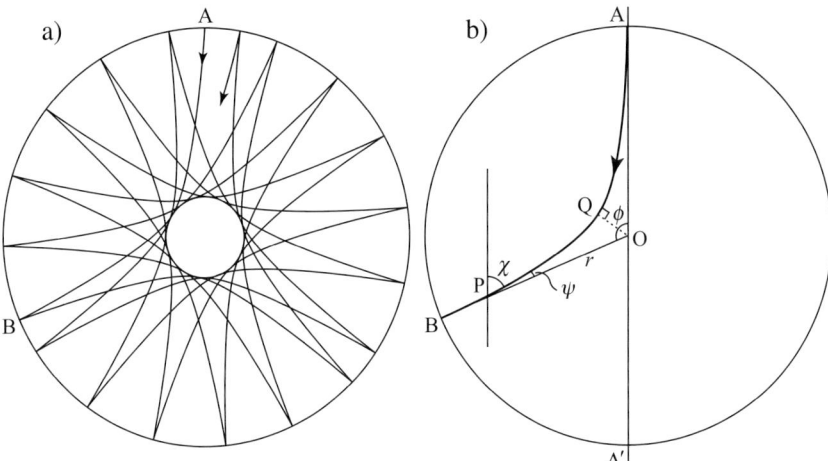

Figure 5. (a): Ray diagram similar to Figure 2(a), in a plane through the centre of a sphere in which sound speed increases inwards. (b): Segment of the ray illustrated in (a): the thin fiducial line through P is parallel to the axis $A'OA$.

incorporate into the analysis is refraction, which is well understood: the rate of rotation of the ray, $d\chi/ds$, where χ is the instantaneous angle subtended by the ray to some fixed direction and s is distance along the ray, is equal to the component of $\nabla \ln c$ perpendicular to the ray, the deflection being such as to turn the ray towards the direction of $-\nabla c$.

A typical stellar ray path is illustrated in Figure (5). Because, on the whole, sound speed decreases outwards (the only exception is in those stellar cores in which nuclear reactions have modified the chemical composition by enough to more than compensate for the non-positive temperature gradient), rays tend to be deflected away from the centre of the sphere.

In Figure 5(b) I show the ray segment joining two points A and B on the surface of the sphere, as in Figure 2(b). The rate of deflection of the ray is $d\chi/ds = \mp(d\ln c/ds)\sin\psi$, the sign being chosen according to whether the point P is on the outwardly or inwardly directed portion of the ray segment, and $ds/dr = \pm\sqrt{1+r^2(d\phi/dr)^2}$. Moreover, $rd\phi/dr = k_\phi/k_r = \tan\psi$. Hence $d\chi/dr = -(d\ln c/dr)\tan\psi$, and therefore, since $\chi = \pi - \psi - \phi$,

$$\frac{d\psi}{dr} = \left(\frac{d\ln c}{dr} - \frac{1}{r}\right)\tan\psi, \qquad (26)$$

which integrates to $\sin\psi = Kc/r$, where K is a constant of integration. Consequently $k_\phi = k\sin\psi = \omega c^{-1}\sin\psi = L/r$, where $L = \omega K$ is constant. This establishes, irrespective of whether or not c varies with r, the r^{-1} dependence of the horizontal component of the wavenumber, and by the waveguide argument in §4 I identify L with $l + \frac{1}{2}$, where l is the degree of the mode. Furthermore, substituting the solution for $\sin \psi$ into the expression for $d\phi/dr$ yields the equation for the ray path:

$$\frac{d\psi}{dr} = \frac{Lc/\omega r}{\sqrt{1 - L^2 c^2/\omega^2 r^2}}. \qquad (27)$$

The standard way to proceed from here is simply to evaluate the travel time T along the ray and the polar angle $\phi(B)$ subtended by A and B about O, and substitute these into the resonance condition

$$\omega T - \alpha \pi - L\phi(B) = 2\hat{n}\pi, \qquad (28)$$

where $\alpha\pi$ is the sum of the phase shifts suffered by the wave on reflection at the surface and on passing through the caustic. If $\Psi = 0$ on $r = R$ and Eq. (6) were valid (asymptotically) everywhere, then α would be $3/2$, as it was in §3 and §4. But in a star, the rapid density variation in the surface layers inhibits propagation and induces reflection at some level typically below the visible surface. In a polytrope of index μ, for example, the outcome is equivalent to assuming that Eq. (6) actually holds right out to $r = R$ (the fiducial radius R being where c^2, extrapolated linearly from well beneath the photosphere, would vanish), but that the total phase shift on reflection is $2\hat{\alpha} \equiv (\mu + \frac{3}{2})\pi$. In more realistic stellar models a similar result holds, where now μ is the value of the effective polytropic index $(d\ln p/d\ln\rho - 1)^{-1}$ in the neighbourhood of the reflecting layer, which is formally the level at which $\omega \simeq \omega_c = c/2H_p$, H_p being the pressure scale height (Gough, 1993). (Note that because the location of the reflecting layer depends on ω, μ is a (generally weakly) varying function of ω.) After the substitution for T and $\phi(B)$, condition (28) becomes the well-known formula

$$\int_{r_t}^{R} \left(1 - \frac{L^2 c^2}{\omega^2 r^2}\right)^{\frac{1}{2}} \frac{dr}{c} = \frac{(n + \mu/2)\pi}{\omega}, \qquad n = 1, 2, \ldots, \qquad (29)$$

where r_t is the distance OQ of closest approach to the origin, and is given by $Lc(r_t)/\omega r_t = 1$, and where once again I have set $n = \hat{n} + 1$. In the limit $n/L \to \infty$, this formula reduces asymptotically to Eqs (1)–(3), once it is accepted that $\mu(\omega)$ can be expanded in inverse powers of ω and written approximately as $2(\varepsilon - \frac{1}{4} + \delta\omega_0^2/\omega)$.

The reduction of Eq. (29) to Eq. (1) is a formal mathematical procedure (Gough, 1986), and it is not straightforward to determine from it the physical origins of the terms. (However, one might guess, correctly, what those origins are from the manner in which I have derived Eq. (29) here, although one cannot do so so readily from previous derivations – in any case, a guess is not a demonstration.) It is expedient instead to work directly with the angle χ defining the orientation of the ray, which satisfies $d\chi/dr = -(d\ln c/dr)\tan\psi$, from which, after substituting for $\tan\psi$, one obtains the total deflection of the ray:

$$\Delta\chi \equiv \chi(B) - \chi(A) = -\frac{2L}{\omega} \int_{r_t}^{R} \left(1 - \frac{L^2 c^2}{\omega^2 r^2}\right)^{-\frac{1}{2}} \frac{c}{r} \frac{d\ln c}{dr} dr. \qquad (30)$$

It is then straightforward from Figure 5(b) to see that $\chi(B) = \pi - \chi(A) - \phi(B)$ and that $\chi(A) = \psi(B)$, and hence that $\phi(B) = \pi - 2\psi(B) - \chi(B) + \chi(A)$. Substituting that, together with the standard formula for T, into condition (28) yields the eigenvalue equation

$$\frac{\omega}{\hat{\omega}_0} = n + \tfrac{1}{2}l + \tfrac{1}{2}\mu + \tfrac{1}{4} - L^2\left[L^{-1}\sin^{-1}\left(\frac{Lc(R)}{\omega R}\right) - \frac{D_L\hat{\omega}_0}{\omega}\right], \quad (31)$$

in which

$$\hat{\omega}_0 = \frac{2\pi}{T} = \left[\frac{1}{\pi}\int_{\tau_t}^{\tau(R)}\left(1 - \frac{L^2c^2}{\omega^2r^2}\right)^{-\frac{1}{2}}d\tau\right]^{-1} \quad (32)$$

and

$$D_L = \frac{1}{\hat{\omega}_0}\int_{\tau_t}^{\tau(R)}\left(1 - \frac{L^2c^2}{\omega^2r^2}\right)^{-\frac{1}{2}}\frac{c}{r}\frac{dc}{dr}d\tau, \quad (33)$$

where

$$\tau = \int_0^r \frac{dr}{c} \quad (34)$$

is acoustical radius, and $\tau_t = \tau(r_t)$. The origins of the terms are now evident: The characteristic frequency $\hat{\omega}_0$ is a measure of the inverse sound travel time along the ray segment, the quantity $n + \tfrac{1}{2}L = n + \tfrac{1}{2}l + \tfrac{1}{4}$ is the characteristic combination of quantum numbers arising from resonance in spherical geometry, a term $\tfrac{1}{2}(\mu + \tfrac{3}{2})$ arises from half the effective phase jump on each reflection at the surface, accounting also for the error in omitting ω_c from the governing Eq. (6), the remaining $\tfrac{1}{4}$ in the combination $\tfrac{1}{2}(\mu + \tfrac{1}{2})$, after accounting for the difference of unity between $\hat{n}$ and n, coming from the encounter with the caustic. The final term in Eq. (31) arises from the phase variation of the interference pattern around the circumference of the sphere (from B to the point A' antipodal to A) with which the phase of the ray must agree. It is in two parts: the first, $L\sin^{-1}(Lc(R)/\omega R)$, is due to the small inclination, $\pi - \psi(A)$ and $\psi(B)$, of the ray to the vertical at the points of reflection (and appears, of course, in the formula for the isothermal sphere); the second arises from the generally much larger deflection of the ray by refraction in the interior.

It is quite evident that in the limit $n/L \to \infty$ equation (31) reduces to equation (1). The frequency $\hat{\omega}_0$ approaches $\omega_0 \equiv \pi/\tau(R)$, because the turning radius τ_t shrinks to zero. (The square-root singularity in the integrand makes a vanishing contribution to the integral in this limit.) Also the term arising from the inclination of the ray at the surface tends to zero. And formally the factor $(1 - L^2c^2/\omega^2r^2)^{-1/2}$ can be omitted from the integrand in the expression (30) for $\Delta\chi$, for the same reason as it can from $\hat{\omega}_0$. However, unlike in the integral for $\hat{\omega}_0$, the resulting

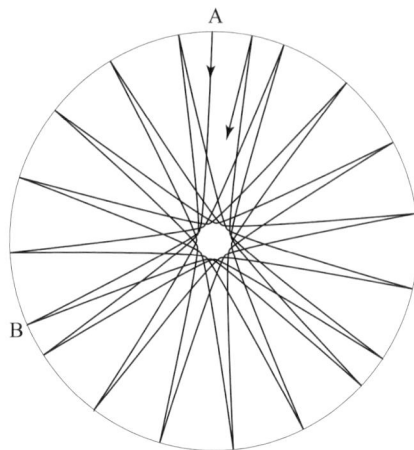

Figure 6. Segment of the ray illustrated in Figure 5(a), but plotted with respect to acoustical radius τ (from Gough, 1984).

integrand (with respect to τ), namely $(c/r)dc/dr$, is much larger near the centre of the sphere than elsewhere, and because in practice n/L is not infinite, the error introduced by this approximation is greater. Accordingly I recommend that the factor be retained. Should that advice be taken, one would then deduce that the large frequency separation is again roughly $\omega_0 = \pi/\tau(R)$, but that the small separation, on retaining the first-order terms in expansions of $D_{1/2}$ and $D_{5/2}$ about $D_{3/2}$ and ignoring the first term in square brackets in Eq. (31), is given approximately by

$$d_{n,l} \simeq \frac{(2l+3)\omega_0}{\omega} \int_{r_t}^{R} \left[1 - \frac{(2l+3)^2 a^2}{4\omega^2}\right]^{-\frac{1}{2}} a^2 \frac{d^2 \ln c}{dr\,da} dr, \qquad (35)$$

in which $a = c/r$ and $d^2\ln c/dr\,da \equiv (d/dr)[(da/dr)^{-1} d\ln c/dr]$; $d_{n,l}$ is a measure principally of the phase change required for resonance as a result of the deflection of the ray by the sound-speed gradient in the core of the star, and it is not related at all to this order of approximation to the modification of the sound travel time T along the ray segment. That the effect of the deflection on T is small is evident from Figure 6, which is identical to Figure 5(a) save that the radial coordinate is now the natural acoustical radius τ: it takes only a casual inspection to appreciate that the length of each ray segment hardly differs from that in Figure 2, which is pertinent to the isothermal sphere.

6. Concluding Remarks

I have demonstrated in simple physical terms why the frequencies ω of high-order acoustic modes of a star take the form they do. The resonant modes of oscillation, of order n and degree l, are formed by the constructive interference of propagating sound waves. In the limit as $n/L \to \infty$, where $L = l + \frac{1}{2}$, the rays of sound pass almost through the centre of the star (leaving an infinitesimal central spherical zone of avoidance). They propagate phase from a point A on the surface of the star directly to almost the antipodal point A' in a time $T \simeq 2 \int_0^R c^{-1} dr$, save for a retardation of $\hat{\alpha}\pi$ ($\hat{\alpha}$ is of order unity) from each reflection at A and A' and of $\frac{1}{2}\pi$ from grazing the zone of avoidance at its bounding caustic surface. This phase must resonate with that of the interference pattern on the surface of the star, whose phase varies as $L\phi$. Thus the phase difference $\omega T - \left(2\hat{\alpha} + \frac{1}{2}\right)\pi - L\Delta\phi$, where $\Delta\phi = \pi$ is the angle subtended by A and A' from the centre of the star, must be an integral multiple of 2π, which leads to the asymptotic relation

$$\omega T/2\pi \sim n + \tfrac{1}{2}l + \hat{\alpha} + \tfrac{1}{2} \quad \text{as } n/L \to \infty. \tag{36}$$

The effective angular wavenumber L must exceed l to take account of the variation of the wave function perpendicular to the azimuthal direction; it exceeds l simply by a constant, $\frac{1}{2}$, (rather than exceeding l^2 by a constant, as would be the case in Cartesian geometry), because the geometry is spherical; n is augmented by $\frac{1}{2}l$ in the equation for ω for essentially the same reason. Analysis of the reflection shows that $\hat{\alpha} = \frac{1}{2}(\mu + \frac{3}{2})$, where μ is an effective polytropic index of the outer layers of the star.

In reality n is not infinite, and the formula (36) should be corrected: because the ray is not exactly radial, whatever the value of l, it meets the surface at a point B that is less than π away from A; consequently both $\Delta\phi$ and T are diminished. The two modifications have opposing effects on ω. If the star were isothermal, as in §4, the augmentation of ω due to the reduction of T would be just half the reduction of ω due to the reduction of $\Delta\phi$, and therefore ω would be decreased. But in a real star, refraction of the waves reduces $\Delta\phi$ much more than the obliquity of propagation from the surface (see Figures 5 and 6) without substantially modifying T, and ω is reduced yet further. When n/L is large, the deflection $\Delta\chi$ of the ray is proportional to the inclination of the ray from the vertical immediately after reflection from the surface, which is proportional to L/n. Hence the phase reduction, $L\Delta\chi$, is proportional to L^2/n, as is the reduction in ω. The deflection $\Delta\chi$ senses the derivative of c, especially near the caustic at the l-dependent radius $r = r_t$; hence the frequency difference $d_{n,l}$ senses a second derivative.

It is worth remarking that the simple limiting formula (36) can never actually be attained. Therefore $\omega_{n-1,l+2}$ is never equal to $\omega_{n,l}$, except under accidental circumstances. The fact that the difference $d_{n,l}$ is not zero (and is proportional to L^2/n when n/L is large) is an intrinsic property of the oscillations of a sphere (although not uniquely so), as is the limiting relation (36). Therefore the small frequency difference is not a deviation, caused by some symmetry-breaking agent, from a degeneracy. It is therefore not a frequency splitting. Generically, the frequencies of all the modes of a sphere (unlike those of a cube, for example) are different, and moreover their ratios are not rational. Consequently, the oscillations of a star are not harmonious. This conclusion is important, for it implies that the information content of a set of N frequencies of a star is likely to be greater than that of a similar set of N frequencies of an harmonious system, which perhaps offers some comfort to we who are bent on asteroseismic inference.

I conclude by addressing $d_{n,l}$ as a diagnostic of age. Broadly speaking, as a star evolves, the mean molecular mass of the stellar material is increased by thermonuclear transmutation preferentially in the inner regions. Consequently the sound speed c is reduced relative to that in the outer regions (there is also a global adjustment of the entire star, which influences the value ω_0 of the large frequency separation), whence dc/dr (which in a chemically homogeneous star is negative) is increased, especially in the core outside the caustic sphere, where $d_{n,l}$ is most sensitive to dc/dr because of the c/r weighting. Consequently $d_{n,l}$ progressively declines with age, and is thereby an age discriminant.

Although my description of the situation is rather simplistic, it does provide the essence of what one is trying to achieve when one seismically calibrates evolved stellar models. But I should point out that there are other phenomena, which I have not described, that also come into play when calibration is being carried out. One is the influence of near-discontinuities in c, or in a derivative of c, at the interfaces between radiative and convection zones, which I mentioned in passing. Another is the fact that the star may not be spherically symmetrical: because modes of different degree l and azimuthal order m sample latitude and longitude differently, their frequencies are influenced differently by horizontal variation. Consequently $d_{n,l}$ is not a measure solely of the sound-speed gradient in the core, and hence of chemical inhomogeneity. Clearly, due caution must be exercised. However, if one could measure the frequencies of all the modes with differing m in the multiplets of like n and l, one could eliminate the influence of asphericity, for the uniformly weighted sum over m of the (singlet) frequencies in a multiplet is determined by only the spherically averaged structure of the star (e.g. Gough, 1993). Moreover, one could also investigate the asphericity from the frequency splitting. In some cases that has already been shown to be possible.

7. Acknowledgements

I am very grateful to Di Sword for preparing the LaTeX file, and to Richard Sword for preparing the diagrams.

References

Brillouin, L.: 1926, *J. Phys. Radium* **7**, 353.

Chaplin, W.J., Elsworth, Y., Howe, R., Isaak, G.R., McLeod, C.P., Miller, B.A., New, R., van der Raay, H.B. and Wheeler S.J.: 1996, *Solar Phys.* **168**, 1.

Christensen-Dalsgaard, J.: 1988, in: J. Christensen-Dalsgaard and S. Frandsen (eds.), *Proc. IAU Symp. 123: Advances in Helio- and Asteroseismology*, Reidel, Dordrecht, 295.

Christensen-Dalsgaard, J.: 1993, in: T.M. Brown (ed.), *Proc. GONG 1992: Seismic Investigation of the Sun and Stars*, A.S.P. Conf. Ser. **42**, 347.

Einstein, A.: 1917, *Verhandl. Deut. Physik Ges.* **19**, No 9-10.

Gough, D.O.: 1984, *Mem. Soc. Astron. Italiana* **55**, 13.

Gough, D.O.: 1986, in: Y. Osaki (ed.), *Hydrodynamic and Hydromagnetic Problems in the Sun and Stars*, Univ. Tokyo, 117.

Gough, D.O.: 1990, *Mat. Fys. R. Dan. Acad. Sci.* **42:4**, 13.

Gough, D.O.: 1993, in: J-P. Zahn and J. Zinn-Justin (eds.), *Astrophysical Fluid Dynamics, Les Houches Session XLVII*, Elsevier, Amsterdam, 399.

Keller, J.B.: 1958, *Ann. Physics* **4**, 180.

Keller, J.B. and Rubinow, S.I.: 1960, *Ann. Physics* **9**, 24.

Tassoul, M.: 1980, *ApJS* **43**, 469.

Tassoul, M.: 1990, *ApJ* **358**, 313.

DIAGNOSTICS OF THE INTERNAL STRUCTURE OF STARS USING THE DIFFERENTIAL RESPONSE TECHNIQUE

IAN ROXBURGH[1,2] and SERGEI VORONTSOV[1,3]

[1]*Astronomy Unit, Queen Mary, University of London, London E1 4NS, UK*
[2]*Observatoire de Paris, Place Jules Janssen, 92195 Meudon, France*
[3]*Institute of Physics of the Earth, B Gruzinsskya 10, Moscow 123810, Russia*

Abstract. We address the problem of the diagnosing the deep interior structure of stars using acoustic p-modes, and investigate the diagnostic capabilities of two complementary approaches both based on the differential response technique (Vorontsov, 1998): (a) direct calibration using a grid of evolutionary stellar models, and (b) linear and non-linear (with consecutive linearisations) inversion of low-degree frequencies. We apply this analysis to the frequencies of a model of an old $0.8 M_\odot$ star, and to the solar frequencies obtained from BiSON measurements, using a 2-D grid of reference models of different mass and age. We explore the convergence and stability of the asteroseismic inversion, performed with the adaptive regularisation technique of Strakhov and Vorontsov (2001).

Keywords: Sun: waves; methods: numerical; stars: oscillations

1. Introduction

The 'differential response technique' (Vorontsov, 1998) is based on the result that in the near surface layers the phases of *p*-modes are a slowly varying function of frequency, $\alpha(\nu)$, independent of degree ℓ. A stellar model is tested by determining the phases of partial waves with the observed frequencies at some truncation radius. (Partial waves are solutions of the oscillation equations, with a given frequency satisfying the surface gravitational boundary condition.) Deviations from a smooth approximation to $\alpha(\nu)$ serve as a measure of the misfit of the model to the input frequencies. We then adjust the model to eliminate the misfit. This technique has been applied to the Sun by Vorontsov (2001, 2002) and to stars by Roxburgh and Vorontsov (2002a,b).

2. Calibration and Inversion of an Evolved $0.8 M_\odot$ Star

The input data in this numerical experiment are low-degree frequencies of a model of a highly evolved star of $0.8 M_\odot$ with central hydrogen abundance $X_c=0.11$. Gaussian noise of 0.1μHz was added to the frequencies. The 2-D reference grid consists of 14×9 evolutionary models with Z=0.02, spaced uniformly in X_c between 0.05 and 0.7, and in mass between 0.8 and $1.2 M_\odot$, computed using the CESAM code

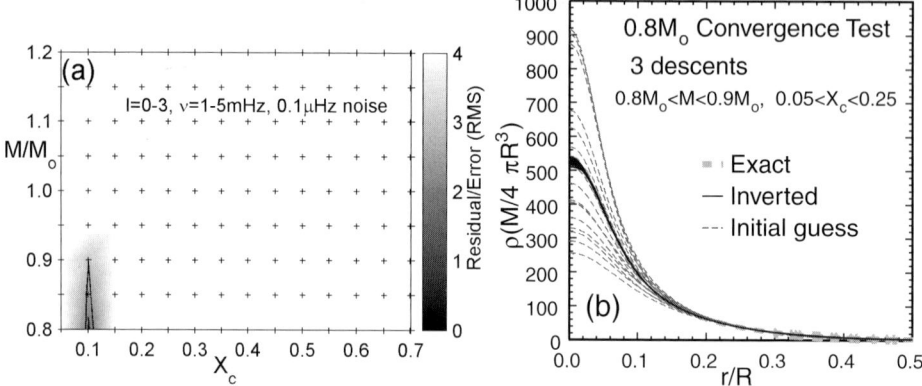

Figure 1. (a) Calibration with frequencies of an old $0.8 M_\odot$ star. Grey scale shows the r.m.s. value of the normalised residuals. Contour levels are in steps of 1.0 (b) Convergence test of the inversion starting from different initial models.

(Morel, 1997) with different input physics than in the input model. Figure 1a illustrates the results with the surface frequency-dependent function approximated by a polynomial of degree $n_p=20$.

Linear (or 'inner') inversions using the conjugate gradient adaptive regularisation technique (Strakhov and Vorontsov, 2001), with $k=20$ iterations, were performed for a range of initial models, each represented by 500 cubic B-splines for the running mean density, and with amplitudes controlled by a polynomial of degree $n_p = 30$. Γ_1 was taken from the model. A corrected hydrostatic model was then constructed and the inner iterations repeated. The quality of the solutions obtained after three outer descents is shown in Figure 1b for twelve initial models.

Figure 2a shows the stability of the solutions to different random realisations of errors in the data. The initial guess was taken as the $1 M_\odot$, $X_c = 0.35$ model – far away from the target. Four outer iterations were needed for convergence. Figure 2b shows the results of a similar inversion, but with data restricted to the frequency range 2–4mHz and with Gaussian noise of 0.2μHz. Here $k = 10$, and $n_p = 10$. The solutions now deviate noticeably from the target model in the central core due to the loss of low-frequency data where the frequencies depart from asymptotic degeneracy. It is important to note that all the solutions which are shown satisfy the 'observational' data set with adequate accuracy. The systematic deviation of the solution from the target model indicates the uncertainty of the asteroseismic solution.

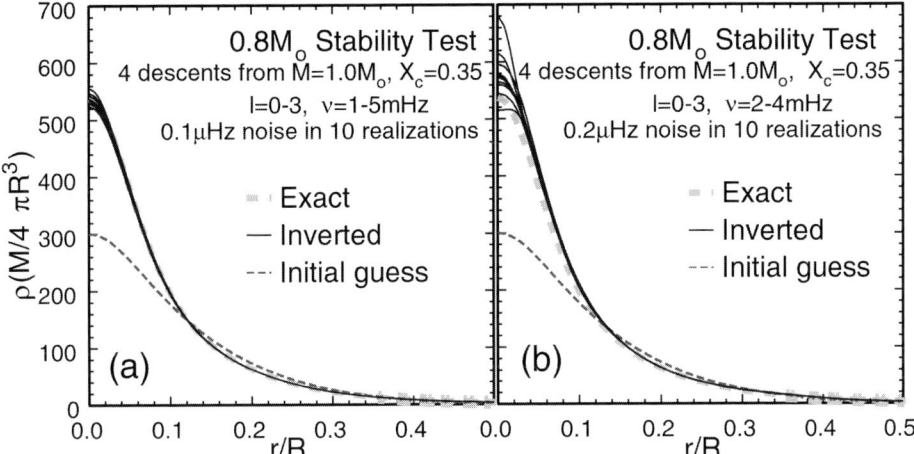

Figure 2. (a) Stability test of the inversion with frequencies of a $0.8 M_\odot$ star with 0.1μHz Gaussian noise. The initial model was close to the present Sun. (b) Stability test with data in a narrow frequency range and with larger noise.

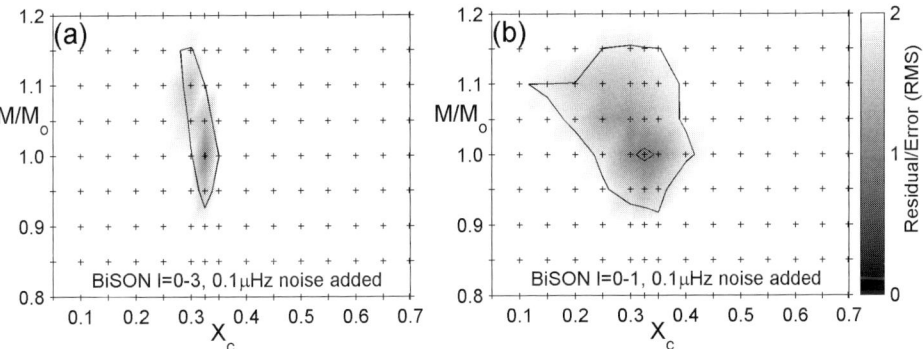

Figure 3. (a) Calibration with BiSON frequencies with 0.1μHz added noise. (b) Results obtained using only modes of degree $\ell = 0, 1$.

3. Calibration and Inversion with BiSON Frequencies

We here used the solar p-mode frequencies infered from 32 months of BiSON measurements (Chaplin et al., 1998). An additional 0.1μHz noise was imposed on the input data. The results of the calibration are shown in Figures 3a,b. The Sun is well localised even when the data are limited to modes of degree $\ell = 0$ and 1 (Fig. 3b).

In the inversion, with data available in the frequency range about 1.5–4 mHz, the polynomial degree for the surface frequency-dependent function was taken to be 15, and the number of inner iterations $k = 15$. The convergence test is illustrated

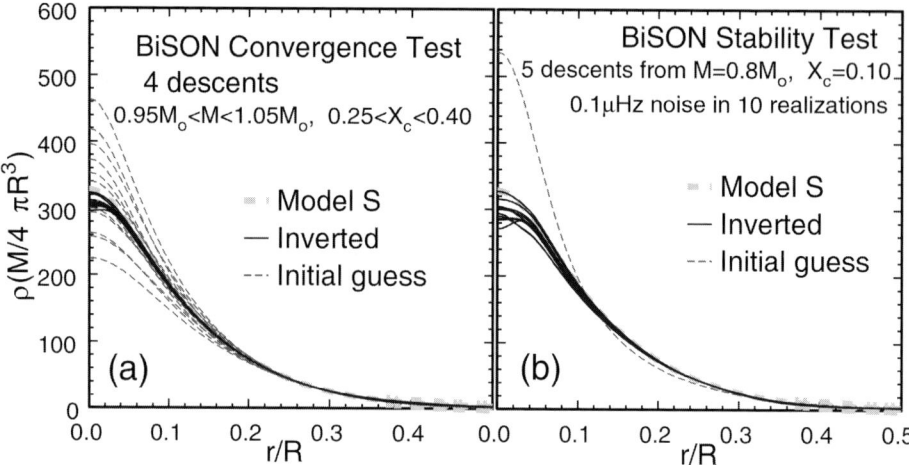

Figure 4. (a) Convergence test of the inversion with BiSON frequencies with $0.1\mu Hz$ added noise (b) Stability test of the inversion with BiSON frequencies with 10 realisations of the Gaussian noise.

in Figure 4a and the stability to random errors is shown in Figure 4b, where a converged model is obtained after 5 outer interations even when starting with an initial model ($0.8M_\odot$, $X_c = 0.10$) which is again very far from the target. As before, all the resulting solutions fit the input data.

4. Discussion

Asteroseismic calibration and inversion, based on the differential response technique, appears to provide an efficient diagnostic tool for probing stellar interiors. The surface function $\alpha(\nu)$, determined by the inversion, can be used for calibrating the He abundance (e.g. Baturin and Vorontsov, 2001), and the depth of convective envelopes (e.g. Roxburgh, 2002).

Acknowledgements

This work was supported in part by the UK Particle Physics and Astronomy Research Council under grant PPA/G/S/1998/00576.

References

Baturin, V.A. and Vorontsov, S.V.: 2001, in: A. Wilson (ed.), *Helio- and Asteroseismology at the Down of the Millenium, ESA SP-***464**, 615.

Chaplin, W.J., Elsworth, Y., Isaak, G.R., Lines, R., McLeod, C.P., Miller, B.A. and New, R.: 1998, *MNRAS* **300**, 1077.

Morel, P.: 1997, *A&A* **124**, 597.

Roxburgh, I.W.: 2002, in: F. Favata, I.W. Roxburgh and D. Galadi (eds.), *Stellar Structure and Habitable Planet Finding*, ESA SP-**485**, 75.

Roxburgh, I.W. and Vorontsov, S.V.: 2002a, ibid, 337.

Roxburgh, I.W. and Vorontsov, S.V.: 2002b, ibid, 349.

Strakhov, V.N. and Vorontsov, S.V.: 2001, in: A. Wilson (ed.), *Helio- and Asteroseismology at the Down of the Millenium*, ESA SP-**464**, 539.

Vorontsov, S.V.: 1998, in: J. Provost and F.-X. Schmieder (eds.), *Sounding Solar and Stellar Interiors*, Obs. Cote d'Azur, Nice, 135.

Vorontsov, S.V.: 2001, in: A. Wilson (ed.), *Helio- and Asteroseismology at the Down of the Millenium*, ESA SP-**464**, 563.

Vorontsov, S.V.: 2002, in: A. Wilson (ed.), *From Solar Min to Max: Half a Solar Cycle with SOHO*, ESA SP-**508**, 107.

STELLAR CONVECTION

PIERRE DEMARQUE and FRANCIS J. ROBINSON
Department of Astronomy, Yale University, Box 208101, New Haven, CT 06520-8101, USA

Abstract. Seismology provides powerful tests of convection deep in stellar interiors. First, the role of convective overshoot and low efficiency convection, two areas of uncertainty with important astrophysical implications, are reviewed briefly. In the rest of the talk, a critical introduction to numerical simulations of radiative hydrodynamics will be given. The basic underlying assumptions and challenges are explained, and some recent results are presented.

Keywords: stars, convection, seismology

1. Introduction

Seismology provides us with a powerful direct probe of convection in stellar interiors. Helioseismology has already yielded a precise estimate of the depth of the solar convection zone, and a sensitive probe of the structure of the outer convective layers of the Sun. Asteroseismology presents greater challenges and opportunities. When used in conjunction with other astrophysical tools, it will provide unique probes of the extent and structure of convective regions in stellar interiors, and stringent new tests of the theory of stellar evolution.

2. Convection in Stars

The importance of convection for transporting energy in stellar interior was recognized early on. Lord Kelvin (1862) conjectured that the Sun's interior was in convective equilibrium, in analogy with the Earth's atmosphere. This means, in contemporary language, that the logarithmic temperature gradient is then given by $\nabla = \nabla_{ad} \equiv (\partial lnT/\partial dlnP)_S$. Lane (1869), in constructing the first solar model, followed Lord Kelvin's suggestion, and assumed convective (adiabatic) equilibrium throughout the interior.

The Schwarzschild (1906) criterion is about the stability of radiative equilibrium against convection. In a radiatively stable layer, $(\nabla - \nabla_{ad}) < 0$. In deep convective regions, $(\nabla - \nabla_{ad}) > 0$ (in order to carry the energy flux), but the superadiabaticity is small, and negligible from the structural point of view. Calculating the structure of deep convective layers is thus straightforward; the assumption of adiabatic equilibrium in the convective regions of the interior made by Lane is an excellent approximation.

Finally, it is interesting to recall the fundamental connection between convection and g-modes. The Brunt-Väisälä frequency of g-modes N is the frequency with which a bubble of gas may oscillate around its equilibrium position under gravity. We have $N^2 = -Ag$, where the Schwarzschild discriminant $A = (dln\rho/dr) - 1/\Gamma_1(dlnp/dr)$, and $\Gamma_1 = (\partial lnp/\partial ln\rho)_S$. As a result, g-modes are damped in convective regions, and we shall discuss only p-modes in the rest of this paper.

3. Transition Regions and Seismology

More interesting from the physics point of view are the transition layers between purely radiative regions and regions of efficient convection. Because the sound speed is sensitive to the structural changes in these regions, their presence is revealed by seismology. For example, one of the most important results of helioseismology has been the determination of the precise depth of the solar convection zone (Christensen-Dalsgaard et al., 1991; Basu and Antia, 1997). Seismology can also tell us about the size of convective cores, and about convective overshoot at the edges of convective cores and above and below convective regions.

In the future, it is also likely that seismology will enable us to probe the structure and dynamics of semi-convective regions in massive stars, horizontal-branch stars, and convective shells in stars on the RGB and AGB. A much better understanding of excitation mechanisms for p- and g-modes in a variety of astrophysical environments will result.

So far, most of the interest has been focused on two problems, the problem of convective core overshoot, and the problem of the physics of the outer convective layers.

3.1. Core overshoot

In addition to stellar physics, core convective overshoot has important implications in the study of the ages of stellar populations in the near-field and at cosmological distances (Yi et al., 2000).

The amount of penetration beyond the edge of a convective core has been considered by many authors (see e.g. Zahn, 1991; Roxburgh, 1998). The apsidal motion test in binary systems provides a direct test of the density distribution in the individual components of the system. This test has been applied with some success to test the mass of the convective core and the extent of overshoot in massive stars (see papers in Giménez et al., 1999).

Another classical test of convective overshoot is derived from the core exhaustion phase during the post-main sequence evolution of massive and intermediate mass stars. This evolutionary phase reveals itself as a gap in the CMD of open star clusters, and as a bump in their luminosity functions. Tests have been performed on cluster CMDs in the Galaxy (e.g. in NGC 2420) by Demarque et al. (1994), and more recently in the LMC (Woo et al., 2002).

Hydrodynamical simulations of convective overshoot have been performed by Singh et al. (1995). In the deep interior, the relaxation times are very long, and because of computational limitations, a spectral approach must be adopted in such calculations (Chan et al., 1994).

It is expected that seismology will provide fundamental new means of detecting the presence of convective cores in stars (Roxburgh and Vorontsov, 1999). For instance, the small frequency spacings, combined with other state-of-the-art astrophysical data, will reveal whether there is a convective core in α Centauri (Guenther and Demarque, 2000). And it may be possible to deduce past histories of overshooting in red giant stars from their mixed (p- and g-mode) oscillation spectra (Audard et al., 1995; Guenther, 1991, 2002). There is already suggestive evidence for mixed modes and 'mode bumping' (Scufflaire, 1974; Unno et al., 1989) in the subgiant η Bootis (Kjeldsen et al., 1995; Christensen-Dalsgaard et al., 1995; Guenther and Demarque, 1996).

3.2. OUTER CONVECTIVE LAYERS

There is a well-known problem in determining the surface boundary conditions in stellar models with a convective envelope. This is because the outer layers (the highly superadiabatic layer (SAL) in particular) are usually described by the mixing length theory (MLT), and the extent of the convection zone depends sensitively on the poorly known mixing length parameter. The mixing length uncertainty remains one the most significant sources of error in deriving ages for globular star clusters (Chaboyer et al., 1998). It also affects the color of giant branches of theoretical isochrones, and is a major uncertainty in the spectral synthesis of distant stellar populations (Yi et al., 1997).

For deep ($v \ll c_s$, where v is the flow velocity and c_s is the sound speed) and efficient ($\nabla - \nabla_{ad}$ just above zero) convection, Chan and Sofia (1989) showed that MLT is a very good approximation to the real situation. However, in the SAL, both the Mach number (v/c_s), and the superadiabacity $\nabla - \nabla_{ad}$, can be of order unity. In this case, MLT is unlikely to apply. The validity of this argument is confirmed by the run of sound speed in the SAL derived by inversion of the helioseismic data, which disagrees with the MLT model (Basu et al., 2000).

The sensitivity of the solar p-mode frequencies to the structure of the SAL has been studied by several authors (see e.g. Monteiro et al., 1996; Demarque et al., 1997, 1999). There is an extensive literature describing the research on 2D and 3D radiative hydrodynamical simulations of the outer layers of the Sun and cool stars. The reader is referred to the reviews by Chan et al. (1991), Nordlund et al. (1996), and to papers in the First Granada Workshop (Giménez et al., 1999).

In the next sections, we describe recent 3D numerical simulations of the outer layers of Sun-like stars, aimed at addressing the SAL structure problem (Robinson, 2000, 2002a,b,c). The mean velocity field from these simulations has then been parameterized for use in 1D stellar models. In the case of the Sun, this approach has yielded significant improvements over the MLT in matching the p-mode frequencies and in calculating the stochastic excitation of the observed p-modes (Li et al., 2002a,b).

4. 3D Radiative Hydrodynamical Simulations of the Outer Layers of Stars

4.1. WHY NOT USE THE BOUSSINESQ OR THE ANELASTIC APPROXIMATION?

Convection in stars is characterized by highly stratified turbulent flows. For example, in the solar convection zone the pressure varies over more than twenty scale heights. Such layers have a very large range of time scales, e.g. the free fall time is about an hour, while the Kelvin-Helmholtz relaxation time is 10^5 years. To simulate even part of such a layer requires very long computations. Chan and Sofia (1989) stressed the importance of thermal relaxation and statistical convergence in modelling turbulent compressible convection. Both of these require very long integration times. To ease the computational burden, simplifications are often made to the governing hydrodynamic equations. The aim is to cover a large time interval in as few computational steps as possible. The problem is that the maximum allowable time step for an explicit computation, is limited by the sound travel time between two adjacent grid points (the C.F.L. condition). To circumvent this restriction, simplifications of the governing equations which filter out sound waves are often used. The two most popular are the Boussinesq (1903) approximation and the anelastic approximation (Batchelor, 1953).

4.1.1. *Boussinesq approximation*
With the exception of thermally induced density fluctuations which produce buoyancy, the density is constant in the Boussinesq approximation. This approximation can only be applied to compressible flows in which the Mach number is small and the total depth is much less than any thermodynamic scale height (Spiegel and Veronis, 1960). In general it cannot be used to model convection in stars.

4.1.2. *Anelastic approximation*
On the other hand, the anelastic approximation (Gough, 1969), allows density stratification. It assumes the relative fluctuations in the thermodynamic variables are small and that the thermodynamic variables fluctuate over time periods of the order of the roll (eddy) turn-over time. Provided the Mach number is small, the turn-over time will be much longer than the acoustic time, and it would appear that sound waves would play a minor role. In the Sun, even at $R = 0.96 R\odot$ the Mach number is less than 10^{-5}, which explains why Toomre and his collaborators (see e.g. Latour et al., 1976) have had so much success in applying the anelastic equations to the solar convection zone. But within a few hundred kilometers of the surface, the Mach number approaches unity, as seen in Figure 1.

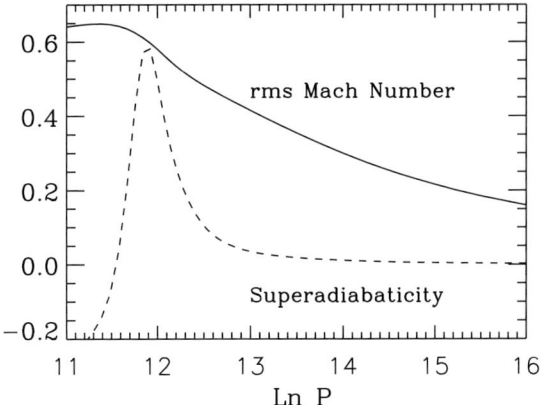

Figure 1. The rms Mach number (solid line), and superadiabaticity in the outer layers of the Sun

5. Description of 3D Radiative Hydrodynamic Simulations

In such an environment, the governing hydrodynamic equations are the fully compressible Navier-Stokes equations (see e.g. Kim et al., 1995, 1996).

$$\partial \rho / \partial t = -\nabla \cdot \rho \mathbf{v} \quad (1)$$
$$\partial \rho \mathbf{v} / \partial t = -\nabla \cdot \rho \mathbf{v}\mathbf{v} - \nabla P + \nabla \cdot \mathbf{\Sigma} + \rho \mathbf{g} \quad (2)$$
$$\partial E / \partial t = -\nabla \cdot [(E + P)\mathbf{v} - \mathbf{v} \cdot \mathbf{\Sigma} + f] \quad (3)$$
$$+ \rho \mathbf{v} \cdot \mathbf{g} + Q_{\text{rad}}$$

where $E = e + \rho v^2 / 2$ is the total energy density and $\rho, \mathbf{v}, P, e$ and $\mathbf{g}$, are the density, velocity, pressure, specific internal energy and acceleration due to gravity, respectively. Ignoring the coefficient of bulk viscosity, the viscous stress tensor for a Newtonian fluid is $\Sigma_{ij} = \mu(\partial v_i / \partial x_j + \partial v_j / \partial x_i) - 2\mu/3 (\nabla \cdot \mathbf{v}) \delta_{ij}$.

5.1. RADIATIVE TRANSFER

In the deeper part of the domain ($\tau > 10^4$), radiative transfer is treated by the diffusion approximation,

$$Q_{\text{rad}} = \nabla \cdot \left[\frac{4acT^3}{3\kappa\rho} \nabla T \right], \quad (4)$$

where κ is the Rosseland mean opacity, a is the Boltzmann constant and c is the speed of light.

In the shallow region, the photon mean free path is at least one tenth of the depth of the atmosphere, so the diffusion approximation may not apply. Instead Q_{rad} is computed as

$$Q_{rad} = 4\kappa\rho(J - B) \tag{5}$$

where the mean intensity J is computed by using the generalized three-dimensional Eddington approximation (Unno and Spiegel, 1966),

$$\nabla \cdot \left(\frac{1}{3\kappa\rho}\nabla J\right) - \kappa\rho J + \kappa\rho B = 0. \tag{6}$$

This is exact for isotropic radiation, and without requiring LTE, the Eddington approximation describes the optically thick and thin regions exactly.

5.2. Sub-grid scale treatment and numerical method

The highly turbulent nature of convection in the sun (Re = 10^{12}), makes it impossible to numerically resolve even 1 % of the dynamical scales. In our 3D simulations we assume most of the energy transport is by the resolved scale motions (large eddies). The dynamic viscosity μ is increased so that it represents the effects of Reynolds stresses on the unresolved (or sub-grid) scales following the prescription of Smagorinsky (1963). Canuto (1997, 2000) has discussed in detail the issues associated with the treatment of sub-grid scales in large eddy simulation (LES) calculations, and has provided an improved sub-grid scale treatment.

The simulation domain is a box of aspect ratio 1.5, which includes the photosphere and the top part of the convection zone. Using a code developed by Chan and Wolff (1982), an implicit scheme (the Alternating Direction Implicit Method on a Staggered grid or ADISM) relaxes the fluid to a self consistent thermal equilibrium. The layer is fully relaxed if: the energy flux leaving the top of the box is within 5 % of the input flux at the base; the horizontally averaged vertical mass flux is less than 0.001 at every vertical level; and the overall thermal structure, i.e. the run of $\nabla - \nabla_{ad}$ (or log T vs. log P), does not change over time. These three criteria must be satisfied, before any useful statistical data can be gathered. The relaxation typically takes about 5 hours of solar surface convection time. A second order explicit method (Adams-Bashforth time integration) gathers the statistics of the time averaged state. The statistical integration time is about 2500 seconds of solar surface convection, and requires about half a million time steps. On an 667 MHz Alpha processor, each integration step on a 80 × 80 × 80 grid, requires about 5 seconds of CPU time. A full simulation on a fine grid (120^3) with a domain depth of 2.5 Mm, takes about 3 months to run.

A standard stellar model, calculated with the Yale stellar evolution code (Guenther and Demarque, 1997), is used to compute the initial stratification i.e. run of pressure, temperature, density, internal energy, for the box. Following Kim and Chan (1997, 1998), the hydrodynamical simulations use the same opacities and equation of state as in the 1D stellar model. The horizontal boundaries are periodic, while the vertical boundaries are stress free. A constant heat flux flows through the base, and the top is a perfect conductor. To ensure that mass, momentum and energy are fully conserved, we use impenetrable (closed) top and bottom boundaries. We solve the conservative form of the hydrodynamics equations.

5.3. COMPARISON WITH PREVIOUS WORK

Although our numerical procedures and our treatments of radiative transfer and sub-grid scale motions differ from those of Stein and Nordlund (1998), we found the velocities and thermal structure to be very close to those found by them. In particular, detailed comparisons of the rms vertical velocity and mean velocity in our solar simulation with those published by Asplund et al. (2000) and of the run of $(\nabla - \nabla_{ad})$ by Rosenthal et al. (1999), are in good agreement in the SAL region and away from the simulation boundaries. This agreement gives us confidence in our ability to simulate stellar convection.

6. Parameterization of 3D Simulations, Implementation in 1D Stellar Models, and P-Mode Frequency Calculations

Implementing the results of 3D simulations into 1D stellar models while retaining as much physical consistency as possible, is an interesting challenge. It is convenient to introduce a 1D turbulent velocity into the equations of stellar structure (Li et al., 2002a).

The turbulent velocity is defined by the velocity variance,

$$v = \left[\sum_{i=1}^{3} (\overline{V_i^2} - \overline{V_i}^2) \right]^{1/2}, \tag{7}$$

where the overbar denotes horizontal and temporal averaging of a 3D hydrodynamical simulation.

By analogy with the work of Lydon and Sofia (1995) on magnetic effects, the 3D turbulence is parameterized in terms of two quantities, the turbulent kinetic energy per unit mass,

$$\chi = \frac{1}{2} v^2 \tag{8}$$

and an anisotropy parameter,

$$\gamma = 1 + 2(v_z/v)^2. \tag{9}$$

The z direction is parallel to the radial direction. As the 3D simulations are in a small box, curvature effects are ignored, hence $v^2 = v_x^2 + v_y^2 + v_z^2$.

The turbulent pressure can be written as

$$P_{turb} = (\gamma - 1)\rho\chi. \tag{10}$$

6.1. A SOLAR MODEL WITH TURBULENT PRESSURE ALONE

The simplest way to take into account turbulence in solar modelling, is to include turbulent pressure (or Reynolds stress) alone, as done by many authors (e.g., Balmforth, 1992). In this case, only the hydrostatic equilibrium equation needs to be modified as follows:

$$\frac{\partial P}{\partial M_r} = -\frac{GM_r}{4\pi r^4}(1+\beta), \qquad (11)$$

where $P = P_{\text{gas}} + P_{\text{rad}}$, and

$$\beta = \left(\frac{2P_{\text{turb}}}{\rho g r} - \frac{\partial P_{\text{turb}}}{\partial P}\right)\left(1 + \frac{\partial P_{\text{turb}}}{\partial P}\right)^{-1}. \qquad (12)$$

6.2. A SOLAR MODEL WITH χ AND γ AS INDEPENDENT VARIABLES

The continuity equation and the equation of transport of energy by radiation remain the same regardless of turbulence. In terms of γ and χ the hydrostatic equilibrium is,

$$\frac{\partial P_T}{\partial M_r} = -\frac{GM_r}{4\pi r^4} - \frac{2(\gamma-1)\chi}{4\pi r^3}, \qquad (13)$$

where the total pressure $P_T = P_{\text{gas}} + P_{\text{rad}} + P_{\text{turb}}$.

The energy conservation equation is also affected by turbulence because the first law of thermodynamics should include the turbulent kinetic energy. The equation of energy transport by convection,

$$\frac{\partial T}{\partial M_r} = -\frac{T}{P_T}\frac{GM_r}{4\pi r^4}\nabla, \qquad (14)$$

does not change in form, but the convection temperature gradient, obtained in the previous section, is different from that without turbulence. The equations that govern envelope integrations also need to be changed accordingly.

6.3. SOLAR P-MODE OSCILLATION FREQUENCIES

Figure 2 shows the frequency differences between a solar model with turbulent pressure included and the standard solar model (single line) and between a solar model with both turbulent pressure and turbulent kinetic energy included and the standard solar model (bunched lines). All are scaled by the mode mass Q_{nl}. The effect on the p-modes of including turbulent kinetic energy is one order of magnitude larger than that of turbulent pressure alone. The inclusion of a pressure term only (Li et al., 2002b) has little effect. Although not described in this way, these conclusions are compatible with the results of Balmforth (1992) (on the effect of

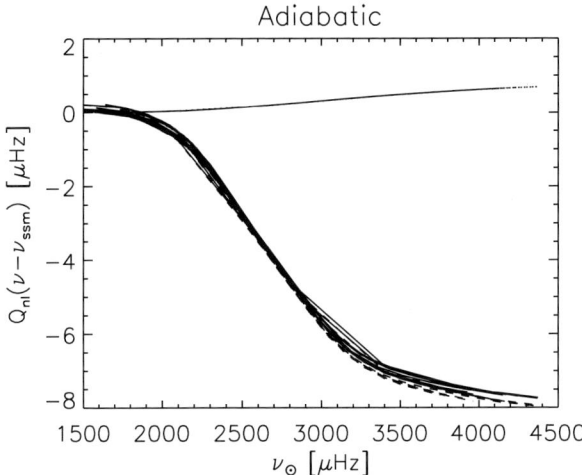

Figure 2. Perturbation of the eigenfrequencies (denoted by v_0) due to the introduction of turbulence into the stellar model, with (bunched lines) and without (single line) turbulent kinetic energy. The symbol v_{ssm} refers to the standard solar model. Each eigenfrequency is scaled by the mode mass Q_{nl}.

turbulent pressure only on p-mode frequencies) and of Rosenthal et al. (1999), who "patched" the 3D simulation of the Sun by Stein and Nordlund (1998) onto a standard solar model.

6.4. P-MODE STOCHASTIC EXCITATION FROM THE 3D SIMULATIONS

The velocity field from the simulation, together with the parameterized solar model yielded the same frequency maximum for the rate of stochastic energy input into solar p-modes as that derived from observation (Roca Cortés, 1999). A similar result was recently obtained by Stein and Nordlund (2001). Note the difference in the Stein-Nordlund approach and ours (Li et al., 2002b). Stein and Nordlund (2001) studied only the acoustic modes present in their simulation, whereas we used the larger set of p-mode frequencies derived from the parameterized solar model.

7. Simulations for Turnoff, Subgiant and Giant Stars

Using the same techniques that we applied to the Sun, we evolved a stellar model of one solar mass beyond the age of the present Sun, to the main sequence turn-off, sub-giant and giant stages and completed 3D simulations for each case.

Preliminary results indicate that the characteristics of the granules change as the star evolves in the $\log g - \log T_{eff}$ plane. The half-width of the two-point velocity correlation distribution decreases from the Sun to the subgiant. This will alter the effective mixing length parameter α_{eff} which is normally assumed to be constant in stellar evolution computations.

The reader is referred to the illustrations of granulation in the companion poster paper presented at this conference, and to the short movies of some of the simulations (Demarque et al., 2002).

8. Future Prospects

We are in the process of extending this analysis to other Sun-like stars for which we have relaxed 3-D simulations. Our purpose is two-fold: (1) to provide improved stellar models, including surface boundary conditions free of the mixing length uncertainty, using a parameterized description of turbulence based on 3-D simulations. (2) to calculate oscillation frequencies and mode excitation information for selected targets stars in upcoming asteroseismic space missions (MOST, then MONS, COROT and Eddington).

It is clear that we have barely scratched the surface in this field. The interaction of asteroseismology and increasingly more sophisticated numerical simulations of hydrodynamics and radiative transfer is likely to yield numerous new insights into stellar physics in the next few years.

Acknowledgements

This research was supported in part by NASA grant NAG5-8406 to Yale University. We thank T. Roca Cortés for sending us his data on stochastic excitation of the solar p-modes by convection. We also wish to acknowledge the contributions of K.L. Chan, D.B. Guenther, Y.-C. Kim, L.H. Li and S. Sofia, our collaborators in the convection simulations.

References

Asplund, M., Ludwig, H.-G., Nordlund, Å. and Stein, R.F.: 2000, *A&A* **359**, 669.
Audard, N., Provost, J. and Christensen-Dalsgaard, J.: 1996, *A&A* **297**, 427.
Balmforth, B.N.: 1992, *MNRAS* **255**, 603.
Basu, S. and Antia, H.M.: 1997, *MNRAS* **287**, 189.
Basu, S., Pinsonneault, M.H. and Bahcall, J.N.: 2000, *ApJ* **529**, 1084.
Batchelor, G.K.: 1953, *Quart. J. Roy. Meteor. Soc.* **79**, 224.
Boussinesq, J.: 1903, *Théorie analytique de la chaleur*, Tome II, Gauthier-Villars, Paris, p. 157.
Canuto, V.: 1997, *ApJ* **478**, 322.
Canuto, V.: 2000, *ApJ* **541**, 79.
Chaboyer, B., Demarque, P., Kernan, P.J. and Krauss, L.M.: 1998, *ApJ* **494**, 96.
Chan, K.L. and Wolff, C.L.: 1982, *J. Comp. Phys.* **47**, 109.
Chan, K.L. and Sofia, S.: 1989, *ApJ* **336**, 1022.

Chan, K.L., Nordlund, Å., Steffen, M. and Stein, R.F.: 1991, in: A.N. Cox, W.C. Livingston and M.S. Matthews (eds.), *Solar Interior & Atmosphere*, U. of Arizona Press, Tucson.
Chan, K.L., Mayr, H.G., Mengel, J.G. and Harris, I.: 1994, *J. Comp. Phys.* **113**, 165.
Christensen-Dalsgaard, J., Gough, D.O. and Thompson, M.J.: 1991, *ApJ*, **378**, 413.
Christensen-Dalsgaard, J., Bedding, T.R. and Kjeldsen, H.: 1995, *ApJ* **443**, L29.
Demarque, P., Sarajedini, A. and Guo, X.-J.: 1994, *ApJ* **426**, 165.
Demarque, P., Guenther, D.B. and Kim, Y.-C.: 1997, *ApJ* **474**, 790.
Demarque, P., Guenther, D.B. and Kim, Y.-C.: 1999, *ApJ* **517**, 510.
Demarque, P., Li, L.H., Robinson, F.J., Sofia, S. and Guenther, D.B.: 2002, in: M.J. Thompson, M.S. Cunha and M.J.P.F.G. Monteiro (eds.), *Asteroseismology Across the HR-diagram*, Kluwer Academic Publishers, Portugal.
Giménez, A., Guinan, E.F. and Montesinos, B.: 1999, *Stellar Structure: Theory and Test of Connective Energy Transport*, ASP Conf. Ser. **173**, San Francisco.
Gough, D.O.: 1969, *J. Atm. Sci.* **26**, 448.
Guenther, D.B.: 1991, *ApJ* **375**, 352.
Guenther, D.B.: 2002, *ApJ* **569**, 911.
Guenther, D.B. and Demarque, P.: 1996, *ApJ* **456**, 798.
Guenther, D.B. and Demarque, P.: 1997, *ApJ* **484**, 937.
Guenther, D.B. and Demarque, P.: 2000, *ApJ* **531**, 503.
Kelvin, Lord: 1862, *Mathematical and Physical Papers* (1911), Vol. 3, p. 255, Cambridge U. Press, Cambridge (original published in 1862).
Kim, Y.-C. and Chan, K.L.: 1997, in: F.P. Pijpers, J. Christen-Dalsgaard and C.S. Rosenthal (eds.), *SCORe'96 Proc.*, *Ap. & Space Science Library* **225**, 131.
Kim, Y.-C. and Chan, K.L.: 1998, *ApJ* **496**, L121.
Kim, Y.-C., Fox, P.A., Sofia, S. and Demarque, P.: 1995, *ApJ* **442**, 422.
Kim, Y.-C., Fox, P.A., Demarque, P. and Sofia, S.: 1996, *ApJ* **461**, 499.
Kjeldsen, H., Bedding, T.R., Viskum, M. and Frandsen, S.: 1995, *AJ* **109**, 1313.
Lane, I.J.H.: 1869, *Amer. J. Sci. 2nd ser.* **50**, 57.
Latour, J., Spiegel, E.A., Toomre, J. and Zahn, J.-P.: 1976, *ApJ* **207**, 233.
Li, L.H., Robinson, F.J., Demarque, P., Sofia, S. and Guenther, D.B.: 2002a, *ApJ* **567**, 1192.
Li, L.H., Robinson, F.J., Sofia, S., Demarque, P. and Guenther, D.B.: 2002b, in preparation.
Lydon, T.J. and Sofia, S.: 1995, *ApJS* **101**, 357.
Monteiro, M.J.P.F.G., Christensen-Dalsgaard, J. and Thompson, M.J.: 1996, *A&A* **307**, 624.
Nordlund, Å., Stein, R.F. and Brandenburg, A.: 1996, in: H.M. Antia and S.M. Chitre (eds.), *Windows on the Sun's Interior*, Bull Astron. Soc. of India **24**, 261.
Robinson, F.J., Demarque, P., Sofia, S., Chan, K.L., Kim, Y.-C. and Guenther, D.B.: 2000, *Helio- and Asteroseismology at the Dawn of the Millenium*, EAS Publ-464, p. 443.
Robinson, F.J., Demarque, P., Li, L.H., Kim, Y.-C., Sofia, S. and Guenther, D.B.: 2002a, *Proceedings of IAU Symp. 210: Modelling Stellar Atmospheres*, June 17–21, 2002, Uppsala, Sweden.
Robinson, F.J., Demarque, P., Li, L.H., Kim, Y.-C., Sofia, S. and Guenther, D.B.: 2002b, in: R. Cavallo, S. Keller and S. Turcotte (eds.), *3D Stellar Evolution*, PASP Conf. Ser., in press.
Robinson, F.J., Demarque, P., Li, L.H., Kim, Y.-C., Sofia, S. and Guenther, D.B.: 2003, *MNRAS*, in press, astro-ph/0212296.
Roca Cortés, T., Montañés, P., Pallé, P.L., Pérez Hernández, F., Jiménez, A., Régulo, C. and the GOLF Team: 1999, in: A. Giménez, E.F. Guinan and B. Montesinos (eds.), *Stellar Structure: Theory and Test of Connective Energy Transport*, ASP Conf. Ser. **173**, 305.
Rosenthal, C.S., Christensen-Dalsgaard, J., Nordlund, Å., Stein, R.F. and Trampedach, R.: 1999, *A&A* **351**, 689.
Roxburgh, I.W.: 1998, in: K.L. Chan, K.S. Cheng and H.P. Singh (eds.), *1997 Pacific Rim Conference on Stellar Astrophysics*, ASP Conf. Ser. **138**, 411.

Roxburgh, I.W. and Vorontsov, S.V.: 1999, in: A. Giménez, E.F. Guinan and B. Montesinos (eds.), *Theory and Tests of Convection in Stellar Structure*, ASP Conf. Ser. **173**, 257.
Schwarzschild, K.: 1906, *Nach. Kön. Ges. Wiss. Göttingen, Math.-phys. Klasse* **195**, 41.
Scufflaire, R.: 1974, *A&A* **36**, 107.
Singh, H.P., Roxburgh, I.W. and Chan, K.L.: 1995, *A&A* **295**, 703.
Smagorinsky, J.: 1963, *Monthly Weather Rev.* **91**, 99.
Spiegel, E.A. and Veronis, G.: 1960, *ApJ* **131**, 442 (correction: 135, 655).
Stein, R.F. and Nordlund, Å.: 1998, *ApJ* **499**, 914.
Stein, R.F and Nordlund, Å.: 2001, *ApJ* **546**, 585.
Unno, W. and Spiegel, E.A.: 1966, *PASJ* **18**, 85.
Unno, W., Osaki, Y., Ando, H., Saio, H. and Shibahashi, H.: 1989, *Non-Radial Oscillations of Stars*, 2nd ed., U. of Tokyo Press, Tokyo.
Woo, J.-H., Gallart, C., Demarque, P., Yi, S. and Zoccali, M.: 2002, *AJ* accepted, astro-ph/0208142.
Yi, S., Demarque, P. and Oemler, A.Jr.: 1997, *ApJ* **486**, 201.
Yi, S. et al.: 2000, *ApJ* **533**, 670.
Zahn, J.-P.: 1991, *A&A* **252**, 179.

DIFFUSION AND MIXING IN MAIN-SEQUENCE STARS

SYLVIE VAUCLAIR

Laboratoire d'Astrophysique, Observatoire Midi-Pyrénées, 14 avenue Edouard Belin, 31400 Toulouse, France

Abstract. The element settling which occurs inside stars, also called 'microscopic diffusion', due to the combined effect of gravity, thermal gradient, radiative acceleration and concentration gradient, leads to abundance variations which cannot be neglected in the computations of stellar structure. These processes where first introduced to account for abundance anomalies in 'peculiar stars', but their importance in the so-called 'normal' stars is now fully acknowledged, specially after the evidence of helium settling in the Sun from helioseismology. The reason why abundance variations as large as predicted by microscopic diffusion are not always observed is due to the influence of macroscopic motions, like rotation-induced mixing, or mass loss, which increase the settling time scales. In the present review, I discuss the theories of element settling and rotation-induced mixing and the importance of their coupling. I also give some comments about the links between diffusion processes and asteroseismology.

Keywords: stars, diffusion, oscillations

1. Element Settling in Stars

1.1. 'Microscopic diffusion' is a 'standard' stellar process

Although the importance of element settling (or 'microscopic diffusion') inside stars was recognized at the very beginning of stellar structure computations (Eddington, 1926), it was long forgotten for a simple reason: the observed abundances in stars did not vary as predicted. Later on, when 'chemically peculiar stars' were observed, the idea that these abundance anomalies could be due to microscopic diffusion emerged again and gave promising results (Michaud, 1970; Michaud et al., 1976; Vauclair et al., 1978a,b). From then on, for more than two decades, it became 'the diffusion hypothesis': element settling was presented as a special process added in some stars to account for their peculiarities. Its fundamental character in stellar structure was not yet back into the minds.

The fact that element settling also occurs in 'normal' stars was recently proved by helioseismic investigations (see Gough et al., 1996). It was known for a long time that helium and metals should have diffused by about 20% down from the solar convection zone since the birth of the Sun up to now (Aller and Chapman, 1960). Recent comparisons between the sound velocity inside the Sun as computed in the models and as deduced from the inversion of seismic modes, confirmed that this settling really occurred.

A star is a self-gravitational gaseous sphere, composed of a mixture of various gases with different partial pressures, masses and atomic spectra. Due to the pressure and thermal gradients and to the selective radiative transfer, individual elements diffuse inside the star, one with respect to the other, leading to a slow but effective restructuration. It is more difficult, in this framework, to understand why the consequences of diffusion processes are not seen in all the stars than to explain abundance anomalies. It is also necessary to account for the fact that, in chemically peculiar stars, the observed anomalies are not as strong as predicted by the theory of pure microscopic diffusion.

These questions are related to the hydrodynamical processes which occur in stellar radiative zones and compete with element settling, thereby increasing the time scale of abundance variations: convection, turbulence, mass loss, rotation-induced mixing...

1.2. Theory of element settling

Inside the convective regions, the rapid macroscopic motions mix the gas components and force their abundance homogenization. The chemical composition observed in the external regions of cool stars is thus affected by the settling which occurs below the outer convective zones, while in hot stars diffusion occurs directly in the atmosphere, which may lead to abundance gradients or 'clouds' in the spectroscopically observed region. As the settling time scales vary, in first approximation, like the inverse of the density, the expected variations are smaller for cooler stars, which have deeper convective zones. While some elements can see their abundances vary by several orders of magnitude in the hottest Ap stars, the abundance variations in the Sun are not larger than a few $\cong 10\%$.

The 'microscopic diffusion' of the chemical elements in stellar plasmas is due to a competition between two kinds of processes: individual ions move under the influence of the local gravity (or pressure gradient), thermal gradient, radiative acceleration and concentration gradient but their motion is slowed down due to collisions. This competition leads to selective element settling.

The computations of abundance variations are based on the Boltzmann equation for dilute collision-dominated plasmas. Because of the stellar gradients, the velocity distribution functions of individual elements deviate from pure maxwellian distribution. Several methods have been used to solve the Boltzmann equation in this framework (see Chapman and Cowling, 1970; Burgers, 1969). The diffusion equation has to be solved simultaneously for all the considered elements. The order of magnitude of the time scales generally implies the computation of many iterations of the diffusion process for a single evolutionary time step in stellar evolution codes. For each computation of a new model along the evolutionary track, the tables of abundances inside the star have to be transferred for every element, as a function of the internal mass. For the model consistency, these abundance profiles must be taken into account in the interpolation of the opacity tables.

For the Sun, the whole process has to be iterated several times from the beginning, with small changes in the original helium mass fraction and mixing length parameter, to obtain the right Sun and the right age (luminosity and radius with a precision of order 10^{-5}).

The diffusion time scales are direct functions of the collision probabilities for the considered species. A good treatment of collisions is thus necessary to obtain the abundance variations with a high degree of precision. Paquette et al. (1986) gave a useful prescription of collision probabilities in a screened coulomb potential, which is generally used in the computations of diffusion in stellar plasmas.

1.3. RADIATIVE ACCELERATIONS

While gravitational settling and thermal diffusion lead to a downward motion of all the elements other than hydrogen (which, in turn, goes up to replace them in global hydrostatic equilibrium), selective radiative transfer may push some elements upwards. This is due to the fact that the elements absorb photons which basically come from the internal layers of the star and re-emit these photons in an isotropic way, thereby gaining a net momentum upwards. This absorption may take place through the continuum (ionization) or through the lines (excitation of the ions). In practice the line process is much more significant than the continuum.

The computations of radiative acceleration on individual ions strongly depend on their atomic characteristics, which have to be precisely known. It also depends on the sharing of the photon flux with the other elements. It represents the same kind of problems as those encountered for the computations of stellar opacities. This is the reason why the most precise computations presently done on radiative accelerations are the result of collaborations between stellar physicists specialists of diffusion processes and atomic physicists specialists of opacity computations.

The importance of the radiative accelerations on individual elements increases with the effective temperature (Michaud et al., 1976). While it is negligible for the Sun (Turcotte et al., 1998b), it becomes larger than the gravity for most elements in hotter stars (Alecian et al., 1993; Turcotte et al., 1998a; Richer et al., 1998; Turcotte et al., 2000). The Montreal models (now called NMM for 'New Montreal Models') are presently the only ones really consistent for diffusion computations, in the sense that the modification of the stellar internal structure induced by the abundance variations are consistently taken into account step by step in the computations. In most other models, the variations of helium are taken into account, as well as a global variation of heavy elements. In the NMM, the opacities are computed according to the real detailed chemical composition.

The radiative accelerations on the elements vary with depth, according to their ionization stage. It may happen to be smaller than gravity at some depth and larger than gravity below. In this case there is an accumulation of the considered element at that special depth, even if it is depleted in the outer layers. This is the case, for example, for iron inside B, A and F stars (Richard et al., 2001; Richard et al.,

2002a). It is then possible that a new convective zone, due to this iron accumulation, take place inside the star, thereby changing the way diffusion processes behave. Note that the same phenomenon occurs in horizontal branch stars and is supposed to be the reason for the oscillations of 'SDB stars' (Charpinet et al., 1997).

We stress however that the NMM, which are very sophisticated in the treatment of element settling, cannot treat in such a precise way the macroscopic motions due to rotation-induced mixing. They can only introduce a very crudely parametrized turbulent diffusion coefficient.

2. Rotation-Induced Mixing

2.1. MERIDIONAL CIRCULATION

The process of meridional circulation and its consequences have been studied by many authors, including Eddington (1926), Sweet (1950), Mestel (1953, 1957), Tassoul and Tassoul (1989), Zahn (1992), Pinsonneault et al. (1992), Maeder and Zahn (1998), Vauclair and Théado (2002) and Théado and Vauclair (2002). The way such processes can slow down microscopic diffusion has also been discussed many times in the literature, beginning with Schatzman (1977), Vauclair et al. (1978a,b), Pinsonneault et al. (1992).

In rotating stars, due to the centrifugal forces, radiative equilibrium is not satisfied. This has to be compensated by a motion of matter with a vertical velocity derived form the equation:

$$\rho T \left(\frac{\partial s}{\partial t} + \mathbf{u}.\nabla s \right) = -\nabla.F + \rho \varepsilon_n \qquad (1)$$

where ε_n is the nuclear energy production, negligible in the outer layers, F the radiative flux, s the entropy density.

The vertical component of the circulation velocity is deduced from Eq. (1). If developed on the second Legendre polynomial, it may be written:

$$u_r = U_r P_2(\cos\theta) \qquad (2)$$

with:

$$U_r = \frac{P}{\rho g T c_P \left(\nabla_{ad} - \nabla + \nabla_\mu \right)} \frac{L}{M_*} (E_\Omega + E_\mu + E_\zeta + E_h) \qquad (3)$$

where:

$$E_\Omega = \qquad (4)$$
$$\frac{8}{3} \left(\frac{\Omega^2 r^3}{GM} \right) \left[1 - \frac{\Omega^2}{2\pi g\rho} \right] - \frac{\rho_m}{\rho} \left[\frac{r}{3} \frac{d}{dr} \left(H_T \frac{d\zeta}{dr} - \chi_T \zeta \right) - \frac{2H_T}{r} \zeta + \frac{2}{3} \zeta \right]$$

$$E_\mu = \frac{\rho_m}{\rho} \left\{ \frac{r}{3} \frac{d}{dr} \left[\left(H_T \frac{d\Lambda}{dr} \right) - (\chi_\mu + \chi_T + 1)\Lambda \right] - \frac{2H_T \Lambda}{r} \right\} \quad (5)$$

$$E_\zeta = \frac{M}{L} T C_p \frac{\partial \zeta}{\partial t} \quad (6)$$

$$E_h = \frac{M}{L} \frac{6}{r^2} C_p T D_h \zeta \quad (7)$$

In these equations, the deviations from perfect gas law are neglected. The thermodynamical parameters are averaged over level surfaces. L is the luminosity at radius r, $M_* = M(1 - \frac{\Omega^2}{2\pi g \rho_m})$ where M is the stellar mass at radius r and ρ_m the mean density inside the sphere of radius r. H_T is the temperature scale height; Λ represents the horizontal μ fluctuations $\frac{\tilde{\mu}}{\mu}$; and ζ the horizontal density fluctuations $\frac{\tilde{\rho}}{\rho}$; χ_μ and χ_T represent the derivatives:

$$\chi_\mu = \left(\frac{\partial \ln \chi}{\partial \ln \mu} \right)_{P,T} \quad ; \quad \chi_T = \left(\frac{\partial \ln \chi}{\partial \ln T} \right)_{P,\mu} \quad (8)$$

Note that E_h has a different behaviour than the other terms, as it is formally proportional to U_r when the horizontal diffusion coefficient D_h is approximated as in Maeder and Zahn (1998), that is $D_h = C_h r U_r$ where C_h is an unknown coefficient.

The relative importance of all these terms is extensively discussed in Vauclair and Théado (2002).

The horizontal component of the meridional circulation, obtained from the equation of mass conservation, is then given by:

$$u_\theta = -\frac{1}{2\rho r} \frac{d}{dr} \left(\rho r^2 U_r \right) \sin \theta \cos \theta \quad (9)$$

Zahn (1992, 1993) suggested that the transport of angular momentum induced by the meridional flow could lead to shear flow instabilities. Due to the density stratification, these instabilities create anisotropic turbulence, more important in the horizontal than in the vertical direction. The coupling between the meridional advection and the horizontal turbulence leads to a special kind of mixing for the chemical species, parametrized as an effective diffusion coefficient. Later on, the transport of chemical species and angular momentum were consistently coupled in the computations, with the assumption that no other process interfered with rotation-induced mixing (Zahn et al., 1997). The results predicted a non negligible differential rotation inside the solar radiative regions, in contradiction with helioseismology. The fact that the Sun rotates as a solid body below the convective zone proves that angular momentum has to be transported by another process,

2.2. THE IMPORTANCE OF μ-GRADIENTS

The importance of the feed-back effect due nuclearly-induced μ-gradients on the meridional circulation has been recognized for a long time. Mestel and Moss (1986) showed how these μ-gradients could slowly stabilize the circulation and expel it from the core towards the external layers. They called this process 'creeping paralysis". On the other hand, the feed-back effect due to the diffusion-induced μ-gradients was not included until recently. Vauclair (1999) showed how, in slowly rotating stars, the resulting terms in the computations of the circulation velocity could become of the same order as the other terms. Théado and Vauclair (2001) computed these terms for the case of Pop II stars and claimed that the induced diffusion-circulation coupling could be the reason for the very small dispersion of the lithium abundances observed in halo stars. More recently, a new analysis of these processes was given by Vauclair and Théado (2002) while Théado and Vauclair (2002) presented a 2D numerical simulation of meridional circulation including μ-gradients (see also Théado and Vauclair, these proceedings).

Due to the meridional circulation, vertical μ- gradients give rise to horizontal μ-gradients as the upward flow brings up matter with a larger μ while the downward flow brings down matter with a smaller μ.

According to Zahn (1992) and Chaboyer and Zahn (1992), the horizontal μ-gradient Λ is given by:

$$\Lambda = -\frac{U_r \cdot r^2}{6D_h} \frac{\partial \ln \mu}{\partial r} \tag{10}$$

With the assumption that: $D_h = C_h U_r \cdot r$, this equation becomes:

$$\Lambda \simeq -\frac{1}{6C_h} \frac{\partial \ln \mu}{\partial \ln r} \tag{11}$$

where C_h should be of order unity.

At the beginning of the stellar evolution on the main sequence, Λ is equal to zero. Then, due to helium settling below the stellar convective zone, a μ-gradient can built so that $|E_\mu|$ becomes very close to $|E_\Omega|$ in a time scale which depends on the local conditions but may be shorter than the main-sequence lifetime. In this case, a kind of 'creeping paralysis' may settle down.

Suppose that the circulation is nearly frozen in a region just below the convective zone. This means that on every level surface, the horizontal μ-gradient Λ is just equal to the critical value Λ_{crit}, for which $|E_\mu|$ is equal to $|E_\Omega|$. However, helium diffuses out of the convective zone where it is homogeneous, that is $\Lambda = 0$. Because of this diffusion, the horizontal μ-gradient is always forced to remain below the critical value in a boundary layer just below the convective zone (in this analysis, no overshooting or tachocline is taken into account, but it would reinforce this conclusion). Under such conditions a new circulation loop may take place, which remains in close connection with the microscopic diffusion process.

3. Diffusion and Asteroseismology in Main-Sequence Stars

3.1. THE SOLAR CASE

Owing to helioseismology, the sound velocity inside the Sun is known with a precision of $\cong$ 0.1% and gives evidence of the occurrence of helium settling as predicted by diffusion computations. Solar models computed in the old 'standard' way, in which the element settling is totally neglected, do not agree with the inversion of the seismic modes. This result has been obtained by many authors, in different ways (see Gough et al., 1996 and references therein). There is a characteristic discrepancy of a few percent, just below the convective zone, between the sound velocity computed in the models and that of the seismic Sun. Introducing the element settling considerably improves the consistency with the seismic Sun, but some discrepancies do remain, particularly below the convective zone where a peak appears in the sound velocity. The reason of this peak is probably due to the steepness of the μ-gradient induced by pure microscopic diffusion (Richard et al., 1996). The helium profiles directly obtained from helioseismology (Basu, 1997, 1998; Antia and Chitre, 1998) show indeed a helium gradient below the convective zone which is smoother than the gradient obtained with pure settling. Introduction of macroscopic motions in competition with the settling slightly smoothes down the helium gradient, and may rub out the peak, although some differences still remain between the models and the seismic inversion results (Brun et al., 1999; Richard et al., 2002b).

Such motions are also needed to reproduce the observed abundances of light elements, namely a lithium depletion by about 140 compared to the protosolar value while beryllium is normal (Balachandran and Bell, 1998). Furthermore, observations of the ^{3}He/^{4}He ratio in the solar wind and in the lunar rocks (Geiss and Gloecker, 1998) show that this ratio may not have increased by more than $\cong$ 10% during the past 3 Gyr in the Sun. While the occurrence of some mild mixing below the solar convective zone is needed to explain the lithium depletion and helps for the conciliation of the models with helioseismological constraints, the ^{3}He/^{4}He observations put a strict constraint on its efficiency. The effect of μ-gradients on the mixing processes has to be invoked to explain these observations in a consistent way.

The question now arises of the possibility to test the effects of diffusion in solar-type stars with future space missions. I discuss below the special case of 'chemically peculiar' stars. For F stars, preliminary computations are underway.

3.2. AM VERSUS δ-SCUTI STARS

Among the main sequence stars which lie inside the instability strip, many chemically peculiar stars are found. The magnetic stars will be discussed below. Here I focus on the so-called Am stars, which are found in the H.R. diagram at the same place as the δ-scuti stars. Generally speaking, the former ones show abundance peculiarities, namely a general overabundance of metals (except calcium and scandium), but no oscillations, while the later ones are pulsating but chemically normal. As discussed by Turcotte et al. (2000), in A-type stars, almost 70% of non chemically peculiar stars are δ-scuti variables at current levels of sensitivity while most non-variable stars are Am stars'. Furthermore, Am stars are slower rotators than δ-scuti stars. This lead Baglin (1972) to suggest a dichotomy between the two kinds of stars. In this region of the H.R. diagram, the stars display two different convective zones in their outer layers: the upper one due to the HI and HeI ionisations and the lower one to the HeII ionisation. The δ-scuti pulsate due to a κ-mechanism which takes place in the second convective zone. When microscopic diffusion occurs, this convective zone disappears due to helium depletion and the κ-mechanism cannot take place anymore (Vauclair et al., 1974).

More recently however, some oscillating Am stars have been discovered (Kurtz, 1989; Kurtz et al., 1995), which challenge the previously accepted theory. Richer et al. (2000) and Turcotte et al. (2000) computed models of Am stars in the framework of the NMM (New Montreal Models). They found that, due to the iron accumulation in the radiative zone below the H and He convective zone, a new convective region appears which increases the diffusion time scales compared to the previous models. In these new models, helium is still substantially present in the helium convective zone at the ages of the considered stars. They claim that it is possible to account for the existence of oscillating Am stars close to the cool boundary of the instability strip.

We must note here that in all these computations, the different convective zones are assumed to be completely connected by overshooting. In the NMM models, the new iron convective zone which appears below the two other ones is also linked to them as soon as it is created. This may not be the case in real stars: even if overshooting exists between the convective zones, the resulting mixing may be incomplete and lead to some slow diffusion, which will modify the results (Bazot and Vauclair, 2003, in preparation).

3.3. RAPIDLY OSCILLATING AP STARS

The observations of rapidly oscillating Ap stars are extensively discussed by Kurtz (this meeting) who shows how complex these objects are. The models which have been proposed up to now are much too simple to be able to account for all their features. Among these models, some are related to the magnetic structure itself (e.g. Shibahashi, 1983) while the others rely on the chemical segregation (Dolez and

Gough, 1982; Dolez et al., 1988; Vauclair et al., 1991; Dziembowski and Goode, 1996; Balmforth et al., 2001).

Dolez et al. (1988) and Vauclair et al. (1991) proposed that the oscillations in roAp stars could be driven by κ-mechanism in the HeII ionisation zone. As helium is always depleted by diffusion, the idea was that a wind could exist at the magnetic poles, creating a helium overabundance. This model was supposed to be the continuation for cool stars of the model proposed by Vauclair (1975) to account for helium rich stars. In cooler stars, the accumulation should not be visible in the atmosphere, but it should still occur at the place where helium becomes neutral (first ionisation zone). As the κ-mechanism is driven by the second ionisation zone, the helium accumulation could be efficient only if it was large enough so that its downward wing could appreciably extent down to this region.

This model for roAp stars has been challenged by Balmforth et al. (2001), who computed the driving of the modes with different helium profiles at the magnetic poles and equator. They found that, contrary to what was assumed before, a helium accumulation induced by diffusion in a wind does not help driving the pulsations. On the contrary, the excitation of the modes is more important when there is less helium in the atmosphere, the basic driving mechanism being induced by hydrogen ionization (this was already suggested by Dziembowski and Goode, 1996). Meanwhile the magnetic equator damps the oscillations due to energy loss by turbulence in the remaining convection zone: this could explain why oscillations aligned with the magnetic axis are preferentially excited in these stars. In this respect, recent spectroscopic observations which show evidences of abundance stratification in roAp stars which are not seen in noAp stars may give interesting clues (Gelbmann et al., 2000; Ryabchikova et al., 2002).

All these models assume that the magnetic field behaves as a dipole, in the framework of the oblique rotator model. However Bigot (this meeting) shows that this may not be the case. The asterosismology of Ap stars is a mine of gold which brings up a large panel of constraints and new information: a real challenge for theoreticians!

Helio- and asteroseismology already renewed our vision of the internal stellar structure, and we may expect new interesting tests and constraints on the diffusion processes which occur in main sequence stars in the very near future.

References

Alecian, G., Michaud, G. and Tully, J.: 1993, *ApJ* **411**, 882.
Aller, L.H. and Chapman, S.: 1960, *ApJ* **132**, 461.
Antia, H.M. and Chitre, S.M.: 1998, *A&A* **339**, 239.
Baglin, A.: 1972, *A&A* **19**, 45.
Balachandran, S. and Bell, R.A., 1998, *Nature* **392**, 791.
Balmforth, N.J., Cunha, M.S., Dolez, N., Gough, D.O. and Vauclair, S.: 2001, *MNRAS* **323**, 362.
Basu, S.: 1997, *MNRAS* **288**, 572.

Basu, S.: 1998, *MNRAS* **298**, 719.
Brun, A.S., Turck-Chièze, S. and Zahn, J.P.: 1999, *ApJ* **525**, 1032.
Burgers, J.M..: 1969, *Flow Equations for Composite Gases*, New York: Academic Press.
Chaboyer, B. and Zahn, J.-P.: 1992, *A&A* **253**, 173.
Chapman, S. and Cowling, T.G.: 1970, *The mathematical Theory of Non-Uniform Gases*, Cambridge University Press, 3rd ed.
Charbonneau, P. and MacGregor, K.B.: 1992, *ApJ* **387**, 639.
Charpinet, S., Fontaine, G., Brassard, P., Chayer, P., Rogers, F.J., Iglesias, C.A. and Dorman, B.: 1997, *ApJ* **483**, L123.
Dolez, N. and Gough, D.O.: 1982, in: J.P. Cox and C.J. Hansen (eds.), *Pulsations in Classical and Cataclysmic Variable Stars*, (JILA, Boulder, CO), p. 248.
Dolez, N., Gough, D.O. and Vauclair, S.: 1988, in: J. Christensen-Dalsgaard and S. Frandsen (eds.), *Advances in Helio- and Asteroseismology*, proc. IAU Symp. 123, 291.
Dziembowski, W.A. and Goode, P.R.: 1996, *ApJ* **458**, 338.
Eddington E., 1926, 'The Internal Constitution of the Stars', Dover Pub.
Geiss, J. and Gloecker, G.: 1998, *Space Science Reviews* **84**, 239.
Gelbmann, M., Ryabchikova, T., Weiss, W.W., Piskunov, N., Kupka, F. and Mathys, G.: 2000, *A&A* **356**, 200.
Gough, D.O., Kosovichev, A.G., Toomre, J., Anderson, E., Antia, H.M., Basu, S., Chaboyer, B., Chitre, S.M., Christensen-Dalsgaard, J., Dziembowski, W.A., Eff-Darwich, A., Elliott, J.R., Giles, P.M., Goode, P.R., Guzik, J.A., Harvey, J.W., Hill, F., Leibacher, J.W., Monteiro, M.J.P.F.G., Richard, O., Sekii, T., Shibahashi; H., Takata, M., Thompson, M.J., Vauclair, S. and Vorontosov, S.V.: 1996, *Science* **272**, 1296.
Gough, D.O. and MacIntyre, M.E.: 1998, *Nature* **394**, 755.
Kurtz, D.W.: 1989, *MNRAS* **238**, 1077.
Kurtz, D.W., Garrison, R.F.,Koen, C., Hofman, G.F. and Viaranna, N.: 1995, *MNRAS* **276**, 199.
Maeder, A. and Zahn, J.-P.: 1998, *A&A* **334**, 1000.
Mestel, L.: 1953, *MNRAS* **113**, 716.
Mestel, L.: 1957, *ApJ* **126**, 550.
Mestel, L. and Moss, D.L.: 1986, *MNRAS* **221**, 25.
Mestel, L., Moss, D.L. and Tayler, R.J.: 1988, *MNRAS* **231**, 873.
Michaud, G.: 1970, *ApJ* **160**, 641.
Michaud, G., Charland, Y., Vauclair, S. and Vauclair, G.: 1976, *ApJ* **210**, 445.
Paquette, C., Pelletier, C., Fontaine, G. and Michaud, G.: 1986, *ApJS* **61**, 177.
Pinsonneault, M.H., Deliyannis, C.P. and Demarque P.: 1992, *ApJS* **78**, 181.
Richard, O., Michaud, G. and Richer, J.: 2001, *ApJ* **558**, 377.
Richard, O., Michaud, G., Richer, J., Turcotte, S., Turck-Chièze, S. and VandenBerg, D.A.: 2002a, *ApJ* **568**, 979.
Richard, O., Théado, S. and Vauclair, S.: 2002b, preprint.
Richard, O., Vauclair, S., Charbonnel, C. and Dziembowski, W.A.: 1996, *A&A* **312**, 1000.
Richer, J., Michaud, G. and Turcotte, S.: 2000, *ApJ* **529**, 338.
Richer, J., Michaud, G., Rogers, F.J., Iglesias, C.A., Turcotte, S. and Leblanc, F.: 1998, *ApJ* **492**, 833.
Ryabchikova, T., Piskunov, N., Kochukhov, O., Tsymbal, V., Mittermayer, P. and Weiss, W.W.: 2002, *A&A* **384**, 545.
Schatzman, E.: 1977, *A&A* **56**, 211.
Sweet, P.A.: 1950, *MNRAS* **110**, 548.
Tassoul, J.L. and Tassoul, M.: 1989, *A&A* **213**, 397.
Théado S. and Vauclair, S.: 2001, *A&A* **375**, 70.
Théado S. and Vauclair, S.: 2002, submitted to *ApJ*.
Turcotte, S., Richer, J. and Michaud, G.: 1998a, *ApJ* **504**, 559.
Turcotte, S., Richer, J., Michaud, G. and Christensen-Dalsgaard, J.: 2000, *A&A* **360**, 603.

Turcotte, S., Richer, J., Michaud, G., Iglesias, C.A. and Rogers, F.J.: 1998b, *ApJ* **504**, 539.
Vauclair, S. and Théado, S.: 2002, submitted to *ApJ*.
Vauclair, S.: 1975, *A&A* **45**, 233.
Vauclair, S.: 1999, *A&A* **351**, 973.
Vauclair, S., Dolez, N. and Gough, D.O.: 1991, *A&A* **252**, 618.
Vauclair, G., Vauclair, S. and Michaud, G.: 1978a, *ApJ* **223**, 920.
Vauclair, S., Vauclair, G., Schatzman, E. and Michaud, G.: 1978b, *ApJ* **223**, 567.
Zahn, J.-P.: 1992, *A&A* **265**, 115.
Zahn, J.-P.: 1993, in: J.-P. Zahn and J. Zinn-Justin (eds.), *Astrophysical Fluid Dynamics*, les Houches, XLVII, 561.
Zahn, J.-P., Talon S. and Matias, J.: 1997, *A&A* **322**, 320.

ARE PULSATION AND MAGNETIC AXES ALIGNED IN roAp STARS?

L. BIGOT[1] and W.A. DZIEMBOWSKI[2]

[1]*N. Bohr Institute for Astronomy, Physics and Geophysics, Copenhagen, Denmark*
[2] *N. Copernicus Center, Warsaw, Poland*

Abstract. It is commonly assumed that in the rapidly oscillating Ap (roAp) stars the mode axis is aligned or nearly aligned with the magnetic field axis. This would be possible if the field is the only important effect causing departure from spherical symmetry. We show that even though these stars are slow rotators, the centrifugal force cannot be neglected. The consequence is that the modes cannot be in general symmetric about the magnetic field. We argue that such a symmetry is not implied by the observed coincidence between the field and pulsation amplitude maxima.

Keywords: pulsations, stars, magnetic field, rotation

1. Introduction

After his discovery of the rapid oscillations of the Ap (roAp) stars, Kurtz (1982) proposed the model of the oblique pulsator. This early model supposed that the magnetic field has a predominant role in the properties of these oscillations so that the eigenmodes and the magnetic field have the same axis of symmetry. Dziembowski and Goode (1985) improved this model by taking into account the effects of the Coriolis force. They found that because of the non-axisymmetry of the rotation, neither of the modes is strictly aligned with the magnetic axis. However, for roAp stars, the effect of the Coriolis force is so weak compared to the magnetic one that the departure from alignment was extremely tiny and people still continue to assume axisymmetry of modes with respect to the magnetic axis (e.g. Kurtz and Shibahashi, 1986; Kurtz, 1990).

In a recent work, Bigot and Dziembowski (2002) revisited the problem of the role of rotation in roAp stars introducing two main improvements: they treated the effect of the magnetic field with a non-perturbative theory (Bigot et al., 2000) and, more importantly, they included the effects of the centrifugal distortion of the star. Even though these stars are quite slow rotators, the higher order effect of rotation is more important for mode alignment. This is due to the fact that they are high overtone pulsators.

In the present paper we would like to stress that this property of aligned magnetic and pulsation axes is no longer valid if we take into account the centrifugal force. We also emphasize that the observed time-coincidence between magnetic and pulsation extrema, generally used as a proof for alignment, can be also interpreted in terms of inclined axes.

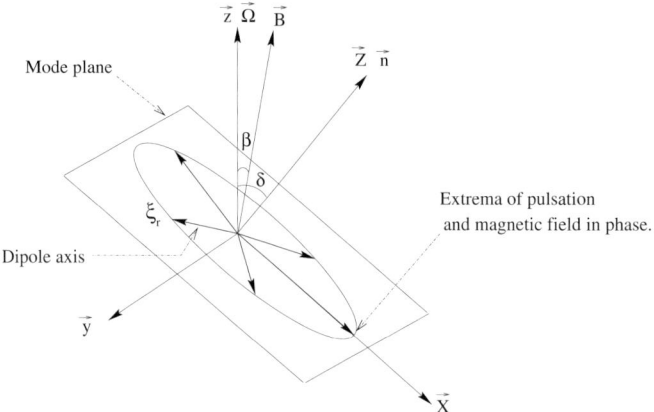

Figure 1. Representation of the dipole mode axis during the pulsation cycle. During the pulsation cycle, each point of the star describes an ellipse in the plane which, in general, is inclined to both the magnetic and rotation axes.

2. Combined Effects of Rotation and Magnetic Field on Nonradial Modes

Since magnetic and rotation axes are inclined, the joint perturbation has no longer axial symmetry and consequently individual modes are no longer described by a single azimuthal number m. The modes are also no longer described by a single degree ℓ, but we ignore this here as it is a less essential effect. We thus consider the displacement vector

$$\vec{\xi} = \sum_{m=-\ell}^{\ell} a_m \vec{\xi}_m, \qquad (1)$$

where the coefficients a_m are solution of an eigenvalue system (Bigot and Dziembowski, 2002). These coefficients depend, among others, on the magnetic field strength, frequency, rotation rate, inclination between magnetic and rotation axes. There is a strong observational evidence that oscillations in roAp stars are dominated by dipole ($\ell = 1$) modes (e.g. Kurtz, 1990). We do not challenge this. The debate concerns the relative orientation of dipole with respect to the magnetic and rotation axes. When an inclined dipole magnetic field is imposed, the three modes represent, in general, three stages of elliptical polarization. Each of modes is characterized by three parameters that are the frequency, the inclination of the mode plane, δ, and the ellipticity angle, ψ. The latter is defined so that the tangent is equal to the ratio of the ellipse axes. We note that when $\psi = 0$ the mode is linearly polarized in the plane formed by the rotation and magnetic axes ($\vec{\Omega}, \vec{B}$) and when $\psi = \pi/2$ the mode axis is orthogonal to this plane. In Figure 1 we show the case of an elliptical mode with the long axis lying in the $(\vec{\Omega}, \vec{B})$ plane. The angle between this axis and the magnetic 'north pole' is $\pi/2 + \delta - \beta$.

The departure from linear polarization arises only from the effect of the Coriolis force and, at typical rotation rates encountered in roAp stars, it is quite small. Consequently, the ellipses described by each point of star are typically very elongated along the longer axis. It is the centrifugal force that is the primary cause of mode perturbation departure from its symmetry about the magnetic field axis. The departure becomes very significant already at rotation period of 3 days for a field of 1 kG at the pole.

3. Observational Constraint

The time-coincidence of the pulsation amplitude and magnetic field maxima has been put forward as the evidence that magnetic and pulsation axes in roAp stars are aligned. In fact, such a coincidence is expected in a whole range of situations. Consider, for instance, a simplified, but not far from realistic, case of linearly polarized mode with an axis in the $(\vec{\Omega}, \vec{B})$ plane. The maximum of pulsation amplitude may occur only when the observer is in this plane. Whether this phase corresponds also to a maximum of the measured field intensity depends on the relative inclinations of the mode axis, $\gamma = \pi/2 - \delta$, magnetic axis, β, and the line of sight, i. During the rotation cycle, both the measured magnetic field intensity and the pulsation amplitude may have one or two maxima. The condition for the time-coincidence of the absolute maxima is that the line of sight is simultaneously the closest to the field and the mode axes. The condition takes form of the following inequality

$$(|\beta - i| - |\pi - \beta - i|)(|\gamma - i| - |\pi - \gamma - i|) > 0,$$

which sets only restrictions on γ but does not imply $\beta = \gamma$. If we have $\beta < i$, the condition reduces to $|\gamma - i| < |\pi - \gamma - i|$.

It is most favourable situation when both the magnetic field and pulsation amplitude exhibit two maxima during one rotation cycle. This is what happens in HR 3831 (Kurtz et al., 1992). Then, from the measured ratios of the maximum values we could, in principle, derive the value of γ. This would provide a valuable test on models of roAp stars. It remains to be seen whether current data allow a meaningful test.

4. Conclusions

In roAp stars the eigenmodes cannot have the same symmetry as the magnetic field. The departure is due to rotation with an axis inclined to the magnetic axis that introduces a non-axisymmetric perturbation to the magnetic modes. The role of the centrifugal distortion of the star is the dominant effect of rotation on oscillations in roAp stars. The distortion may cause a significant departure of mode symmetry from that of the magnetic field. We have also shown, on a simple example, that the usual interpretation of the coincidence between maxima of the magnetic versus pulsation variation in terms of aligned magnetic and pulsation axes is not a proof for alignment. The inclined dipole modes that we found are then not in contradiction with the observations of the magnetic data.

Acknowledgements

The work was partially supported by KBN grant 5P03D.

References

Bigot, L. et al.: 2000, *A&A* **356**, 218.
Bigot, L. and Dziembowski, W.A.: 2002, *A&A* **391**, 235.
Dziembowski, W.A. and Goode, P.R.: 1985, *ApJL* **296**, L27.
Kurtz, D.W.: 1982, *MNRAS* **200**, 807.
Kurtz, D.W.: 1990, *ARA&A* **28**, 607.
Kurtz, D.W. and Shibahashi, H.: 1986, *MNRAS* **223**, 557.
Kurtz, D.W. et al.: 1992, *MNRAS* **255**, 289.

SEISMIC DIAGNOSTICS OF STELLAR CONVECTION TREATMENT FROM OSCILLATION AMPLITUDES OF P-MODES

R. SAMADI[1,2], M.J. GOUPIL[2], Y. LEBRETON[2], Å. NORDLUND[3] and F. BAUDIN[4]

[1] Astronomy Unit, Queen Mary, University of London, London, UK
[2] Observatoire de Paris-Meudon, Meudon, France
[3] Niels Bohr Institute for Astronomy Physics and Geophysics, Copenhagen, Denmark
[4] Institut d'Astrophysique Spatiale, Orsay, France

Abstract. The excitation rate P of solar p-modes is computed with a model of stochastic excitation which involves constraints on the averaged properties of the solar turbulence. These constraints are obtained from a 3D simulation. Resulting values for P are found ~ 4.5 times larger than when the calculation assumes properties of turbulent convection which are derived from an 1D solar model based on Gough (1977)'s formulation of the mixing-length theory (GMLT). This difference is mainly due to the assumed values for the mean anisotropy of the velocity field in each case. Calculations based on 3D constraints bring the P maximum closer to the observational one. We also compute P for several models of intermediate mass stars ($1 \lesssim M \lesssim 2\,M_\odot$). Differences in the values of $P_{\max}$ between models computed with the classical mixing-length theory and GMLT models are found large enough for main sequence stars to suggest that measurements of P in this mass range will be able to discriminate between different models of turbulent convection.

Keywords: convection, turbulence, oscillations, Sun

1. Introduction

Excitation of solar-type oscillations is attributed to turbulent movements in the outer convective zone of intermediate mass stars.

Accurate measurement of the rate P at which acoustic energy is injected into such oscillations will be possible with the future seismic missions (e.g. COROT and Eddington). Comparison between measured and theoretical values of P obtained with different models of turbulent convection will then provide valuable information about the properties of stellar convection zones.

Models for stochastic excitation have been proposed by several authors (e.g. Goldreich and Keeley, 1977; Balmforth, 1992; Samadi and Goupil, 2001). In the present work, we consider the formulation of Samadi and Goupil (2001, Paper I hereafter). Constraints on the time averaged properties of the solar turbulent medium are obtained from a 3D simulation. They allow us to compute P and compare it with solar seismic observations and results obtained with Gough (1977)'s formulation of the mixing-length theory (GMLT hereafter).

2. The Model of Stochastic Excitation

According to Paper I, the rate at which a given mode with frequency ω_0 is excited can be written as:

$$P(\omega_0) \propto \int_0^M dm \, \frac{\Phi}{3} \rho_0 w^4 \left\{ \frac{16}{15} \frac{\Phi}{3} \left(\frac{d\xi_r}{dr} \right)^2 S_R + \frac{4}{3} \left(\frac{\alpha_s \tilde{s}}{\rho_0 w} \right)^2 \frac{g_r}{\omega_0^2} S_S \right\} \quad (1)$$

In Eq. (1), ρ_0 is the mean density, ξ_r is the radial component of the fluid displacement adiabatic eigenfunction ξ, $\alpha_s = (\partial p/\partial s)_\rho$ where p denotes the pressure and s the entropy, $\tilde{s}^2$ is the rms value of the entropy fluctuations, $g_r(\xi_r, m)$ involves the first and the second derivatives of ξ with respect to r, $S_R(\omega_0, m)$ and $S_S(\omega_0, m)$ are driving sources inferred from the Reynolds stress and the entropy fluctuations respectively, Φ is Gough (1977)'s mean anisotropy factor defined as $\Phi(m) \equiv <\mathbf{u}^2> - <\mathbf{u}>^2/w^2$ where $\mathbf{u}$ is the velocity field, $<.>$ denotes time and horizontal average and w is the mean vertical velocity ($w^2 \equiv <u_z^2> - <u_z>^2$). Expressions for $S_R(\omega_0, m)$, $S_S(\omega_0, m)$ and $g_r(\xi_r, m)$ are given in Paper I.

The driving sources S_R and S_S include the turbulent kinetic energy spectrum $E(k, m)$, the turbulent spectrum of the entropy fluctuations $E_s(k, m)$ and $\chi_k(\omega)$ the frequency-dependent part of the turbulent spectra which models the correlation time-scale of an eddy with wavenumber k. The quantity $\chi_k(\omega)$ is modeled here with a non-gaussian function constraint from the 3D simulation.

3. Numerical Constraints and Computation in the Solar Case

We consider a 3D simulation of the upper part of the solar convective zone as obtained by Stein and Nordlund (1998). The simulated domain is 3.2 Mm deep and its surface is 6×6 Mm^2. The grid of mesh points is $256 \times 256 \times 163$, the total duration 27 min and the sampling time 30s.

The simulation data are used to determine the quantities $E(k, m)$, $E_s(k, m)$, w, $\tilde{s}$ and Φ involved in the theoretical expressions for S_R, S_S and P. More details will be given in forthcoming papers.

We compute P according to Eq. (1): the eigenfunctions (ξ) and their frequencies ($\nu = \omega_0/2\pi$) are calculated with Balmforth (1992)'s non-adiabatic code for a solar 1D model built with the GMLT approach. The k-dependency of $E(k, z)$, is modeled as $(k/k_0)^{+1}$ for $k < k_0$ and as $(k/k_0)^{-5/3}$ for $k > k_0$ where $k_0 = 2\pi/\beta\Lambda$, $\Lambda = \alpha H_p$ is the mixing-length, H_p the pressure scale height. The value of the mixing-length parameter, α, is imposed by a solar calibration of the 1D GMLT model. The value of k_0 – hence of β – is obtained from the 3D simulation. The above analytical k-dependency of $E(k, z)$ and $E_s(k, z)$ reproduce the global features of $E(k, z)$ and E_s derived from the 3D simulation.

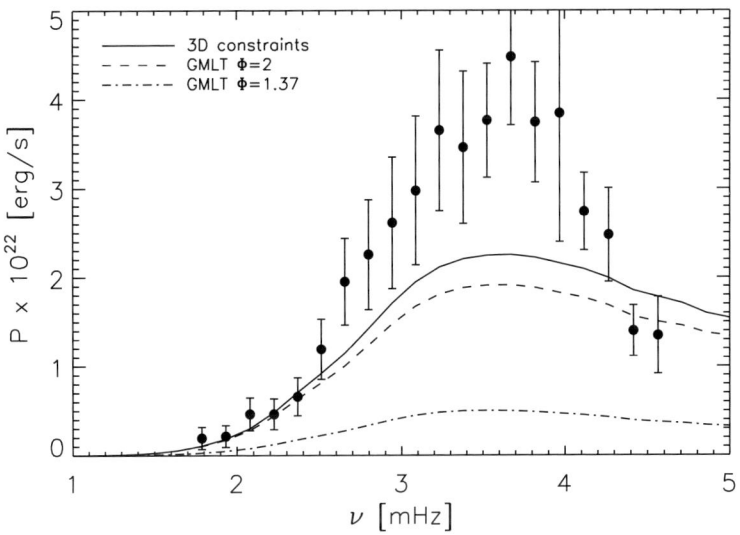

Figure 1. Solar p-modes excitation rate $P(\nu)$ (See text for details).

Results of P computations are presented in Figure 1 (**solid curve**) and compared with values of P (**filled dots**) derived from solar seismic measurement by Chaplin et al. (1998).

We also compute P in the case when the quantities involved in Eq. (1) are all obtained from the GMLT model. In the mixing-length approach, the anisotropy factor, Φ, is a free parameter. We then assume two different values for Φ: $\Phi = 1.3745$ (**dot dashed curves**) and $\Phi = 2$ (**dashed curves**). The value $\Phi = 1.3745$ used in the GMLT model provides the best fit between computed solar damping rates and the measured ones by Chaplin et al. (1998).

Without any adjustement of scaling parameters and instead using all the constraints inferred from the 3D simulation, we find a maximum of P much larger ($\sim$ 4.5 times larger) than is obtained using a 1D GMLT solar model with $\Phi = 1.37$. This difference is mainly due to the assumed value for Φ in the excitation region. Indeed, analysis of the 3D simulation suggests that $\Phi \sim 2$ in the excitation region. When $\Phi = 2$ is used to compute the excitation rate with the GMLT model, P comes close to P calculated with the 3D simulation constraints.

Our result then shows that the values of Φ found for the solar GMLT model when adjusted on the damping rates is not compatible with the actual properties of turbulent medium in the excitation region. An improvement could come from a consistent calculation which would assume a depth dependent $\Phi(m)$, as suggested by the simulation, in both damping and excitation rates computation.

Our calculations using the 3D constraints bring P_{max}, the P maximum, closer to the solar seismic measurements but still under-estimates them by a factor ~ 2. More sophisticated assumptions for $k_0(z)$ will likely lead to a better agreement with the observations.

4. Scanning the HR Diagram

We consider two sets of stellar models previously investigated by Samadi et al. (2001): Models in the first set are computed with the classical MLT whereas those belonging to the the second set are computed with GMLT. We find that the maximum of P, P_{max}, scales as $3.2 \log(L/L_\odot \times M_\odot/M)$ for the first set and scales as $3.5 \log(L/L_\odot \times M_\odot/M)$ for the second one (L, $L_\odot$, M and $M_\odot$ have their usual meaning). This result suggests that measurements of P in several differents intermediate mass stars ($1 \lesssim M \lesssim 2\,M_\odot$) will enable one to discriminate between different models of turbulent convection.

Acknowledgements

RS acknowledges support by the Particle Physics and Astronomy Research Council of the UK under the grant PPA/G/O/1998/00576. We thank Guenter Houdek for providing us the solar model.

References

Balmforth, N.J.: 1992, *MNRAS* **255**, 639.
Chaplin, W.J., Elsworth, Y., Isaak, G.R., Lines, R., McLeod, C.P., Miller, B.A. and New, R.: 1998, *MNRAS* **298**, L7.
Goldreich, P. and Keeley, D.A.: 1977, *ApJ* **212**, 243.
Gough, D.O.: 1977, *ApJ* **214**, 196.
Samadi, R. and Goupil, M.-J.: 2001, *A&A* **370**, 136. (**Paper I**)
Samadi, R., Houdek, G., Goupil, M.-J., Lebreton, Y. and Baglin, A.: 2001, *1st Eddington Workshop: Stellar Structure and Habitable Planet Finding*, ESA SP-485, 87.
Stein, R.F. and Nordlund, A.: 1998, *ApJ* **499**, 914.

OF VARIABILITY, OR ITS ABSENCE, IN HgMn STARS

SYLVAIN TURCOTTE
Lawrence Livermore National Laboratory, L-413, P.O. Box 808, Livermore, CA 94551, USA

OLIVIER RICHARD
Département de physique, Université de Montréal, Succ. Centre-Ville, Montréal, Québec, Canada

Abstract. Current models and observations of variability in HgMn stars disagree. We present here the models that argue for pulsating HgMn stars with properties similar to those of Slowly Pulsating B Stars. The lack of observed variable HgMn stars suggests that some physical process is missing from the models. Some possibilities are discussed.

Keywords: abundance anomalies, pulsations

1. Introduction

HgMn stars are late B type stars featuring large and varied abundance anomalies. They are slowly rotating, non-magnetic and mostly young stars and are thought to be the purest examples of stars undergoing microscopic diffusion without the complications posed by mixing processes in cooler stars with convective upper layers (Vauclair and Vauclair, 1982). From a pulsation point of view they are as of yet entirely constant with a maximum photometric variability estimated at less than 5 mmag (Adelman, 1998). There has only been a few studies of spectral variability in HgMn stars leading to suggestions of some variability in Hg lines (Adelman et al., 2002). There are no serious suggestions of pulsations in line profile variations at this time (see e.g. Turcotte et al., these proceedings). There have been claims of rotational variability linked to possible magnetic fields or abundance spots (Adelman et al., 2002).

HgMn stars share a part of the HR diagram in which pulsating stars are present as shown in Figure 1. Those stars, the Slowly Pulsating B stars (SPBs), are in many ways similar to HgMn stars, many seem to be slowly rotating, but differ in that they are chemically normal. In addition, there is a theoretical expectation that pulsations should be driven in HgMn stars at least as much as in SPBs, suggesting that detailed studies of the stability of HgMn stars, both observationally and theoretically, are likely to yield new information on the structure and dynamics of B stars.

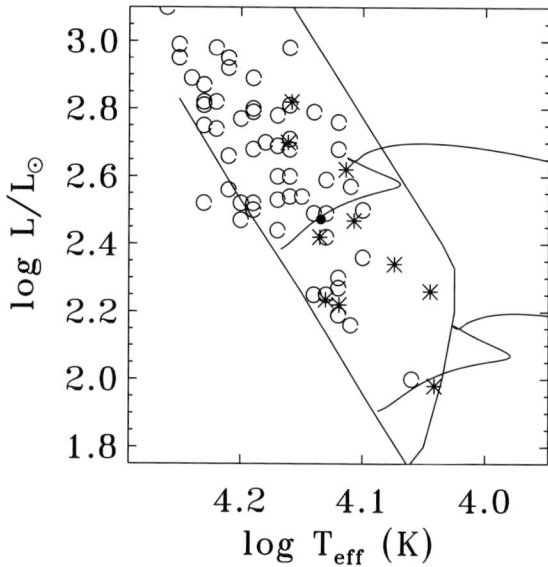

Figure 1. HR Diagram showing some HgMn stars (asterisks), some SPB stars (open circles), and the theoretical limit of the SPB instability region (Pamyatnykh, 1999). The evolutionary tracks for models of 3 and 4 $M_\odot$ are shown and the location of the model which will be discussed here is indicated by the filled circle.

2. Theoretical Expectations

Sophisticated models of diffusion in stars can now be made for cool B stars following the work of (Richer et al., 2000) and (Richard et al., 2001). In those papers, it has been demonstrated that diffusion in A and B stars can lead to a substantial increase in opacity in the region where heavy elements contribute the most to it, at a temperature around 200 000 K. In some cases, when metals are allowed to accumulate enough, a convection zone can form in that region. This is also the same region in which pulsations in B stars (both SPBs and hotter β Cephei stars) are driven (Pamyatnykh, 1999).

The increase in opacity in the model with diffusion is illustrated on the left-hand side of Figure 2 and the resulting changes in the integrand to the work integral are shown on the right-hand side. Though the opacity differs in these models throughout the upper envelope, the work integral for this and other high order g-modes depends very little on the regions with $\log T < 5.0$. This may imply that these modes may not be damped efficiently at low temperature. That will be discussed in the next section.

Adding mixing to prevent the formation of abundance stratification in the models, would ensure that solar metallicities would be retained and one would therefore

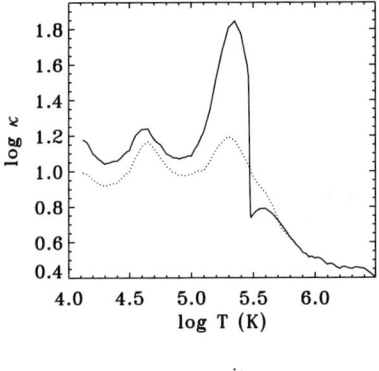 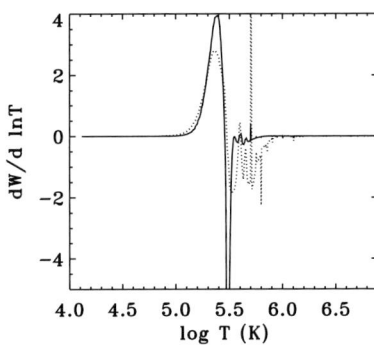

a. opacity b. integrand of the work integral

Figure 2. The left-hand side shows the logarithm of the Rosseland opacity with respect to temperature. The solid line is for a 4 $M_\odot$ model at 100 Myr with diffusion while the dotted curve is a similar model with practically no diffusion. The right-hand side shows the integrand of the work integral for the same models for a $\ell = 1, n = 26$ gravity mode. Pressure modes and low order radial modes are damped while high order gravity modes are excited. In this case the mode is overstable in both models but with a slightly larger growth rate in the model with diffusion.

expect driving as it occurs in SPB stars. This would lead to a normal fraction of variable HgMn stars, i.e. roughly 50% of HgMn stars in the instability region would be variable with amplitudes of approximately 10 mmag.

3. Reconciling the Models with the Observations

As the models are contradicted by observations in B stars it is obvious that some additional physical process or processes need to be included in the models either to reduce the metallicity in the driving region or to increase damping elsewhere.

It is fair to ask if the models with diffusion are representative of HgMn stars. In fact they do not reproduce the surface anomalies of typical HgMn stars very well. The models require that the entire upper envelope, from the iron convection zone to the photosphere, be mixed to avoid numerical problems. This is unlikely to be the case in real stars. The very large and very varied anomalies found in HgMn stars most likely show that diffusion occurs in the atmosphere. In fact, it is difficult to argue that surface abundances tell us much about the internal composition of HgMn stars. It is therefore unlikely that the abundance profiles in our models reproduce well those in real HgMn stars. As a result the opacity profile can also be expected to be different.

Models with varying homogeneous chemical abundances in the superficial mixed zone have been examined to infer the effect of abundance variations in possibly damping pulsations. No hints that it may be the case has been found but a more detailed analysis is needed to rule out this possibility.

Mass loss could also be invoked to explain the differences between SPBs and HgMn stars. Mass loss can remove surface abundance anomalies and after some time could possibly empty the reservoir of iron-peak elements in the driving region for pulsations. However, a metallic wind as described by Babel (1996) could in principle remove some elements while others would still be left to accumulate in the photosphere. This scenario requires a conspiracy where most or all of the elements that contribute most to the metal opacity bump would be removed on a short time scale, and as a result be roughly solar or slightly depleted at the surface. The other elements observed to be very anomalous should not contribute significantly to the opacity at $T \sim 2000\,000$ K.

Acknowledgements

This work was performed in part under the auspices of the U.S. Department of Energy, National Nuclear Security Administration by the University of California, Lawrence Livermore National Laboratory under contract No.W-7405-Eng-48. Olivier Richard thanks G. Michaud for his financial support. We thank the Réseau Québecois de Calcul Haute Performance (RQCHP) for providing us with the computational resources required to compute the stellar models.

References

Adelman, S.J.: 1998, *A&AS* **132**, 93.
Adelman, S.J., Gulliver, A.F., Kochukov, O.P. and Ryabchikova, T.A.: 2002, *ApJ* **575**, 449.
Babel, J.: 1996, *A&A* **309**, 867.
Pamyatnykh, A.A.: 1999, *Acta Astronomica* **49**, 119.
Richard, O., Michaud, G. and Richer, J.: 2001, *ApJ* **558**, 377.
Richer, J., Michaud, G. and Turcotte, S.: 2000, *ApJ* **529**, 338.
Vauclair, S. and Vauclair, G.: 1982 *ARA&A* **20**, 37.

A STUDY OF THE SOLAR-LIKE PROPERTIES OF β HYDRI

MARIA PIA DI MAURO
INAF-Osservatorio Astrofisico di Catania, Via S. Sofia 78, 95123 Catania, Italy

JØRGEN CHRISTENSEN-DALSGAARD
Institut for Fysik og Astronomi, and Teoretisk Astrofysik Center, Aarhus, Denmark

LUCIO PATERNÒ
Dipartimento di Fisica e Astronomia dell'Università di Catania, Catania, Italy

Abstract. We investigate properties of the internal structure of HR2021, better known as β Hydri, a G2 IV subgiant with mass close to solar and for which observations by Bedding et al. (2001) have shown the presence of solar-like oscillations. We have computed models of β Hydri, based on updated global parameters, and compared the computed frequencies for the models with the observed oscillation spectrum.

Keywords: Stars: oscillations, stars: internal structure

1. Introduction

β Hydri is one of the best candidates for asteroseismic studies since it is one of the better-observed individual stars other than the Sun, providing accurate estimates of the basic parameters. It is a G2 IV subgiant, with a luminosity of $L/L_\odot = 4.01 \pm 0.25$ (Pijpers, private communication) and an effective temperature of $T_{\text{eff}} = (5860 \pm 70)$ K (Castro et al., 1999). The mass and metallicity ([Fe/H] $=-0.11 \pm 0.07$, Castro et al., 1999) are close to solar, so that it might be taken as representative of the future evolutive state of our star.

Recently Bedding et al. (2001) obtained radial-velocity data of very high quality, showing clear evidence for solar-like oscillations in β Hydri. Furthermore, for the purpose of the present study Kjeldsen (private communication) made a first attempt at mode identification, resulting in 13 p-mode frequencies in the range $(790 - 1300)$ μHz, with harmonic degrees $l = 0 - 2$, characterized by a large separation of $\Delta \nu = (55.77 \pm 0.17)$ μHz and a small separation of $\delta \nu = (4.7 \pm 0.5)$ μHz.

2. Evolution Models for β Hydri

We computed evolution models of β Hydri with the code of Christensen-Dalsgaard (1982), by employing OPAL96 opacities (Iglesias and Rogers, 1996), the EFF

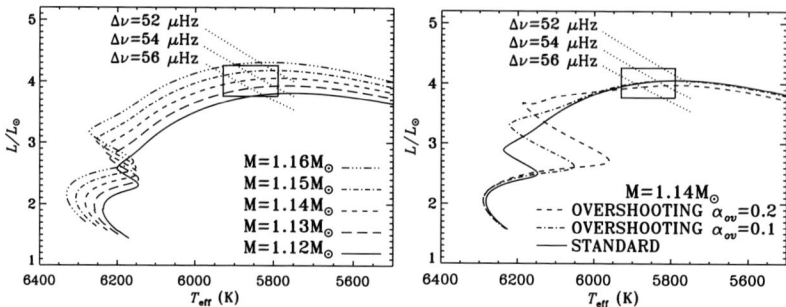

Figure 1. Computed evolutive tracks plotted in H-R diagrams for several masses (left panel) and at fixed mass for different extents of the overshooting from the convective stellar core (right panel). The plotted models assume the same metallicity $Z = 0.015$. The rectangle defines the one-sigma error box for the observed luminosity and effective temperature. The slanted dotted lines indicate models with constant large separation $\Delta \nu$.

equation of state (Eggleton et al., 1973), as well as Bahcall and Pinsonneault (1992) nuclear reaction rates and the mixing-length formalism (MLT) for convection. Evolution started from homogeneous zero-age main-sequence models with an initial hydrogen abundance of $X = 0.7$ and $Z = 0.015 \pm 0.005$. A set of evolutionary tracks is shown in Figure 1, chosen to match the observed effective temperature and luminosity. Models were also calculated by including, during the main-sequence phase, convective overshooting extending from the convective core over a distance $\ell_{ov} = \alpha_{ov} \min(r_c, H_p)$, where α_{ov} is a parameter, r_c the radius of the convective core and H_p the pressure scale height at the edge of the core.

We confirm (Dravins et al., 1998) that β Hydri is in the post-main-sequence phase of evolution, with a helium core and a hydrogen-burning shell. Assuming the input parameters, the mass seems to be limited to the range $(1.07 - 1.20)\,M_\odot$, the age is about $(5.2 - 6.1)$ Gyr and the radius is $R \simeq 2\,R_\odot$. Moreover, the models at the actual location of β Hydri appear to not be affected by variations of the convective overshooting parameter (see also Fernandes and Monteiro, these proceedings), probably because they have a small convective core at the beginning of evolution, which disappears after roughly 5 Gyr.

3. Adiabatic Oscillation Frequencies

We have computed adiabatic oscillation frequencies for models which match the observed properties of β Hydri.

The detailed properties of the computed frequencies $\nu_{n,l}$ are conveniently illustrated in an *echelle diagram*, where the spectrum is divided into segments of fixed length, corresponding to the large frequency separation $\Delta \nu = \nu_{n,l} - \nu_{n-1,l}$, reflecting the asymptotic behaviour of the frequencies. The results for a $1.14 M_\odot$

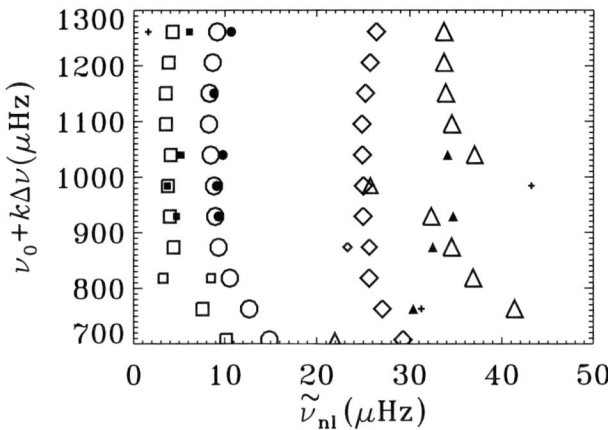

Figure 2. Echelle diagram in which frequencies are expressed as $\nu_{n,l} = \nu_0 + k\Delta\nu + \tilde{\nu}$, where ν_0 is an arbitrary reference frequency, k is an integer and $\tilde{\nu}$ is a reduced frequency in the range $(0 - \Delta\nu)$. In this figure $\Delta\nu = 55.4\,\mu\text{Hz}$ and $\nu_0 = 652\,\mu\text{Hz}$. Open symbols represent frequencies of a model with mass $M = 1.14\,M_\odot$, $Z = 0.015$ and without overshooting. The filled symbols show observed frequencies (Kjeldsen, private communication). Circles are used for modes with $l = 0$, triangles for $l = 1$, squares for $l = 2$, diamonds for $l = 3$. The size of the open symbols indicates the relative surface amplitude of oscillation of the modes. Crosses are employed for modes with too small amplitude (e.g. g modes). Here $\delta\nu_0 = 4.8\,\mu\text{Hz}$.

model of about the observed $\Delta\nu$ are shown in Fig. 2. It is evident that for the modes of degree $l = 0$ and 2 the asymptotic structure is indeed approximately satisfied, with a well-defined small separation $\delta\nu_0 = \nu_{n,0} - \nu_{n-1,2}$. However, the modes with $l = 1$ show a far less regular structure. This is a consequence of the avoided crossings, which occur when a g mode interacts with a p mode of similar frequency and same harmonic degree, giving rise to a mode with mixed character (a very similar behaviour has been found for models of η Bootis; e.g., Christensen-Dalsgaard et al., 1995). In fact, as the star evolves the core contracts and the buoyancy frequency increases with the consequence that g modes with higher frequencies are allowed to propagate. The effect of the coupling becomes much weaker for modes with $l \geq 2$, since in this case the gravity waves are more efficiently trapped in the stellar interior and widely separated from the region of propagation of acoustic waves.

A detailed comparison between observed and computed frequencies (right panel of Figure 3) shows that the behaviour differs substantially from the solar case with, as expected, more scatter for the $l = 1$ modes than for the other degrees. We conclude that the theoretical models can reasonably reproduce the observed spectrum. In particular, we found that the theoretical large separation (left panel of Figure 3) is about $\Delta\nu \simeq (52 - 57)\,\mu\text{Hz}$, while the theoretical small separation is about $\delta\nu_0 \simeq 5\,\mu\text{Hz}$, values which are certainly consistent with the observed ones. However, while the evolutionary scenario of β Hydri looks already quite well

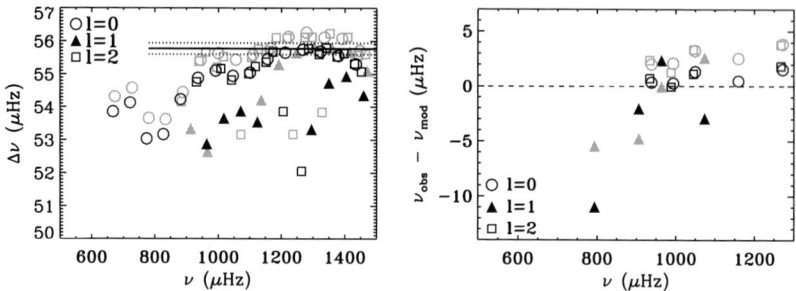

Figure 3. On the left panel the large separation calculated for two models with mass $M = 1.14 M_\odot$ and $Z = 0.015$, without overshooting (grey symbols) and with overshooting with $\alpha_{ov} = 0.2$ (black symbols), is shown as function of the frequency. The solid line indicate the observed $\Delta \nu$ and the dotted lines fix the observed error limits. On the right panel the differences between observed and theoretical frequencies calculated for the same models are plotted as function of frequency.

constrained thanks to the existing spectroscopic and photometric data, it is evident that only more accurate asteroseismic observations will allow further investigation of the properties of the interior of this star.

Acknowledgements

This work was supported in part by the Danish National Research Foundation through its establishment of the Theoretical Astrophysics Center. The authors are grateful to H. Kjeldsen for providing them with the first attempt of identification of observed frequencies and to F. Pijpers for obtaining the basic input parameters.

References

Bahcall, J.N. and Pinsonneault, M.H.: 1992, *ApJ* **395**, L119.
Bedding, T.R. et al.: 2001, *ApJ* **549**, 105.
Castro, S., Porto de Mello, G.F. and da Silva, L.: 1999, *MNRAS* **305**, 693.
Christensen-Dalsgaard, J.: 1982, *MNRAS* **199**, 735.
Christensen-Dalsgaard, J., Bedding, T.R. and Kjeldsen, H.: 1995, *ApJ* **443**, L29.
Dravins, D., Lindegren, L. and VandenBerg, D.A.: 1998, *A&A* **330**, 1077.
Eggleton, P.P., Faulkner, J. and Flannery, B.P.: 1973, *A&A* **23**, 325.
Iglesias, C.A. and Rogers, F.J.: 1996 *ApJ* **464**, 943.

GIANT VIBRATIONS IN DIP*

T.C. TEIXEIRA[1,2], J. CHRISTENSEN-DALSGAARD[1,3], F. CARRIER[4], C. AERTS[5],
S. FRANDSEN[1], D. STELLO[1], T. MAAS[5], M. BURNET[4], H. BRUNTT[1],
J.R. DE MEDEIROS[6], F. BOUCHY[4], H. KJELDSEN[1,3] and F. PIJPERS[1,3]

[1] *Institut for Fysik og Astronomi, Aarhus Universitet, Denmark*
[2] *Centro de Astrofísica da Universidade do Porto, Portugal*
[3] *Teoretisk Astrofysik Center, Danmarks Grundforskningsfond, Denmark*
[4] *Observatoire de Genève, Switzerland*
[5] *Katholieke Universiteit Leuven, Instituut voor Sterrekunde, Belgium*
[6] *Dep. de Física, Universidade Federal do Rio Grande do Norte, Brazil*

Abstract. This work reports the discovery of solar-type oscillations in the giant star ξ Hydrae.

Keywords: Techniques: asteroseismology – Stars: evolution – Stars: giants – Stars: individual: ξ Hydrae

1. Introduction

Recent years have been the stage for the exciting start of observational asteroseismology. The development of high-resolution, high-precision spectrographs like CORALIE, and the unexpected access to the WIRE satellite, have resulted in a breakthrough in the observations of solar-type oscillations. However, the detection of solar-type oscillations has so far been confined to stars that lie close to the Sun in mass. There have been reports of detections in two giant stars (α UMa: Buzasi et al., 2000; Arcturus: Merline, 1999; Retter et al., these proceedings), but they suffer from severe problems of theoretical interpretation, in the case of α UMa, and from observational uncertainties, in the case of Arcturus.

This work reports the discovery of solar-type oscillations in the giant star ξ Hydrae. The evidence supporting this claim is presented, as well as a precursory discussion of the modelling within the parameter space defined by the uncertainties in the observational parameters.

2. The observational ξ Hya

ξ Hya is a bright, slowly rotating, G7 III star located in the southern hemisphere. Since giants are expected to show high-amplitude oscillations, it was a perfect candidate for observations with the CORALIE spectrograph at the 1.2-m Swiss telescope at La Silla. For a month high-precision spectra were taken every night,

---

* Based on observations obtained with the CORALIE spectrograph on the 1.2-m Swiss Euler telescope at La Silla, Chile.

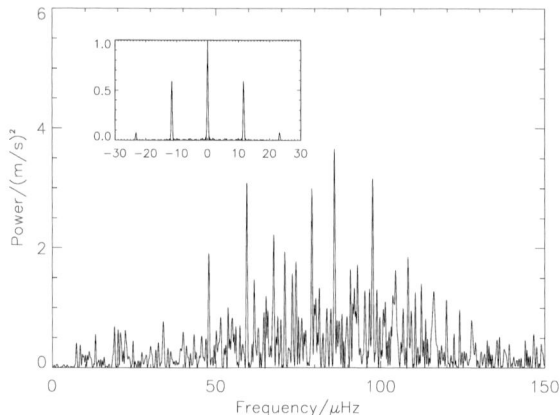

Figure 1. Power spectrum of ξ Hya, with the window function shown in the inset.

on an hourly basis, in the wavelength range 3875 − 6820 Å. The optical spectra were then converted to a radial velocity time series, which resulted in the power spectrum shown in Figure 1.

Even from visual inspection of the power spectrum it is possible to identify an excess power hump in the range ∼ 50 − 130 μHz with clear individual peaks. Such a spectrum is clearly reminiscent of that of the Sun and of other (near-) main-sequence stars (e.g. α Cen A: Bouchy and Carrier, 2002). In fact, it can be shown (Frandsen et al., 2002) that the power spectrum of ξ Hya displays a dominant frequency spacing of $\Delta\nu \sim 6.8\,\mu$Hz. Such near-uniform spacing of the strongest peaks is a unique signature of solar-type, convection driven oscillations, since no other known mechanism present in the interior of cool stars could produce such a spectrum.

As a by-product of the observations, it was possible to (re)determine the atmospheric parameters of ξ Hya: $T_{\rm eff}$ = 5000 ± 100 K, L = 61.1 ± 6.2 $L_\odot$, $\log g$ = 2.85, [M/H] = 0.06±0.07 (Z = 0.019±0.006), and $v \sin i$ = 1.8±1.0 km s^{-1}.

3. The Model ξ Hya

In order to study the nature of the oscillations detected in ξ Hya, it is necessary to compare the observed frequency spectrum with model predictions. Using the evolution code of Christensen-Dalsgaard (1982), evolutionary tracks were produced spanning the error box defined by the uncertainties in $T_{\rm eff}$, L and Z, given above. Taking a value of Z = 0.019, the evolutionary track that passes through the observed ($T_{\rm eff}$, L) corresponds to a mass of 3.07$M_\odot$. ξ Hya is a H-shell burning star, located in a dip of its evolutionary track in the HR diagram (Figure 2). It is acknowledged that another solution could lie in the He-core burning stage of

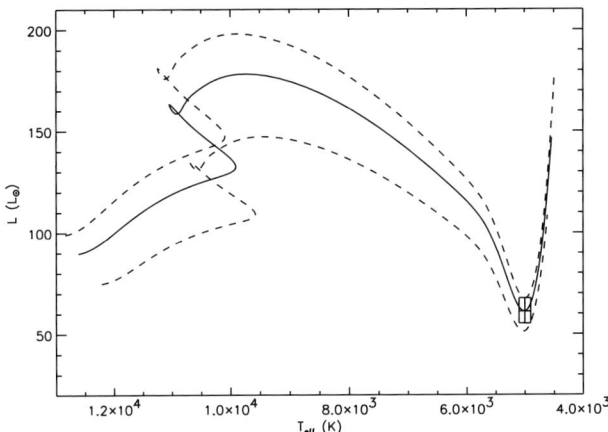

Figure 2. HR diagram with evolutionary tracks for $Z = 0.019$ for masses 3.07 $M_\odot$ (solid line), 2.93 and 3.15 $M_\odot$ (lower and upper dashed lines, respectively). The location of ξ Hya in the diagram is given by the error box.

a lower mass track, but currently our code does not run that far, and no other appropriate tracks were available for this work. For that reason, only the H-shell burning solution is being analysed here.

Oscillation frequencies were calculated for several models within the error cube defined by the uncertainties in L, T_{eff} and Z, taking two values for the mixing length parameter: $\alpha = \alpha_\odot$ and $\alpha = 1.2\alpha_\odot$, where $\alpha_\odot$ is the solar value. The differential error between the model results and the observations, $(\langle \Delta \nu \rangle_{\text{model}} - \Delta \nu_{\text{obs}})/\sigma_{\text{obs}}$, is shown in Figure 3. Despite the large amount of information contained in that figure, it is clear that, even taking the most cautious estimate of the observed large separation, some models deviate from the observed values by more than 3σ and can therefore be discarded.

The most important conclusion to draw from this work is that asteroseismology can be done across the HR diagram and can deliver a significant improvement in the accuracy to which stellar parameters can be determined.

Acknowledgements

The authors wish to thank the financial support of the following institutions: the Fund for Scientific Research of Flanders (project G.0178.02 to CA and TM); the Swiss National Science Foundation; the Danish National Science Foundation, through the establishment of the Theoretical Astrophysics Center; the Aarhus University; the Danish Natural Science Research Council; the Fundação para a Ciência e a Tecnologia, Portugal (research grant SFRH/BPD/3545/2000 to TCT).

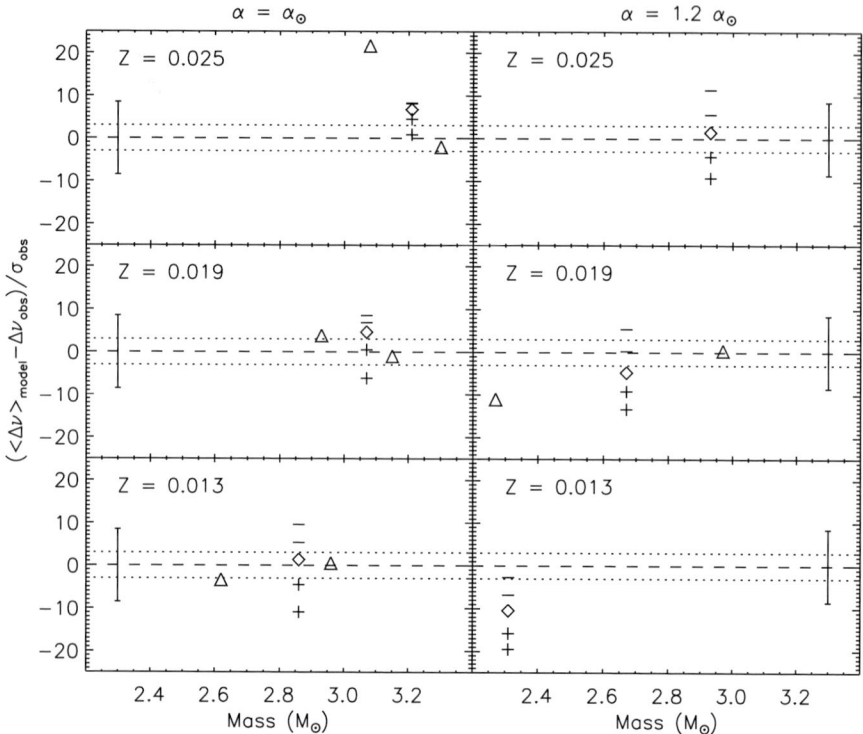

Figure 3. Differential error between the model and observed large separation, $(\langle \Delta \nu \rangle_{\mathrm{model}} - \Delta \nu_{\mathrm{obs}})/\sigma_{\mathrm{obs}}$, plotted as a function of the mass of the model. $\langle \Delta \nu \rangle_{\mathrm{model}}$ is the average large separation in the model in the range $50-100\,\mu$Hz, $\Delta \nu_{\mathrm{obs}}$ is the large separation from the observations and σ_{obs} is the corresponding uncertainty. The dashed lines correspond to a null deviation, and the dotted lines to 3σ deviations. For each subplot, the 3 values of the mass correspond to 3 evolution tracks: the lower and upper values (triangles) graze the error box in Figure 2 at the bottom and top, respectively, and the middle value (diamond, −'s and +'s) goes through the observed (T_{eff}, L). The diamond corresponds to the evolutionary time where T_{eff} and L equal the observed values, while the minus (−) and plus (+) symbols correspond to respectively earlier and later times in the same track, but still within the error box. The error bar in each subplot corresponds to three times the actual uncertainty in the estimate of the observational value of the large separation, $\Delta \nu_{\mathrm{obs}}$, if different methods are used to obtain that value.

References

Bouchy, F. and Carrier, F.: 2002, *A&A* **390**, 205.
Buzasi, D., Catanzarite, J., Laher, R. et al.: 2000, *ApJ* **532**, L133.
Christensen-Dalsgaard, J.: 1982, *MNRAS* **199**, 735.
Frandsen, S., Carrier, F., Aerts, C. et al.: 2002, *A&A*, accepted.
Merline W.J.: 1999, in: J.B. Hearnshaw and C.D. Scarfe (eds.), *Precise Stellar Radial Velocities*, ASP Conf. Ser. **185**, 187.

STOCHASTIC EXCITATION OF GRAVITY WAVES BY OVERSHOOTING CONVECTION IN SOLAR-TYPE STARS

BORIS DINTRANS[1], AXEL BRANDENBURG[2], ÅKE NORDLUND[3] and
ROBERT F. STEIN[4]

[1] *Laboratoire d'Astrophysique, Observatoire Midi-Pyrénées, 14 avenue Edouard Belin, 31400 Toulouse, France*
[2] *Nordic Institute for Theoretical Physics, Blegdamsvej 17, 2100 Copenhagen Ø, Denmark*
[3] *Theoretical Astrophysics Center, and Astronomical Observatory, Juliane Maries Vej 30, 2100 Copenhagen Ø, Denmark*
[4] *Department of Physics and Astronomy, Michigan State University, East Lansing, MI 48824, USA*

Abstract. The excitation of gravity waves by penetrative convective plumes is investigated using 2D direct simulations of compressible convection. The oscillation field is measured by a new technique based on the projection of our simulation data onto the theoretical g-modes solutions of the associated linear eigenvalue problem. This allows us to determine both the excited modes and their corresponding amplitudes accurately.

Keywords: Stars: oscillations – Convection

1. Introduction

Two-dimensional simulations of compressible convection have shown that it is possible to excite internal gravity waves (IGW) in radiative zones of solar-type stars from the downward penetrating plumes (Hurlburt et al., 1986, hereafter HTM86; Hurlburt et al., 1994; Kiraga and Jahn, 1995). However, detecting IGW with confidence is challenging given the stochastic nature of the excitation mechanism.

We propose here a new detection method which allows us to measure rigorously both the spectrum and amplitude of excited g-modes. This method is first applied to the g-mode oscillations of an isothermal atmosphere and then, to IGW generated in a 2D-simulation of a convective zone embedded between two stable ones.

2. Detecting g-Modes using the Anelastic Subspace

In hydrosimulations, wave fields are commonly measured using two main methods: *(i)* the simplest one consists in recording the vertical velocity at a fixed point and then performing the Fourier transform of the sequence (see, e.g., HTM86); *(ii)* a more complicated method consists in taking two Fourier transforms, in space and time, of the vertical mass flux (Stein and Nordlund, 1990). However, these two methods are not well adapted to detect IGW in our problem because the Fourier

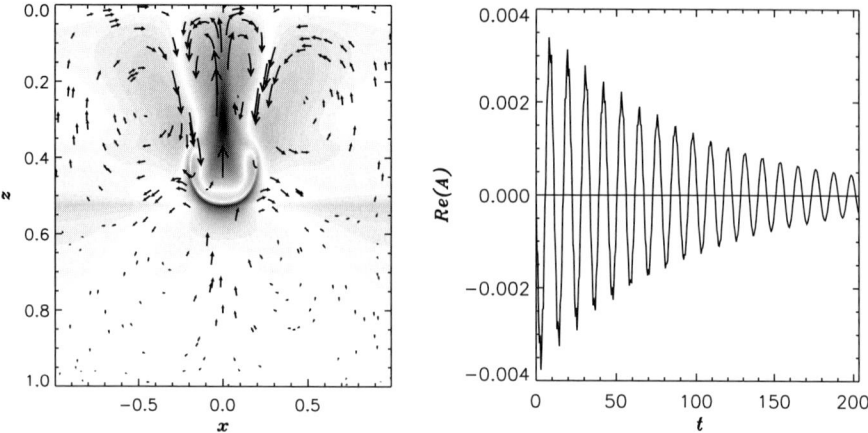

Figure 1. Left: slide of the velocity field superimposed on the contours of entropy perturbations for a 2D-simulation of an oscillating entropy bubble embedded in an isothermal atmosphere. Right: time-evolution of the real part of its projection coefficient A_{10} (here t is in units of d/c_s).

transforms are calculated over the *whole* simulation while IGW are *stochastically* excited by penetrating plumes.

Our new detection method takes into account the random nature of this excitation. Indeed, it is based first, on projections of the simulated velocity field $\vec{v}(k, z, t)$ onto the anelastic eigenvectors $\vec{\psi}_{kn}(z)$ as

$$\vec{v}(k, z, t) = \sum_{n=0}^{\infty} <\vec{\psi}_{kn}, \vec{v}> \vec{\psi}_{kn}(z) = \sum_{n=0}^{\infty} A_{kn}(t)\vec{\psi}_{kn}(z), \quad (1)$$

and second, on time-frequency diagrams of the complex coefficients $A_{kn}(t)$. As a consequence, the immediate spectrum (the set of frequencies ω) and amplitudes (defined as $|A_{kn}|$) of stochastically excited g-modes are reached and not only their 'mean' values over the whole simulation. It is instructive to consider the simplest possible case, that is, the propagation of a single gravity mode with horizontal wavenumber k through the computation domain: applying Eq. (1) leads in this case to a projection coefficient $A_{kn}(t) \propto \exp(i\omega_{kn}t)$, where ω_{kn} denotes the frequency of the anelastic eigenmode of degree k and order n.

3. Results

3.1. Oscillations of an isothermal atmosphere

As a first test, we apply our method to detect IGW excited by an oscillating entropy bubble embedded in an isothermal atmosphere of depth d (see Fig. 1, left panel). In this case, the building of the anelastic subspace is simplified since we found analytic solutions for the eigenfrequencies and their associated eigenvectors (Dintrans et al., 2002).

In Figure 1 (right panel), we show the real part of the projection coefficient $A_{10}(t)$, i.e. we projected the left panel velocity field onto the first anelastic eigenmode of the isothermal atmosphere at $k = 1$ and $n = 0$. As expected, we found that $A_{10}(t)$ behaves like $\exp(i\omega_{10}t)$ (with $\omega_{10} \simeq 0.569 c_s/d$, where c_s is the constant adiabatic sound speed) whereas the mode amplitude $|A_{10}|$ decreases as $\exp(-t/t_\nu)$ with $t_\nu \propto \nu^{-1}$ (ν being the constant kinematic viscosity of the simulation).

3.2. 2D-SIMULATIONS OF PENETRATIVE CONVECTION

Once our method validated, we study the excitation of IGW by overshooting convection using high-resolution two-dimensional simulations of a three-layer polytropic model. That is, we solve the following equations:

$$\begin{cases} \dfrac{D \ln \rho}{Dt} = -\operatorname{div} \vec{u}, \\ \dfrac{D\vec{u}}{Dt} = -(\gamma - 1)\left(\vec{\nabla} e + e\vec{\nabla} \ln \rho\right) + \vec{g} + \dfrac{1}{\rho}\vec{\nabla} \cdot (\rho \nu \vec{\mathcal{S}}), \\ \dfrac{De}{Dt} = -(\gamma - 1)e \operatorname{div} \vec{u} + \dfrac{1}{\rho}\vec{\nabla} \cdot (\mathcal{K}\vec{\nabla} e) + \nu \vec{\mathcal{S}}^2 - \dfrac{e - e_0}{\tau(z)}, \end{cases} \quad (2)$$

where $\vec{u}$ denotes the velocity, e the internal energy, ρ the density, $\mathcal{K} = K/c_v$ where K is the radiative conductivity and c_v the specific heat at constant volume, $\vec{\mathcal{S}}$ the stress tensor and, finally, $\tau(z)$ is a cooling time (see Brandenburg et al., 1996 for more details).

Figure 2 (left panel) shows an example of such a simulation of a convective zone of depth $d = 1$ ($0 < z < 1$) embedded between two stable ones ($-0.15 < z < 0$ and $1 < z < 3$). IGW are excited in the bottom radiative zone by penetrating downward plumes and the evolution of the projection coefficient $A_{10}(t)$ is now more chaotic (see right panel). However, by applying a time-frequency diagram on this sequence, we extracted three IGW events (with $\omega_{10} \simeq 0.251\sqrt{g/d}$), emphasized as thick lines in the figure.

Acknowledgements

This work has been supported by the European Commission under Marie-Curie grant no. HPMF-CT-1999-00411.

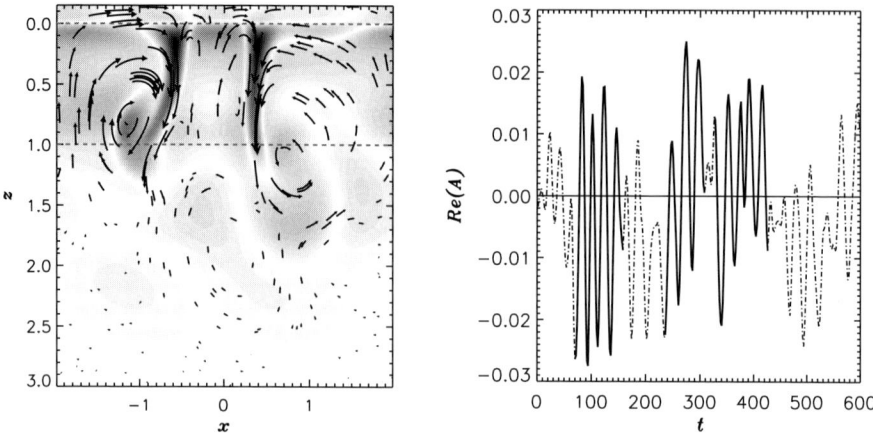

Figure 2. Left: slide of the velocity field superimposed on the contours of entropy perturbations for a two-dimensional convection simulation. Two strong downwards plumes penetrate into the bottom radiative zone and excite IGW. Right: time-evolution of the real part of the projection coefficient A_{10} (here t is in units of $\sqrt{d/g}$). Thick lines emphasize IGW events detected using a time-frequency diagram.

References

Dintrans, B., Brandenburg, A., Nordlund, Å. and Stein, R.F.: 2002, *A&A*, in preparation.
Brandenburg, A., Jennings, R.L., Nordlund, Å., Rieutord, M., Stein, R.F. and Tuominen, I.: 1996, *J. Fluid Mech.* **306**, 325.
Hurlburt, N.E., Toomre, J. and Massaguer, J.M.: 1986, *ApJ* **311**, 563. **[HTM86]**
Hurlburt, N.E., Toomre, J., Massaguer, J.M. and Zahn, J.-P.: 1994, *ApJ* **421**, 245.
Kiraga, M. and Jahn, K.: 1995, *Acta Astron.* **45**, 685.
Stein, R.F. and Nordlund, Å.: 1990, in: Y. Osaki and H. Shibahashi (eds.), *Progress in Seismology of the Sun and Stars*, Lecture Notes in Physics, Springer-Verlag, Berlin, Vol. 367, 93

P-MODE OSCILLATIONS OF α CEN A

A. THOUL*, R. SCUFLAIRE, B. VATOVEZ, A. NOELS, P. MAGAIN, M. BRIQUET
and M.-A. DUPRET

Department of Astrophysics and Geophysics, University of Liège, Belgium

Abstract. Models of α Cen A & B have been computed using the masses determined by Pourbaix et al. (2002) and the data derived from the spectroscopic analysis of Neuforge and Magain (1997). The seismological data obtained by Bouchy and Carrier (2001, 2002) do help improve our knowledge of the evolutionary status of the system. All the constraints are satisfied with a model which gives an age of about 6 Gyr for the binary.

Keywords: stars: binaries: visual – stars: individual: α Cen – stars: oscillations – stars: evolution

1. Introduction

The binary system Alpha Centauri offers a unique opportunity to test our knowledge of stellar physics in solar-type stars other than the Sun. Its proximity allows very good determinations not only of its parallax but also of its orbital parameters so that the masses and the luminosities of both components can be determined. In addition, effective temperatures and chemical compositions can be obtained through spectroscopic analyses. Recently, solar-like p-mode oscillations have been discovered in α Cen A from ground-based analysis (Bouchy and Carrier, 2001; Carrier et al., 2002; Bouchy and Carrier, 2002).

In a very recent calibration and stability analysis, Thévenin et al. (2002) have found that the asteroseismological constraints, namely the large and small frequency spacings, required a slight decrease in the mass assigned by Pourbaix et al. (2002) to α Cen A.

Using an entirely new numerical code and a slightly different physics, we find models for the α Cen binary system which agree not only with all the non asteroseismic constraints, including the masses and the requirement of the same age for both components, but also with the asteroseismic constraints, including the frequencies themselves.

---

* Chercheur Qualifié au Fonds National de la Recherche Scientifique, Belgium.

2. Observational Constraints

We adopt the masses determined by Pourbaix et al. (2002), i.e., $M_A/M_\odot = 1.105$ and $M_B/M_\odot = 0.934$. From their spectroscopic analysis of the α Centauri system, Neuforge and Magain (1997) have determined the effective temperature, the gravity and the metallicity of both components. The effective temperatures are given by $T_{\text{eff},A} = 5830 \pm 30$K and $T_{\text{eff},B} = 5255 \pm 50$K. The gravities are given by $\log g_A = 4.34 \pm 0.05$ and $\log g_B = 4.51 \pm 0.08$. For the metallicity, they obtained $[\text{Fe/H}]_A = 0.25 \pm 0.02$ and $[\text{Fe/H}]_B = 0.24 \pm 0.03$. To deduce Z/X, we assume that this ratio is proportional to the abundance ratio Fe/H and we adopt the solar value $(Z/X)_\odot = 0.023$ with an uncertainty of 10 %, given by Grevesse and Sauval (1998). We then have the same value for both components of α Cen: $Z/X = 0.040 \pm 0.005$. However, Grevesse and Sauval (private communication) now favour a lower value $(Z/X)_\odot = 0.0209$ with the same uncertainty. With this value, we would have $Z/X = 0.037 \pm 0.004$. It seems safe to say that Z/X is between 0.033 and 0.045.

3. The Models

A number of evolutionary sequences have been computed from the main sequence with CLÉS (Code Liégeois d'Évolution Stellaire).

We have computed a number of models for both components, with different initial chemical compositions, defined by (X, Z), and convection parameter α. We require that both components of the binary system reach their respective positions in the HR-diagram at the same age. This requirement of simultaneity determines a line in the (X, Z)-diagram to the right of which the B component reaches its observed position in the HR-diagram after the A component has already left its own observed position. This constraint, together with the constraint on the Z/X value, delimits a permitted area in the (X, Z) diagram.

4. Asteroseismology of α-Centauri

We have calculated the oscillation frequencies of α Cen A using a standard adiabatic code, for a grid of models where the mass M_A is fixed at the value determined by Pourbaix et al. (2002), i.e., $M_A = 1.105 M_\odot$. The only free parameters are α, X, and Z, and the last two can only vary within the permitted area.

The oscillation mode (l=0, n=21) is the one with the highest amplitude in the Bouchy and Carrier (2002) spectrum, and we assumed therefore that it is the best observationally determined mode. Therefore, for each sequence of evolution, and within the error boxes on $(T_{\text{eff}}, L/L_\odot)$, we determined the model for which the oscillation frequency of the mode (l=0, n=21) fits exactly the observed frequency.

TABLE I
Mode frequencies (in μHz) of α Cen A.

	Observations[a]			Our model		
	$l = 0$	$l = 1$	$l = 2$	$l = 0$	$l = 1$	$l = 2$
n = 15			1833.1	1730.5	1778.5	1828.0
n = 16	1841.3	1887.4	1934.9	1834.9	1882.8	1932.6
n = 17		1991.7	2041.5	1939.3	1987.7	2038.1
n = 18		2095.6	2146.0	2044.4	2093.7	2144.7
n = 19	2152.9	2202.8	2251.4	2150.8	2200.4	2251.7
n = 20	2258.4	2309.1	2358.4	2257.5	2307.5	2358.7
n = 21	**2364.2**	2414.3	2464.1	**2364.2**	2414.4	2465.9
n = 22	2470.0	2519.3	2568.5	2471.0	2521.5	2573.1
n = 23	2573.1	2625.6		2578.0	2629.0	2681.0
n = 24	2679.8	2733.2	2782.9	2685.5	2736.9	2789.2
n = 25	2786.2	2837.6	2887.7	2793.4	2845.2	2897.6

[a] Bouchy and Carrier (2002).

We calculated the modes $l = 0$, $l = 1$ and $l = 2$ for n=15 to 25. The results are summarized in Table I. We then calculated the large frequency spacings $\Delta\nu_{l,n} = \nu_{l,n} - \nu_{l,n-1}$ and the small frequency spacings $\delta\nu_{0,n} = \nu_{l,n} - \nu_{l+2,n-1}$ and their averages over n=15 to 25. We have also calculated the small spacings $\delta\nu(n) = \nu_{n+1,0} + \nu_{n,0} - 2\nu_{n,1}$.

5. Discussion and Conclusions

Our best model satisfies all the observational constraints and corresponds to $X = 0.70$ and $Z = 0.0275$ for the chemical composition, $\alpha_A = \alpha_B = 1.8$ for the mixing-length parameter (the value for the sun is 1.77), and yields an age of 6.41Gyr for the α Cen system.

The average of the large spacings is slightly larger than the one obtained from the observations, but those large spacings remain within the error bars, as shown in Figure 1. The values for the small spacings fall within the error bars as well, as shown in Figure 1. The small frequency spacings $\delta\nu(n)$ are slightly larger than the observed ones, but fall within the error bars, except for the largest frequencies. Finally, we note that this solution was found for a value of the mixing-length parameter α very close to the solar value.

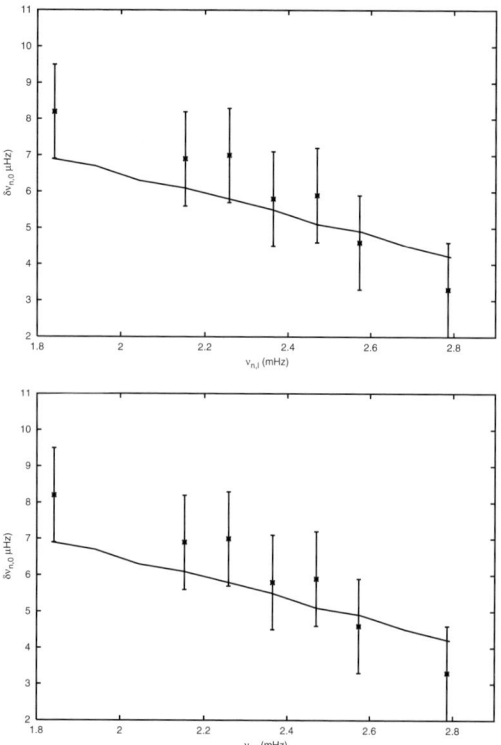

Figure 1. Large and small frequency spacings as a function of the frequency. The symbols indicate the values obtained by Bouchy and Carrier (2002), with error bars of $\pm 1.3\mu$Hz. The three lines correspond to our model for $l = 0, l = 1$ and $l = 2$.

References

Bouchy, F. and Carrier, F.: 2001, *A&A* **374**, L5.
Bouchy, F. and Carrier, F.: 2002, *A&A* (in press).
Carrier, F., Bouchy, F., Meynet, G., Maeder, A., Provost, J., Berthomieu, G. and Thévenin, G.: 2002, in: C. Aerts, T.R. Bedding and J. Christensen-Dalsgaard (eds.) *Radial and Nonradial Pulsations as Probes of Stellar Physics*, A.S.P. Conf. Ser. **259**, 460.
Grevesse, N. and Sauval, A.J.: 1998, *Space Sci. Rev.* **85**, 161.
Neuforge-Verheecke, C. and Magain, P.: 1997, *A&A* **328**, 261.
Pourbaix, D., Nidever, D., McCarthy, C., Butler, R.P., Tinney, C.G., Marcy, G.W., Jones, H.R.A, Penny, A.J., Carter, B.D., Bouchy, F., Pepe, F., Hearnshaw, J.B., Skuljan, J., Ramm, D. and Kent, D.: 2002, *A&A* **386**, 280.
Thévenin, F., Provost, J., Morel, P., Berthomieu, G., Bouchy, F. and Carrier, F.: 2002, *A&A* (in press).

OUTSTANDING ISSUES ON PULSATIONS IN YOUNG STARS

M. MARCONI
INAF – Osservatorio Astronomico di Capodimonte, Naples, Italy; E-mail: marcella@na.astro.it

F. PALLA
INAF – Osservatorio Astrofisico di Arcetri, Florence, Italy; E-mail: palla@arcetri.astro.it

Abstract. We briefly summarize the impact of protostellar evolution on the initial conditions of pre-main-sequence (PMS) contraction and identify two different kinds of instability in young stars closely connected to the PMS physical conditions. These are the δ Scuti type pulsations in intermediate-mass PMS stars (Herbig Ae stars) and the pulsation instability of brown dwarfs and very low mass stars. For the former, we review recent observational and theoretical work and provide the location in the H-R diagram of known candidates. For the latter, the possibility that deuterium can be an efficient driving mechanism of the instability is explored.

Keywords: stars: δ Scuti – stars: pre-main sequence

1. The Impact of Star Formation on PMS Evolution

In order to answer to the basic question of *where can pulsating young stars be found in the classical H-R diagram*, a brief discussion of the impact of protostellar evolution on the initial conditions for PMS contraction is necessary. Schematically, our current understanding of protostellar evolution can be summarized as follows:

- Protostellar radii are small, few $R_\odot$ for solar-mass protostars. This is much lower (by about a factor of 10) than the typical initial radii assumed in standard PMS models, implying a shorter contraction time before reaching the conditions for H-burning.
- Deuterium burning occurs during the accretion phase for protostars of mass $M_p > 0.2$-0.3 $M_\odot$ for typical mass accretion rates of $\dot{M}_{acc} \sim 10^{-5}$ $M_\odot$ yr^{-1}. For lower mass accretion rates, the onset of D-burning begins at smaller values of the mass, close to the standard substellar limit at 0.08 $M_\odot$.
- The energy release from the small amounts of interstellar deuterium ([D/H]$\sim 1 - 2 \times 10^{-5}$) that is burnt at the centre of low-mass protostars is sufficient to turn the internal structure convectively unstable. For such objects, a mass-radius relation is readily established, as a consequence of the thermostatic nature of D-burning. During these phases, the accretion time scale (t_{acc}) is shorter than the contraction time (or Kelvin-Helmoltz time, t_{KH}). Thus, the protostar can keep accreting as long as there is enough circumstellar material (which is generally the case for low-mass stars). Most of the total luminosity is due to accretion, with a non-negligible contribution from nuclear burning.

- For more massive protostars, the energy release of D-burning cannot balance the increased strength of the gravitational pull and the protostar begins contraction while accreting (i.e., $t_{KH} < t_{acc}$). The internal structure also changes since convection is no longer the main transport mechanism throughout the protostar. Objects in the mass range $\sim$2–4 $M_\odot$ undergo a thermal relaxation and a dramatic swelling of the radius due to D-burning in a subsurface shell. Then, contraction rapidly ensues while still accreting.

- Protostars with mass higher than $\sim$6 $M_\odot$ have central temperatures high enough to start the ignition of the standard pp-reactions in the core. Then, most of the luminosity generated by the protostar is due to nuclear energy and not by accretion. In these conditions, the protostar is no longer contracting and has basically settled on the zero-age MS. The main result of this phase is that high mass stars do *not* have a PMS phase at all: once they become optically visible, they are already on the ZAMS.

Once the main accretion phase is over (for whatever reason: lack of molecular gas in the parent cloud, or because of the dispersal circumstellar matter by jets and outflows), a protostar of low and intermediate mass can start its contraction phase towards the ZAMS, following the well known evolutionary paths along convective and radiative tracks. However, the physical conditions inherited in the accretion phase are drastically different from those assumed in standard PMS calculations that ignore the previous history. As described above, the main differences are: the role of D-burning for very low- and low-mass stars (from the brown dwarf regime up to $\sim$1 $M_\odot$), and the anomalous internal conditions of intermediate-mass objects ($\sim$2–4 $M_\odot$) that start the quasi-static contraction phase with an inert, radiative core and a small convective external layer.

The PMS evolution of this two class of objects is quite different. Brown dwarfs and very low-mass stars descend the convective tracks burning the full content of the accreted deuterium when the central temperature exceeds $\sim 10^6$ K. On the contrary, low-mass stars have already exhausted most of the available deuterium while accreting. Intermediate mass stars start immediately their evolution on almost fully radiative tracks with surface temperatures of about 6000 K, not too far from the red edge of the classical instability strip for evolved stars.

Since PMS contraction lasts longer than the previous, optically invisible protostellar phase, it is easier to identify possible pulsation and oscillation mechanisms during this phase. In addition, young stars have dispersed most of the circumstellar material and are thus easier to observe when searching for periodic photometric variability. It is well known that very young stars show a wealth of both long- and short-term variability, most likely due to the interaction of the central object(s) with the circumstellar material (disk and envelope). These processes make the identification of the intrinsic variability much harder to find. Thus, for searches of pulsations it is better to concentrate on more mature stars that are no longer in the main accretion phase.

We identify two areas of study of instabilities in young stars closely related to the physical conditions described above: pulsations induced by D-burning in brown dwarfs and very-low mass stars (the ϵ mechanism), and the instability strip for intermediate mass stars (the κ mechanism). Since the latter has been worked out theoretically and confirmed observationally by a variety of groups, we will begin our presentation with a summary of the recent work on the field of pulsating Herbig Ae stars. We will then introduce the possibility that deuterium can be an efficient driving mechanism for pulsation in young brown dwarfs that are currently discovered in large numbers in nearby star forming regions.

2. δ Scuti-Type Pulsation in PMS Intermediate Mass Stars

PMS stars with $M > 1.5 M_\odot$ cross the pulsation instability strip during their contraction toward the MS, with a crossing time ranging from about 0.05 t_{KH} to 0.1 t_{KH}. The existence of pulsating stars among PMS intermediate-mass stars was originally suggested by Breger (1972), who identified the first two candidates, V588 Mon and V589 Mon in the NGC2264 cluster. About 20 years later, new empirical evidences for δ Scuti-like pulsations were found by Kurtz and Marang (1995) and Donati et al. (1997) for the Herbig Ae stars HR 5999 and HD 104237, respectively.

Stimulated by these observations, a theoretical study of the modal stability of PMS stars with mass $1.5 \leq M/M_\odot \leq 4.0$ was carried out by Marconi and Palla (1998) using nonlinear, convective radial pulsation models. By following the evolution along PMS tracks, models were selected at different effective temperatures and their stability investigated for the first three radial modes. As a result, the locus of the theoretical instability strip for young stars was characterized. The shaded region in Figure 1 represents the location in the HR diagram of the instability strip between the second overtone blue edge and the fundamental red edge, whereas lines are (solid) PMS (Palla and Stahler, 1993) and (dotted) post-MS (Castellani et al., 1999) tracks, respectively.

In turn, a growing observational interest for PMS δ Scuti-like pulsation has developed and several new members of the class have been identified in the last few years (see Marconi et al., 2002, for details). The list of the currently 12 known PMS pulsators is reported in Table I. Column 1 gives the object's name; columns 2 to 5 list the measured frequencies; the visual amplitude and mean magnitude are given in columns 6 and 7, and column 8 reports the spectral type.

2.1. COMPARISON OF THE MEASURED FREQUENCIES WITH THE PREDICTIONS OF PULSATING MODELS

The pulsation periods can be predicted by radial linear nonadiabatic pulsation models for each selected mode (Marconi and Palla, 1998). It is well known that even on the basis of the simplest linear adiabatic theory the period is expected to be a function of the mean stellar density and then of mass, luminosity and effective temperature. In addition, the evolutionary prescriptions provide constraints on the luminosity level for each mass and effective temperature. This implies that the comparison between empirical and theoretical periods, for a given set of evolutionary tracks, in principle allows to estimate the luminosity and the effective temperature of the pulsators. Obviously, if only one period is observed, different combinations of luminosity and effective temperature can simultaneously satisfy the pulsation relation for the period and the evolutionary prescription. In this case, independent information (e.g. empirical values from the literature) is needed in order to remove the degeneracy. This is, for instance, the case of the pulsator H254 in the IC 348 cluster where only one frequency at 7.4 c/d is measured (see Ripepi et al., 2002a). In order to discriminate between models that reproduce this period, the empirical values of (L_*, T_{eff}) provided by Luhman et al. (1998) are adopted. By varying the intrinsic model parameters within the observational uncertainties, two solutions are obtained corresponding to pulsation in the fundamental or in the first overtone.

If more than one period is observed and the accuracy is high enough, the comparison with model predictions is able to provide a unique solution for the position of the star in the HR diagram. In this case, we have at least two pulsation relations for the periods of different modes to be combined with the evolutionary constraints, and three unknown quantities (mass, luminosity and effective temperature), can thus be derived. This is the case of the variables BL50 and HP57, for which Pigulski et al. (2000a,b) derived very accurate frequencies (see Table I).

On the contrary, if the frequencies are less accurately determined, the procedure can lead to different solutions even in the case of multiperiodic variables. This is the case of V346 Ori and V351 Ori where only two of the four measured frequencies have been confirmed by subsequent and/or simultaneous observations. For both stars, a solution matching the two most reliable frequencies has been obtained, as described in Pinheiro et al. (these proceedings; see also Pinheiro et al., 2002; Ripepi et al., 2002b; Ripepi, these proceedings). Moreover, while a radial pulsation model is able to reproduce all the observed periodicities in V346 Ori, no single solution is found for V351 Ori (as well as for variable 4 in NGC 6383), indicating that nonradial modes may also be important for these stars. With these caveats, the solution matching the two confirmed frequencies has been adopted for V351 Ori (Ripepi et al., 2002b), whereas that matching the two highest amplitude frequencies has been chosen for variable 4 in NGC 6383 (Zwintz and Weiss, these proceedings).

By applying this analysis to all of the PMS δ Scuti stars reported in Table I, we can derive their position in the H-R diagram and make a comparison with the predictions of the nonlinear instability strip (see Figure 2). Even if this result is heavily dependent on the assumption of radial pulsation, the agreement is quite satisfactory: only three pulsators are found to be bluer than the second overtone blue edge and indeed they are predicted to pulsate in higher overtones. The derived stellar parameters of all the variables are reported in Table II.

TABLE I
Properties of the 12 known PMS δ Scuti stars.

Name	F1 (c/d)	F2 (c/d)	F3 (c/d)	F4 (c/d)	ΔV (mag)	V (mag)	S.T.
V588 Mon	7.1865±0.0006	?	?		0.04	9.7	A7
V589 Mon	7.4385±0.0006	?			0.04	10.3	F2
HR5999	4.812±0.010				0.02	7.0	A7
HD104237	33±0.2				0.02	6.6	A7
HD35929	5.10±0.13				0.02	8.1	A5
V351 Ori	15.687±0.002	13.337±0.002	16.868±0.002	11.780±0.002	0.1	8.9	A7
BL 50	13.9175±0.0005	9.8878±0.0009			0.02	14.5	–
HP 57	12.72557±0.0002	15.52437±0.0003			0.03	14.6	–
HD142666	21.43±3				0.01	8.8	A8
V346 Ori	35.3±2.3	22.6±2.7	45.5±2.5	18.3±2.5	0.015	10.1	A5
H254	7.406±0.008				0.02	10.6	F0
NGC6383-4	14.376	19.436	13.766	8.295	0.014	12.61	A7

TABLE II

Stellar parameters and mode identification of the known PMS δ Scuti stars. The modes are: F (Fundamental), FO (First Overtone), SO (Second Overtone), TO (Third Overtone), FoO (Fourth Over tone), FiO (Fifth Overtone)

Star	$M/M_\odot$	$\log L/L_\odot$	$\log T_e$	Source	mode(s)
V351Ori	1.80	1.142	3.866	Ripepi et al. 2002b	F,FO
V346Ori	1.55	0.965	3.863	Pinheiro et al. 2002	F,SO
V346Ori	1.70	0.740	3.863	Pinheiro et al. 2002	F,FO,TO,FiO
HD104237	2.20	1.500	3.886	van den Ancker et al. 1998	FiO
HR5999	4.00	2.120	3.845	Marconi and Palla 1998	SO
HP57	2.00	1.250	3.857	Marconi et al. 2002	FO,SO
BL50	2.50	1.600	3.860	Marconi et al. 2002	SO,FoO
V588 Mon	3.50	2.050	3.903	this paper	TO
V589 Mon	2.50	1.540	3.845	this paper	FO
HD142666	1.70	1.030	3.880	Natta et al. 1997	FO
HD35929	3.40	1.920	3.857	Marconi et al. 2000	FO
HD35929	3.80	2.060	3.851	Marconi et al. 2000	SO
H254	2.60	1.620	3.854	Ripepi et al. 2002a	FO
H254	2.30	1.450	3.857	Ripepi et al. 2002a	F
NGC6383-4	2.50	1.680	3.908	this paper	TO,FiO

The main limitations of this method are: 1) the uncertainties still affecting many of the observed frequencies, due to poor data quality and/or the aliasing problem; 2) the difficulty to discriminate between PMS and post-MS evolutionary phases on the basis of radial models, in particular for pulsators that are predicted to be located close to the MS; 3) the fact that the likely presence of nonradial modes is not taken into account. Concerning the first point, significant improvements can be obtained by means of multisite campaigns and are expected from future satellite missions (e.g. COROT, MONS). As for the last two issues, it is clear that both radial and nonradial models should be computed in order to better understand the intrinsic properties of the rather unexplored class of young variable stars.

3. Pulsational Instability of Brown Dwarfs and Very-Low Mass Stars

As we have described in Sect. 1, stars with a mass of ~ 0.2 $M_\odot$ and below are expected to begin their PMS phase with the full amount of interstellar deuterium available for burning as soon as the central conditions become appropriate. Since these objects are fully convective even prior to the onset of D-burning, it is thus possible that they become pulsationally unstable, especially at the beginning of the burning phase, when the coefficients of vibrational instability are the largest

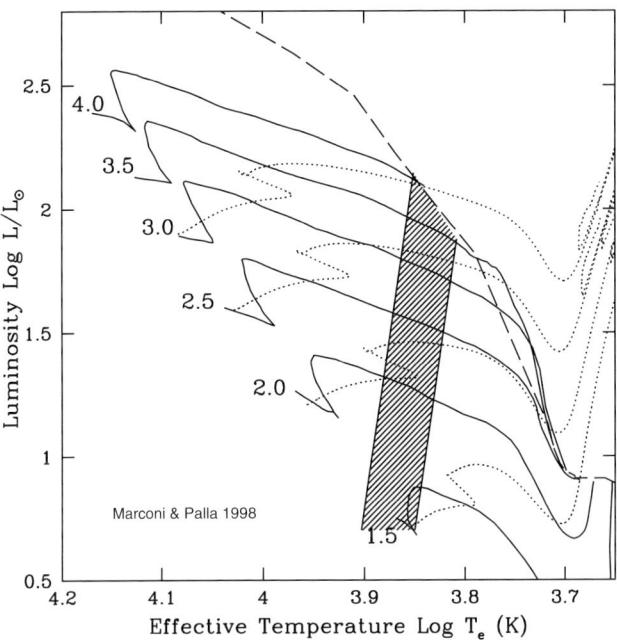

Figure 1. The predicted instability strip of PMS δ Scuti stars (shaded region), along with PMS (solid) and post-MS (dotted) evolutionary tracks. The dashed line represents the birthline (see Marconi and Palla, 1998 for details).

(e.g. Toma, 1972). If detected, it would be the first time that the effects of the ϵ mechanism are observed in the H-R diagram.

The interest for brown dwarfs is motivated by the recent discovery of large populations of these objects both in the field and in nearby clouds (e.g. Luhman et al., 2000; Dougados et al., 2001; Béjar et al., 2001; Delfosse et al., 1999; Kirkpatrick et al., 2000). Brown dwarfs in the field are too old for the ϵ mechanism related to D-burning to be operative. On the other hand, the most active star forming regions in the solar neighbourhood are few Myr old, exactly the age when brown dwarfs are expected to undergo active D-burning at their centres (e.g. Chabricr et al., 2000).

Brown dwarfs exist in large numbers, with current estimates of the substellar mass function indicating that they are nearly as numerous as stars in the solar neighbourhood (Bouvier et al., 1998; Reid et al., 1999). The ubiquity of brown dwarfs in different astrophysical contexts (star forming regions, open clusters, field) suggests that their formation is not restricted to specific (and restricted) initial conditions, but is a natural outcome of the collapse of dense cores: wherever stars form, brown dwarfs form as well (e.g. Luhman et al., 2000). However, the main formation mechanism is still debated. In one scenario, brown dwarfs form like stars by accretion from the collapse of a dense core. Another view supports the idea that they are 'failed stars', the result of the dynamical ejection of very low mass,

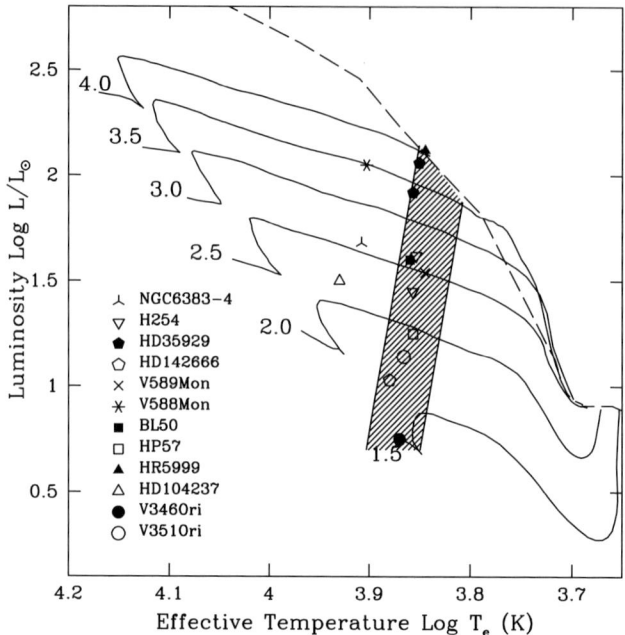

Figure 2. Position in the HR diagram of known PMS δ Scuti stars.

accreting in multiple systems (Reipurth and Clarke, 2001). Notwithstanding this divergence of interpretation, both models suggest that the initial seed has formed by accretion of dense gas and therefore has not had time to start burning deuterium at the centre.

3.1. D-BURNING AND PULSATIONAL INSTABILITY

We are interested in the excitation of non-adiabatic oscillations of the type $\xi(x, t) = \xi(x) e^{\sigma t}$, where σ is the coefficient of instability. From the conservation of mass, momentum and energy, we can write

$$\sigma \sim \int \frac{\delta T}{T} \left(\delta \epsilon_n - \frac{\partial \delta L_R}{\partial m} \right) dm - \int \frac{\delta T}{T} \left(\frac{\partial \delta L_C}{\partial m} \right) dm + \int (\Gamma - 5/3) \, \alpha \, L_C \, dm$$

in this expression, ϵ_n is the destabilizing term due to the strong sensitivity of D-burning on temperature: $\epsilon \sim \rho T^{16}$. Thus, a small perturbation on temperature induces a variation of the energy generation rate which is an order of magnitude bigger. In general, the quantity σ is a complex number: if the real part is positive, the perturbation is unstable; otherwise, the oscillation is damped around the equilibrium solution. The time scale for the development of the instability, τ_{inst}, is proportional to the inverse of σ. For the effectiveness of the instability, it is important that this time scale be shorter than the deuterium burning time (τ_D), which is typically of the order of several Myr in stars of mass lower than ~ 0.2 M$_\odot$. The instability is measured by the value of the parameter $\kappa = \tau_D/\tau_{inst}$, that should be $\gg 1$.

Estimates of the pulsations periods due to D-burning in low-mass stars have been computed by Gahm et al. (1989) who considered the case for stars in the mass range 0.2 to 3.0 M$_\odot$, assuming polytropic models of index $n = 1.5$. They found that the typical periods vary from ~ 2.5 days for a star of 3 M$_\odot$ to ~ 10 hr for a 0.2 M$_\odot$ object. Our extension of the calculations to stars of lower mass and brown dwarfs indicates that the predicted pulsation period is ~ 6 hr for 0.08 M$_\odot$ and 3 hr for 0.04 M$_\odot$. We have used simplified (polytropic) models of the internal structure of a brown dwarf and therefore the derived periods must be considered preliminary. The typical value of κ varies between 50 and 200. Thus, the onset of active D-burning can efficiently excite pulsations in these interesting objects.

The expected periods of the pulsations can be readily detected with short term monitoring of likely candidates. The optimal wavelength regions is the infrared where young brown dwarfs are brightest. The remaining question is where to find the good candidates. In our view, one of the best targets is the ρ Ophiuchi cloud, a ~ 1 Myr old cluster forming region at a distance of 140 pc from the Sun. Recent surveys have identified a conspicuous population of objects with mass below 0.2–0.1 M$_\odot$ (e.g. Natta et al., 2002). As shown in Figure 3, their location in the H-R diagram is very close to the theoretical instability strip for D-burning, with little dependence on the specific set of evolutionary tracks considered. Thus, these systems are prime targets for a monitoring campaign. Alternatively, one could consider the case of the four brown dwarfs discovered in the Taurus-Auriga association, again at 140 pc distance and of similar age. However, the determination of the surface parameters (luminosity and effective temperature) of these objects is less secure than in the case of the ρ Ophiuchi dwarfs. In any case, the search for pulsating brown dwarfs holds promising results.

4. Summary

The recent progress in the field of PMS δ Scuti pulsators has been discussed and the results based on the comparison between predicted and observed pulsations frequencies have been critically presented. The inferred position in the H-R diagram of the twelve known PMS δ Scuti is in agreement with the theoretical instability strip for the first three radial modes. However, it is clear that a study of nonradial modes is needed to understand better the nature of the observed pulsations and to constrain the intrinsic properties of this rather unexplored class of variables. In the case of very low-mass stars and brown dwarfs, we have pointed out the efficiency of the ϵ mechanism as a driving source for the possible pulsations of these structure

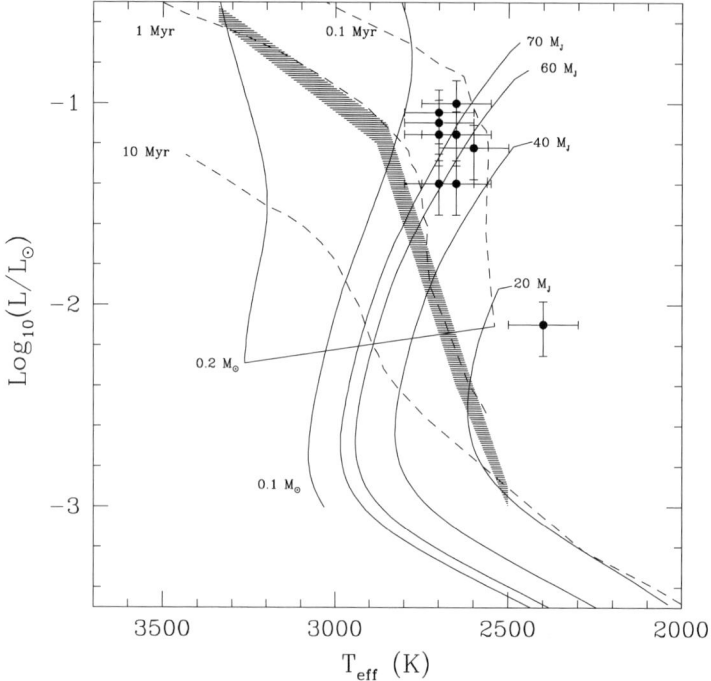

Figure 3. Position in the HR diagram of the brown dwarfs in the ρ Oph cluster. The shaded region is the instability strip due to D-burning. The evolutionary tracks (solid lines) are from D'Antona and Mazzitelli (1997). The dashed lines are isochrones at selected times.

during the short lived D-burning phase in the earliest phases of the their evolution. A first estimate of the expected periods is in the range of several hours and less. The detection of pulsating brown dwarfs in nearby star forming regions could represent an important result of variability studies.

References

Béjar, V.J.S., Martín, E.L., Zapatero Osorio, M.R., Rebolo, R., Barrado y Navascués, D., Bailer-Jones, C.A.L., Mundt, R., Baraffe, I., Chabrier, C. and Allard, F.: 2001, *ApJ* **556**, 830,
Bouvier, J., Stauffer, J.R., Martin, E.L., Barrado y Navascues, D., Wallace, B. and Béjar, V.J.S.: 1998, *A&A* **336**, 490.
Breger, M.: 1972, *ApJ* **171**, 539.
Castellani, V., Degl'Innocenti, S. and Marconi, M.: 1999, *MNRAS* **303**, 265.
Chabrier, G., Baraffe, I., Allard, F. and Hauschildt, P.: 2000, *ApJ* **542**, 119.
D'Antona, F. and Mazzitelli, I.: 1997, *Mem. S.A.It.* **68**, 807.
Delfosse, X., Tinney, C.G., Forveille, T., Epchtein, N., Borsenberger, J., Fouqué, P., Kimeswenger, S. and Tiphène, D.: 1999, *A&AS* **135**, 41.
Donati, J.-F., Semel, M., Carter, B.D., Rees, D.E. and Cameron, A.C.: 1997, *MNRAS* **291**, 658.
Dougados, C., Menard, F., Magnier, E., Cuillandre, J.-C., Lai, O., Manset, N., Fahlman, G., Martin, E. L., Bouvier, J., Veillet, C., Martin, P. and Forveille, T.: 2001, *Bull. CFH* **43**, 12.

Gahm, G.F., Fischerstrom, C., Lindroos, K.P. and Liseau, R.: 1989, *A&A* **211**, 115.
Kirkpatrick, J.D., Reid, I.N., Liebert, J., Gizis, J.E., Burgasser, A.J., Monet, D.G., Dahn, C.C., Nelson, B. and Williams, R.J.: 2000, *AJ* **120**, 447.
Kurtz, D.W. and Marang, F.: 1995, *MNRAS* **276**, 191.
Luhman, K.L., Rieke, G.H., Lada, C.J. and Lada, E.A.: 1998, *ApJ* **508**, 347.
Luhman, K.L., Rieke, G.H., Young, E.T., Cotera, A.S., Chen, H., Rieke, M.J., Schneider, G. and Thompson, R.I.: 2000, *ApJ* **540**, 1016.
Marconi, M. and Palla, F.: 1998, *ApJ* **507**, L141.
Marconi, M., Palla, F. and Ripepi, V.: 2002, *Comm. in Asteroseismology* **141**, 13.
Marconi, M., Ripepi, V., Alcalá, J.M., Covino, E., Palla, F. and Terranegra, L. 2000: *A&A* **355**, L35.
Natta, A., Grinin, V.P., Mannings, V. and Ungerechts, H.: 1997, *ApJ* **491**, 885.
Natta, A., Testi, L., Comeron, F. et al.: 2002, *A&A*, in press.
Palla, F., Stahler, S.W.: 1993, *ApJ* **418**, 414.
Pigulski, A., Kołaczkowski, Z. and Kopacki, G.: 2000a, *AcA* **50**, 113.
Pigulski, A., Kołaczkowski, Z. and Kopacki, G.: 2000b, in: L. Szabados and D. Kurtz (eds.), *The Impact of Large-Scale Surveys on Pulsating Star Research*, ASP Conf. Ser. **203**, 499.
Pinheiro, F.J.G., Folha, D.F.M., Marconi, M., Ripepi, V., Palla, F., Monteiro, M.J.P.F.G. and Bernabei, S.: 2002, *A&A* (submitted).
Reid, I.N., Kirkpatrick, J.D., Gizis, J.E. and Liebert, J.: 1999, *ApJ* **527**, 105.
Reipurth, B. and Clarke, C.: 2001, *AJ* **122**, 432.
Ripepi, V., Marconi, M., Bernabei, S., Palla, F., Pinheiro, F.J.G., Folha, D.F.M., Terranegra, L., Arellano Ferro, A., Jiang, X.J., Alcalá, J.M. and Oswalt, T.D.: 2002b, *A&A* (submitted).
Ripepi, V., Palla, F., Marconi, M., Bernabei, S., Arellano Ferro, A., Terranegra, L. and Alcalá, J.M.: 2002a, *A&A* **391**, 587.
van den Ancker, M.E., de Winter, D. and Tjin A Djie, H.R.E.: 1998, *A&A* **330**, 145.
Toma, E.: 1972, *A&A* **19**, 76.

OUTSTANDING ISSUES FOR POST-MAIN SEQUENCE EVOLUTION:

Instability Strips and Excitation of Modes in Compact Pulsators

GILLES FONTAINE and PIERRE BRASSARD
*Département de Physique, Université de Montréal,
C.P. 6128, Succ. Centre-Ville, Montréal, Québec H3C 3J7, Canada*

STÉPHANE CHARPINET
Observatoire Midi-Pyrénées, 14 Avenue Edouard Belin, 31400 Toulouse, France

Abstract. We review the question of the empirical and theoretical instability regions in the HR diagram for evolved, compact stars. These include the three families of pulsating white dwarfs (g-mode pulsators excited through mechanisms associated with partial ionization and convection in the stellar envelope), the pulsating subdwarf B stars (p-mode variables excited through a classic kappa mechanism associated with the radiative levitation of iron in the stellar envelope), and the 'Betsy stars', the brand new class of long-period, g-mode pulsators of the subdwarf B type discovered recently.

Keywords: compact pulsators, instability strips

1. The Compact Pulsators

The compact pulsators are defined as those pulsating stars with surface gravities larger than 10^5 cm s^{-2}. Figure 1 provides a bird's eye view of that portion of the surface gravity-effective temperature plane where they are found[*]. With the recent discovery of a new type, as reported by Betsy Green at this meeting, we now know of *five distinct families of compact pulsators*. Three of those belong to the general category of white dwarf stars, while the other two belong to the class of hot B subdwarf (sdB) stars.

About 75% of the white dwarfs descend from post-AGB remnants which have retained a thin H envelope. The dashed curve shown in the figure is representative of the evolution of such H-atmosphere white dwarfs. This channel leads to the formation of a first family of compact pulsators, the pulsating DA (or ZZ Ceti) white dwarfs, by the time a star evolves through a narrow range of effective temperature centred around $T_{\rm eff} \sim 11{,}500$ K in which H recombines in the envelope of the star.

A minority of white dwarfs descend from post-AGB remnants which have managed to get rid of their residual H through the so-called 'born-again scenario'. This

---

[*] The use of this diagram instead of the more standard HR diagram is more convenient for making the connection with the results of quantitative spectroscopic analyses which lead to the determination of the atmospheric parameters log g and $T_{\rm eff}$.

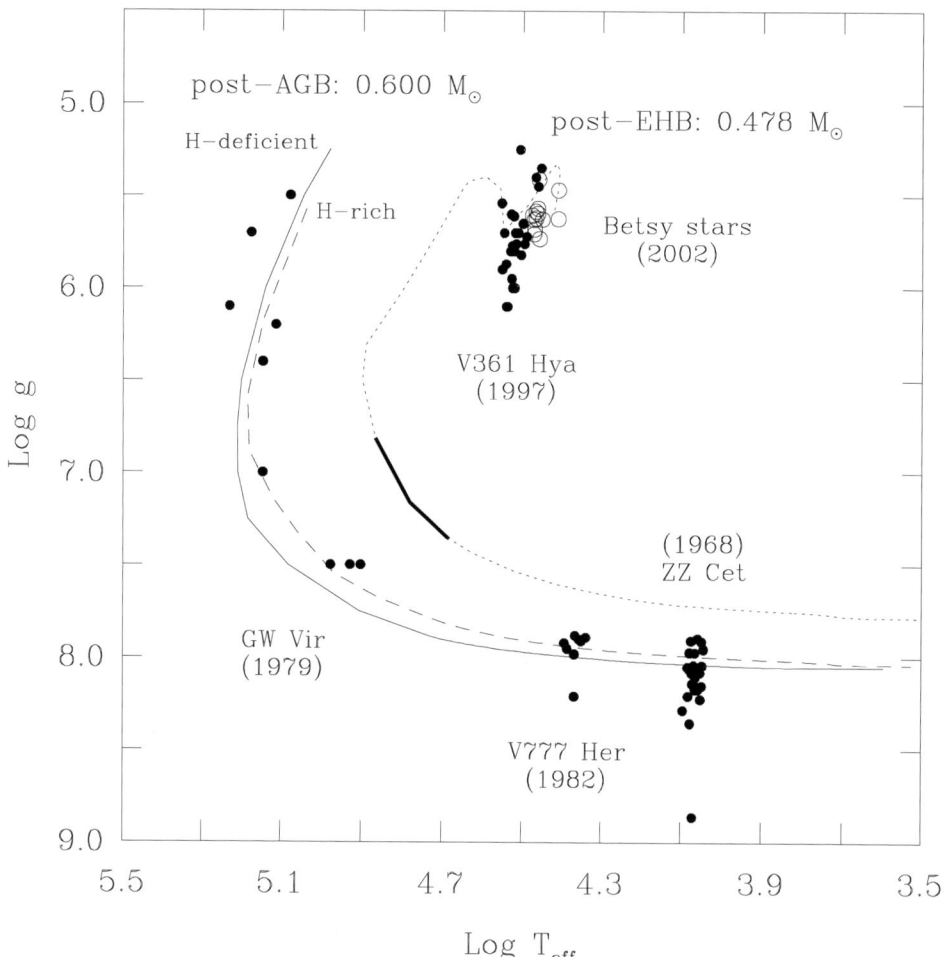

Figure 1. Region of the log g-$T_{\rm eff}$ plane where the compact pulsators are found. Each of the five distinct families is identified by its official IAU name (except for the newly-reported "Betsy stars"), and the year of the report of the discovery of the prototype of each class is also indicated. Typical evolutionary tracks are plotted showing 1) the track followed by a 0.6 $M_\odot$ post-AGB, H-rich star which becomes a H-atmosphere white dwarf (dashed curve), 2) the path followed by a 0.6 $M_\odot$ post-AGB, H-deficient star which becomes a He-atmosphere white dwarf (solid curve), and 3) the path followed by a 0.478 $M_\odot$ post-EHB model which leads to the formation of a low-mass H-atmosphere white dwarf (dotted curve).

second channel gives rise to *two* other families of compact pulsators: the extremely hot ($T_{\rm eff} \sim$ 100,000 K) pulsating PG1159 white dwarfs (or GW Vir stars), and the much cooler ($T_{\rm eff} \sim$ 25,000 K) pulsating DB white dwarfs (or V777 Her stars). There is a natural evolutionary link between these two types of pulsators along the solid curve shown in Figure 1 (see, e.g., Brassard and Fontaine, these proceedings), and a GW Vir star is bound to pulsate again later on in its lifetime, but, this time, as a DB white dwarf.

The first short-period pulsating sdB stars were discovered at the South African Astronomical Observatory and their existence was reported in 1997 (Kilkenny et al., 1997). Officially named V361 Hya stars, they are commonly referred to as the EC 14026 pulsators, after the prototype, a star discovered during the course of the Edinburgh-Cape Survey. In the same general area of the log g-T_{eff} diagram, but distinctly cooler, one finds the second family of pulsating sdB stars, the long-period variables discovered quite recently by Betsy Green and her students. The evolution of sdB stars in the diagram is typified by the dotted curve which shows the track followed by a 0.478 $M_\odot$ post-EHB model. This channel contributes to a small fraction, $\sim 2\%$, of the white dwarf population in the form of low-mass objects. Note that the heavy part of the dotted track corresponds to a theoretical instability region predicted by the models of Charpinet et al. (1997a). In Nature, these would correspond to low-mass DAO white dwarfs – relatively rare objects – undergoing g-mode oscillations excited through the epsilon mechanism, but this sixth family of compact pulsators, if indeed it exists, has yet to be discovered.

2. The Instability Regions

For the purposes of this review, we have compiled the latest available data and used only estimates of atmospheric parameters based on reliable quantitative spectroscopy. The picture that emerges, zooming in each region, is as follows.

2.1. PG1159 (GW VIR) STARS

Figure 2 shows the part of the log g-T_{eff} plane where the GW Vir stars are found. There are currently 9 known GW stars with published estimates of their atmospheric parameters (Dreizler and Heber, 1999; Miksa et al., 2002). Their positions are indicated by filled circles. Note that in this extreme regime of effective temperature, NLTE effects are quite important and model atmosphere analyses are involved and difficult to carry out. Note further that 4 of the 9 objects shown here are also central stars of planetary nebulae (CSPN), and they have been identified by larger open circles. There are 6 other pulsating CSPN known, but their atmospheric parameters are not known to any degree of accuracy and they are not considered here.

There are also nonvariable stars of the PG1159 spectral type – 14 of them – which occupy the same general area of the log g-T_{eff} diagram as the pulsators. Their coexistence with the GW Vir stars is currently a puzzle, although there are a number of potential explanations such as a very strong blue edge dependence on the mass, the effects of metallicity, and the effects of winds and diffusion. Those possibilities remain to be investigated in details.

The periods observed in GW Vir stars are found in a wide range from 400 to upward of 2000 s and correspond to low-degree g-modes. The rather broad range of

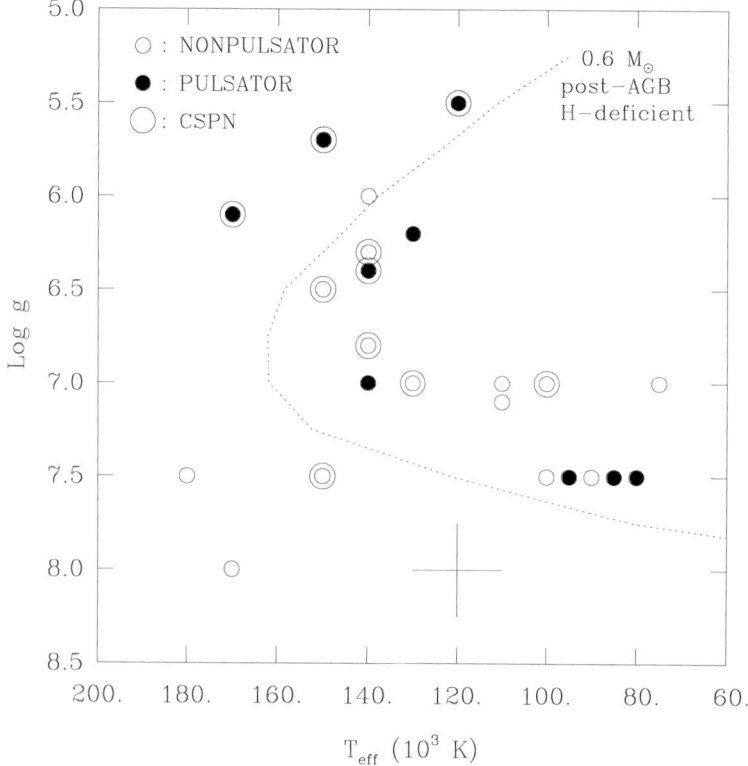

Figure 2. Instability region for the GW Vir stars. The known pulsators are indicated by small filled circles, and the nonvariable PG1159 stars are indicated by small open circles. The CSPN among those objects are identified by the larger open circles. The cross illustrates typical uncertainties on the atmospheric parameters. The dotted curve is the evolutionary track of a typical post-AGB, H-deficient model.

periods is readily explained in terms of the wide range of surface gravities exhibited in Figure 2. As would be expected, the more evolved, denser, and higher-gravity GW Vir stars show the shorter periods. From the point of view of driving, the mechanism is the same along the evolutionary track shown in the figure, and it is then inappropriate to divide the GW Vir stars in two categories (the so-called PNNV and DOV stars) as has been done very often in the literature. This driving process has been identified many years ago and is a kappa mechanism associated with the ionization of the K-shell electrons in C and O which, together with He, are the main envelope chemical constituents. Convection plays no role in the pulsations of PG1159 stars.

The most comprehensive analyses of the properties of GW Vir star models remain those of Saio (1996) and Gautschy (1997). Both authors have been able to explain, at the broad qualitative level, the ranges of periods actually observed in these pulsators. However, more work remains to be done at the quantitative level. The current major weakness in the field resides with the fact that more realistic

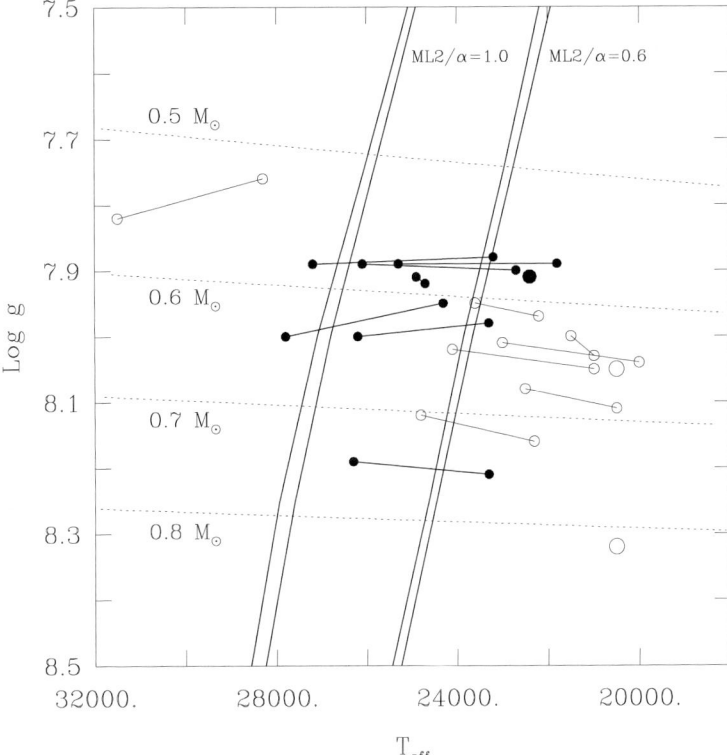

Figure 3. Instability region for the V777 Her stars. The positions of the pulsators are indicated by small filled circles while those of the nonvariable stars by small open circles. The positions are uncertain, depending on the presence or not of small traces of hydrogen in the atmospheres. Those 3 objects with *known* H/He atmospheric abundances are indicated by larger symbols. The solid curves represent theoretical blue edges, sensitive to the assumed efficiency of convection and also to small traces of hydrogen in the envelope. The dotted curves illustrate evolutionary tracks for He-atmosphere white dwarfs of different masses.

models of post-AGB, born-again stars are badly needed. The work of Herwig et al. (1999) in this area shows very interesting promises.

2.2. PULSATING DB (V777 HER) WHITE DWARFS

There are currently 9 known pulsating DB white dwarfs, but one of which is not considered here because it is a southern hemisphere object with very poor spectroscopy. Otherwise, the positions of the 8 others in the log g-$T_{\rm eff}$ diagram are indicated by small filled circles in Figure 3. There are also 9 nonvariable DB stars in the same general area of the diagram but, as indicated by the small open circles, those are either hotter or cooler than the pulsators.

One difficulty in the interpretation of optical spectra of DB white dwarfs is that there is a sensitivity to small traces of H which might be present in the otherwise

He-dominated atmospheres (see Beauchamp et al., 1999 for a detailed discussion of this). These traces are generally invisible in the optical domain, but may be detected in the UV range. For those objects with no UV data available, optical spectroscopy gives two solutions for each star: a set of log g-T_{eff} values for an assumed pure He atmosphere (the points to the left in the figure), and another set assuming a H trace at the limit of detection (the points to the right in the figure). Despite the 'fuzziness' introduced by this uncertainty, the data available in Figure 3 are compatible with the idea of a narrow instability strip which would contain only pulsators.

Observationally, the pulsating DB white dwarfs show multiperiodic luminosity variations with periods in the range 150–1100 s. Those are interpreted as low-degree, low-order ($k \sim 1$–20) g-modes. An examination of models reveals that the ultimate cause of instability is due to the recombination of HeIII and HeII in the envelope of a DB star. This recombination process occurs around $T_{\mathrm{eff}} \sim 25{,}000$ K in a typical He-atmosphere white dwarf. It leads to the formation of a superficial convection zone which grows in size (the bottom of the convection zone moves downward while the top remains close to the photosphere) with further cooling. Such a convection zone can carry a very significant fraction of the total energy flux. It is found, in fact, that the location of maximum driving does not correspond to the maximum of the opacity profile (as in a classic kappa mechanism), but is confined in a narrow region near the base of the convection zone. Hence, contrary to PG1159 pulsators, convection-pulsation interactions cannot be neglected in DB stars. The usual neglect of such interactions in nonadiabatic calculations of DB models – due to the lack of a general credible prescription – implies that our general understanding of driving in these stars remains quite incomplete. This is especially true for the modelling of the red edge of the instability strip where convection-pulsation interactions are expected to be particularly important. And indeed, to our knowledge, no one has been able yet to model satisfactorily the empirical DB instability strip shown in Figure 3. While the theoretical blue edge can be adjusted to the empirical blue edge by selecting properly the convective efficiency in the models, the theoretical red edge, under the neglect of convection-pulsation interactions, is always much redder than the empirical red edge.

After the initial exploratory work of Winget et al. (1983) and the study of Bradley and Winget (1994), we have been the only ones to carry out detailed nonadiabatic calculations of the pulsation properties of models of DB white dwarfs using modern opacities and detailed model atmospheres. Some of our work has been reported in Beauchamp et al. (1999). For the purposes of this review, we have carried out additional calculations using our most recent evolutionary models of DB stars as briefly described in Brassard and Fontaine (these proceedings). Our theoretical blue edges are shown by the solid curves in Figure 3 where we used two flavors of the mixing-length convective efficiency. The double curves show the (small) effects of including small traces of hydrogen in the envelopes of our equilibrium models. Since, as in the previous studies, we made the so-called frozen-in convection approximation, our red edges are not very realistic (they are too cool) and are not shown in the figure.

2.3. PULSATING DA (ZZ CETI) STARS

The available data show a clear example of a *pure* instability strip in the case of the ZZ Ceti stars. This is illustrated in Figure 4 where the results of an analysis of 28 out of the 32 known ZZ Ceti stars have been reported. This analysis is an extension of the work of Bergeron et al. (1995) carried out by Pierre Bergeron and using the same homegeneous methods and tools as before. In particular, the 4 known ZZ Ceti stars with rather poor optical spectra have not been included here to preserve the homogeneity of the sample. The 28 filled circles shown in the figure and compared to the known constant stars (the open circles), define an empirical instability strip which is consistent with the idea of a pure strip. This result is at odds with the many (erroneous!) claims which have been made over the years that the ZZ Ceti strip contains both constant and variable stars (note that theory has no provision for a 'hidden' parameter that could explain such a mixed population) but is quite consistent with the position that one of us (G.F.) has maintained over more than two decades.

From an observational point of view, the ZZ Ceti stars share many properties in common with the pulsating DB white dwarfs. Their observed periods are found in the range 110–1200 s and are, again, associated with low-degree, low-order g-modes. Likewise, the instabilities can be traced back ultimately to the recombination of the main envelope constituent, hydrogen in this case, which occurs around $T_{\rm eff} \sim 12{,}000$ K. This recombination of HII leads to the formation and development of a superficial convection zone with further cooling. The region of maximum driving is found just at the base of that convection zone, and that zone can carry even more flux than that found in the DB white dwarfs. Historically, there has been quite a bit of confusion as how to name the driving mechanism found in models of DA (and DB) white dwarfs. Terms such as 'kappa mechanism' (which is incorrect as we have seen), 'generalized kappa mechanism', 'partial ionization mechanism', 'convective blocking', and, more recently, 'convective driving' have all been used in the literature. Whatever the correct name, it is clear that convection plays an essential role in the pulsation properties of the ZZ Ceti stars.

There have been a few attempts to include pulsation/convection interactions in models of DA white dwarfs, notably by Brickhill (1991) and, more recently, by Goldreich and Wu (1999). We show, in Figure 4, the results of our own recent efforts at including a more realistic treatment of convection. Specifically, we assumed that the convective flux in a DA adjusts instantly to the perturbations, we perturbed the optically-thin layers, and we included thermal imbalance terms. The dashed lines (to be compared with the solid lines which define the empirical boundaries) show our results for the theoretical instability strip. While very preliminary, these

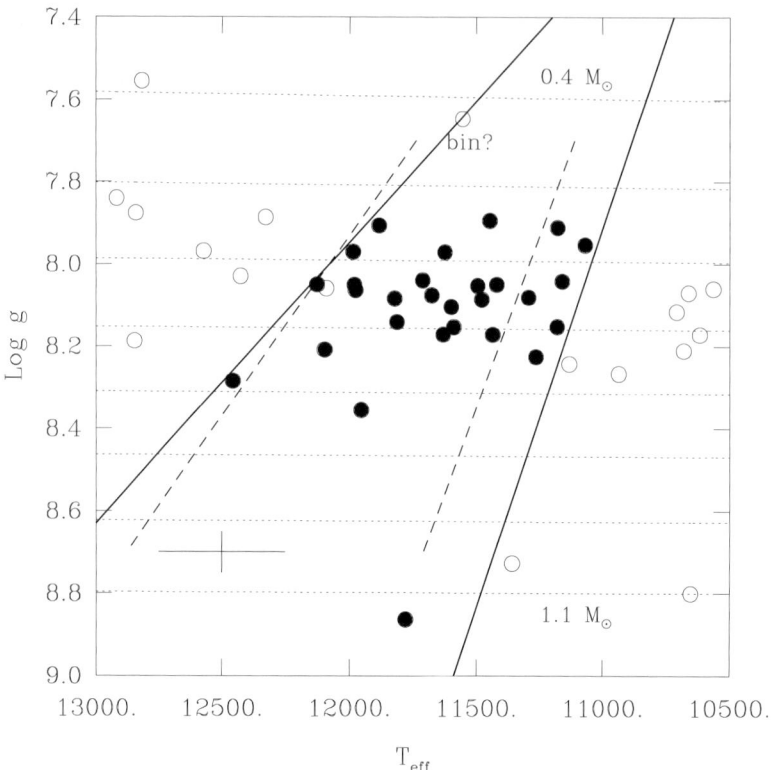

Figure 4. Instability region for the ZZ Ceti stars. The positions of the pulsators are indicated by the filled circles while those of the nonvariable stars by the open circles. The solid (dashed) lines represent the empirical (theoretical) boundaries of the ZZ Ceti instability strip. The dotted curves illustrate evolutionary tracks for H-atmosphere white dwarfs of different masses.

results are encouraging in that both the blue and red edges from theory are not too different from the empirical data.

2.4. EC 14026 (V361 HYA) STARS

There are currently 30 known short-period sdB pulsators, and the distribution of 26 of them in the log g-$T_{\rm eff}$ diagram is shown in Figure 5. As a comparison, the open circles in the figure give the positions of 70 sdB stars which have been searched for variability and found to be constant (see Billères et al., 2002). The solid curve shows the contour of maximum instability as predicted by the second-generation models of Charpinet et al. (1997b). It is clear here that both variable and nonvariable stars coexist in the same region of the surface gravity-effective temperature plane. This coexistence has yet to be explained satisfactorily and constitutes one of the outstanding issues in our understanding of EC 14026 pulsators. The suggestion has been that weak stellar winds as well as differences in age are the likely explanations (see, e.g., Charpinet et al., 2001).

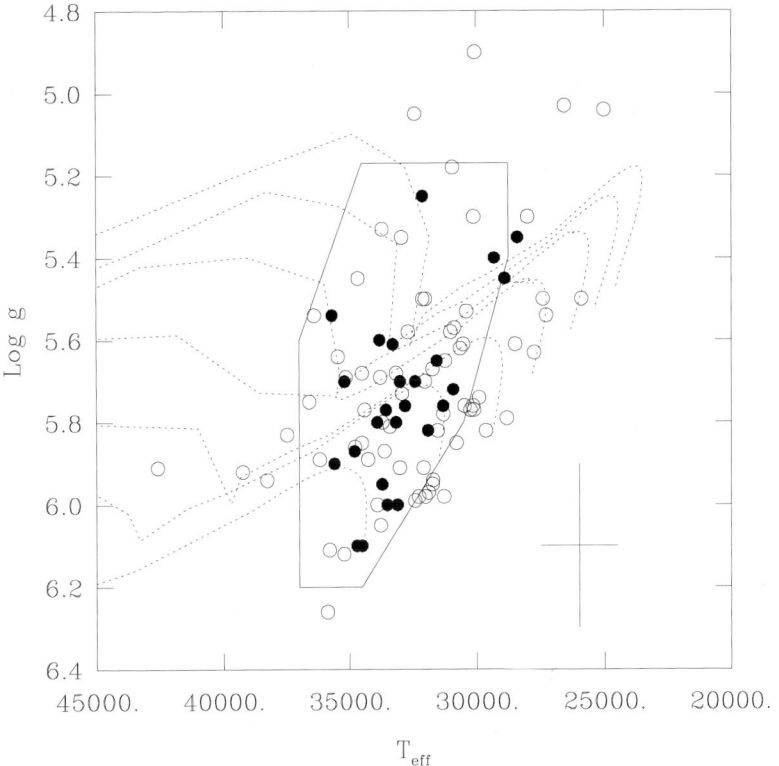

Figure 5. Instability region for the EC 14026 stars. The positions of the pulsators are indicated by the filled circles while those of the nonvariable stars by the open circles. The solid curve defines the region of maximum instability as derived from nonadiabatic calculations. The dotted curves illustrate evolutionary tracks for typical models of EHB stars with different envelope masses.

The EC 14026's are, like the other compact pulsators, multiperiodic variables. The observed periods are found in the range 80-600 s, but, more typically, in the narrow range 100–200 s. These variations correspond to low-degree, low-order p-modes. Values of $l = 0$, 1, 2, and 3 and $k = 1$–7 have been assigned to the observed periods in the sdB star PG 0014+067 (Brassard et al., 2001). Unlike the other types of compact pulsators which are g-mode variables, however, the EC 14026 stars are the only known p-mode objects in the lower part of the HR diagram. They also stand out among the compact pulsators in that nonadiabatic calculations have been extremely successful at explaining their properties (see Charpinet et al., 2001 for a review of this). In this context, it is worthwhile to point out that, back in 1997, when one of us (S.C.) computed the instability contour shown in Figure 5, only 4 EC 14026's were known. Over the years, the newly-found pulsators have kept falling within that region, to the point where we can claim that the driving mechanism is now well identified. The excitation mechanism is a classic kappa mechanism associated with a maximum in the opacity profile due to a local overabundance of iron. This local overabundance in the envelopes of EC 14026 stars is caused by diffusion processes (specifically, the competition between radiative levitation and gravitational settling), which are known to be active in sdB stars. To be complete, let us mention that pulsation/convection interactions are completely negligible in models of sdB stars.

2.5. THE BETSY STARS

We show, in Figure 6, the region of the log g-$T_{\rm eff}$ diagram where the newly-discovered Betsy stars are found. As compared to the EC 14026 stars (open circles), the Betsy stars (heavy crosses) occupy a domain which is definitely cooler. The two regions of instability appear to touch, but do not overlap. The locations of the Betsy stars correspond to EHB stars which have thicker H-envelopes than the EC 14026 pulsators. It is not clear at this stage how this difference can be interpreted, but it presumably means something.

The discovery and global properties of these new pulsators have been reported by Betsy Green at this meeting. Among others, the observed periods of the order of an hour automatically imply that those objects are g-mode pulsators. For a typical sdB model with parameters compatible with the location of the Betsy stars in the log g-$T_{\rm eff}$ diagram, the observed periods imply that $k \sim$ 15–20, assuming that $l = 1$.

The most outstanding issue concerning the Betsy stars is currently their very existence. What is (are?) the mechanism (mechanisms?) responsible for the excitation of g-modes in these cooler sdB stars? Since more than half of the sdB stars are part of close binary systems, it is possible than tidal excitation of g-modes are of relevance for at least some of the Betsy stars. A preliminary exploration of this possibility has been presented by Fontaine et al. (these proceedings) at this meeting and they concluded that this avenue is well worthy of further study. However, because some of the Betsy stars appear not to be members of close binary systems, and, in addition, their actual distribution in the log g-$T_{\rm eff}$ plane would be difficult to explain on the basis of binarity alone, another driving mechanism must be sought. Currently, the iron driving mechanism which explains so successfully the existence of EC 14026 stars does not appear to be able to drive pulsation modes (not even the p-modes and, a fortiori, the deeper g-modes) in the cooler Betsy stars. However, that issue remains to be studied in some more details. The possibility, although apparently remote at the outset, that H-burning at the base of the H-rich envelopes in *nonstandard* models of sdB stars could drive g-modes through the epsilon mechanism is also worthy of investigation. If such 'boosted' H-burning makes any sense at all in models of sdB stars, one would expect indeed to find the more 'active' stars in the region of the surface gravity-effective temperature plane where the H envelope is the more massive, i.e., on the cooler side as observed. The Betsy stars offer us a very interesting challenge!

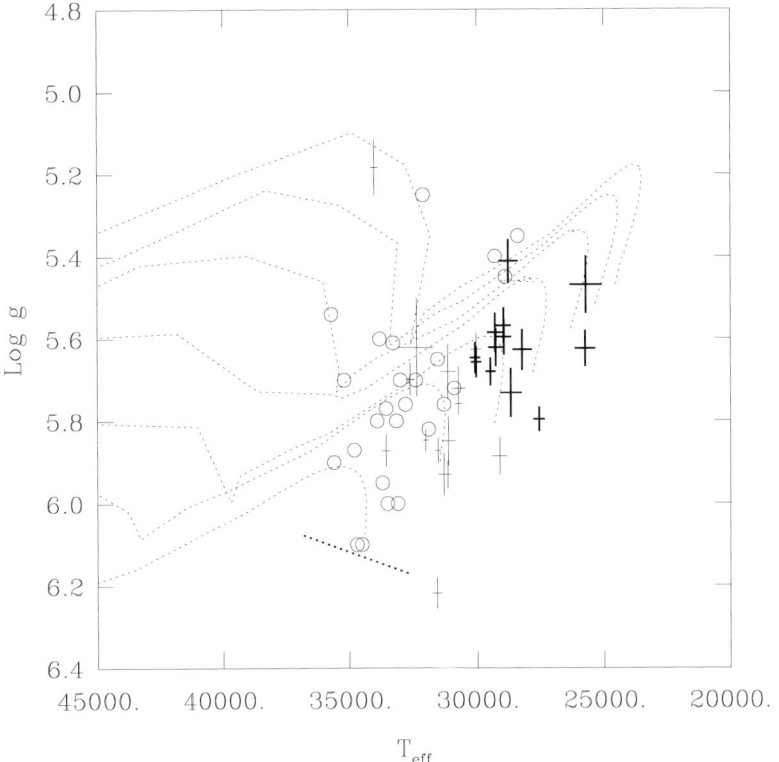

Figure 6. Instability region for the Betsy stars. The heavy crosses give the positions of 13 long-period sdB pulsators, while the light crosses show the positions of 15 constant stars which have been searched for long-period photometric activity. In comparison, the open circles indicate the positions of the 26 short-period sdB (EC 14016) pulsators. The dotted curves illustrate evolutionary tracks for typical models of EHB stars with different envelope masses increasing from left to right. The heavy dotted curve defines the ZAHEMS.

References

Beauchamp, A., Wesemael, F., Bergeron, P., Fontaine, G., Saffer, R.A., Liebert, J. and Brassard, P.: 1999, *ApJ* **516**, 887.
Bergeron, P., Wesemael, F., Lamontagne, R., Fontaine, G., Saffer, R.A. and Allard, N.: 1995, *ApJ* **449**, 258.
Billères, M., Fontaine, G., Brassard, P. and Liebert, J.: 2002, *ApJ*, in press.
Bradley, P.A. and Winget, D.E.: 1994, *ApJ* **421**, 236.
Brassard, P., Fontaine, G., Billères, M., Charpinet, S., Liebert, J. and Saffer, R.A.: 2001, *ApJ* **563**, 1013.
Brickhill, A.J.: 1991, *MNRAS* **251**, 673.
Charpinet, S., Fontaine, G. and Brassard, P.: 2001, *PASP* **113**, 775.
Charpinet, S., Fontaine, G., Brassard, P. and Dorman B.: 1997a, *ApJ* **489**, L149.
Charpinet, S., Fontaine, G., Brassard, P., Chayer, P., Rogers, F.J., Iglesias, C.A. and Dorman, B.: 1997b, *ApJ* **483**, L123.
Dreizler, S. and Heber, U.: 1999, *A&A* **334**, 618.

Gautschy, A.: 1997, *A&A* **320**, 811.
Goldreich, P. and Wu, Y.: 1999, *ApJ* **511**, 904.
Herwig, F., Blöcker, T., Langer, N. and Driebe, T.: 1999, *A&A* **349**, L5.
Kilkenny, D., Koen, C., O'Donoghue, D. and Stobie, R.S.: 1997, *MNRAS* **285**, 640.
Miksa, S., Deetjen, J.L., Dreizler, S., Krurk, J.W., Rauch, T. and Werner, K.: 2002, *A&A*, in press.
Saio, H.: 1996, in: C.S. Jeffery and U. Heber (eds.), *Hydrogen-Deficient Stars*, ASP Conf. Ser. **96**, 361.
Winget, D.E., Van Horn, H.M., Tassoul, M., Hansen, C.J. and Fontaine, G.: 1983, *ApJ* **268**, L33.

HOT SUBDWARFS: MAGNETIC, OSCILLATORY AND OTHER PHYSICAL PROPERTIES

R. OREIRO[1,2,3], F. PÉREZ HERNÁNDEZ[3,1], M. MANTEIGA[4], A. ULLA[2],
J.M. GONZÁLEZ PÉREZ[5], M.R. ZAPATERO OSORIO[6], R. GARCÍA LÓPEZ[3,1],
J. MACDONALD[7], P. THEJLL[8], A. FERRIZ-MAS[2], R.A. SAFFER[9] and V. ELKIN[10]

[1] *Departamento de Astrofísica, Universidad de La Laguna, E-38200 Spain*
[2] *Departamento Física Aplicada, Universidade Vigo, E-36200 Vigo, Spain*
[3] *Instituto de Astrofísica de Canarias E-38200 La Laguna, Spain*
[4] *Depart. Ciencias de Navegación y la Tierra, Univ. da Coruña, Spain*
[5] *Nordlysobservatoriet, Universitetet i Tromsø, N-9037 Norway*
[6] *Laborat. Astrofísica Espacial y Física Fundamental, E-28080 Madrid, Spain*
[7] *Department of Physics and Astronomy, Delaware University, USA*
[8] *Danmarks Meteorologiske Institut, DK-2100 København Ø, Denmark*
[9] *Department of Astronomy & Astrophysics, Villanova University, USA*
[10] *Spec. Astroph. Observat., Russ. Acad. Sciences, 357 147 Nizhnij Arkhyz, Russia*

Abstract. Hot subdwarf stars (hot sds) are blue subluminous objects. Only a few determinations are available to date regarding the study of such aspects as rotation, microturbulent velocities or the magnetic nature of these objects. Over 26 sdBs are known to date to be multiperiodic rapid oscillators. This project presents preliminary results of new observations and models of a sample of pulsating and non-pulsating hot sds, including considerations on mass loss and eventual magnetic properties.

1. Introduction

Hot sds split into two well separated spectroscopic sequences: the O, named sdO subclass, and the B, named sdB subclass. Their location in the log $T_{\rm eff}$–log g plane is different: the sdB stars, rich in Hydrogen, are in the region with $T_{\rm eff}$ between 20000 K and 35000 K, and log g around 5.25-6.50; on the other hand the sdO stars are He-rich, with $T_{\rm eff}$ ranging from 35000 K to 90000 K and log g from 4.0 to 6.5.

They are common and abundant stars. According to Jeffery and Pollacco (1998), about 50% in any searching for blue objects are sdB stars. Moreover, they dominate the old population of giant elliptical galaxies, as reflected by their UV excess (Bica et al., 1996).

The evolutionary state of hot sds corresponds to the final Horizontal Branch phase (HB) in which stars derive their energy from the conversion of He to C. Depending on the mass of the H envelope, sds could be identified with stars evolved directly from the Giant Branch to the Extended-HB or with descendants of AGB stars (Saffer et al., 1998). It is supposed that they evolve directly to white dwarfs, but only a 25% of WD have a hot sds as progenitor (Heber, 1986).

The metallic abundances show anomalies partially caused by the balance between gravitational settling and radiative levitation of heavy elements, although this process alone does not explain satisfactorily the observations. As possible mechanisms proposed in an attempt for explaining the He, C and Si anomalies are: mass loss, although it should be sufficiently low that it would be undetectable, rotation and magnetic fields (Michaud et al., 1985; Fontaine and Chayer, 1998; Elkin, 1996).

A few years ago, Kilkenny et al. (1997) found the first pulsating subdwarf. Nowadays over 26 oscillating hot sds are known. All of them are multiperiodic sdB pulsators, located in a region between 33000 K and 35000 K and $\log g$ from 5.6 to 5.96, with periods around 5 minutes. There are currently no known pulsating sdO. In any case, the mechanism of excitation of sdBs is not valid for the sdOs. The frequency peaks observed seem to correspond mainly to p modes of low radial orders, as follows from a comparison with stellar models. This is in agreement with the excitation mechanism proposed so far (e.g. Fontaine et al., 1998).

Out of this range of instability, PG0856+121, discovered by Piccioni et al. (2000), has $T_{eff} \simeq 26400$ K (Saffer et al., 1994). We have observed it in our 2002 campaign during 3 hours at the IAC80 telescope, and we can confirm its oscillatory nature as shown in the power spectrum of Figure 1 (left panel). However, the main frequency peak shown in Figure 1 does not correspond to that found by Ulla et al. (2001) and some caution is required to interpret the nature of the variability of this star.

2. Our Program for Analyzing a Sample of Hot Subdwarfs

There are more than 1200 cataloged hot sds. In particular, we have selected 15 sdOs and 15 sdBs from the Palomar Green Catalogue trying to cover a wide range in gravities, T_{eff} and metallicities. We have omitted those of binary nature and with V greater than 14.

We are performing a photometric study of the selected stars to find new pulsating sds, using the 3-channel Tromsø-Texas Photometer. One of the observed stars is BS16545-49. Figure 1 (right panel) shows its power spectrum during one night at the IAC80 telescope. The peaks with greater amplitudes could correspond to modes with periods around 500 and 700 seconds. We have also found evidence of oscillations in the star BD+254655 (González et al., 2002).

The next step in the investigation is a high resolution spectroscopic study to obtain accurate values of $\log g$, T_{eff} and composition. In Table I we show the physical parameters we have calculated for the sdOs of our sample. We are working to calculate the same parameters for our sdBs. With these stellar parameters we can compute precise structure and evolution models with the code of J. MacDonald (see Mullan and MacDonald, 2001). His code includes element diffusion and mass loss, and we are currently working to include the magnetic and rotational effects.

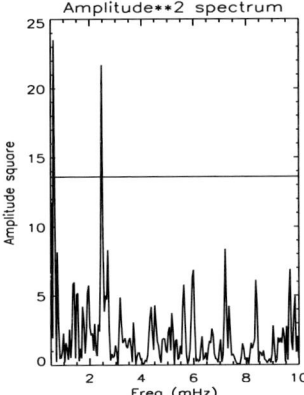 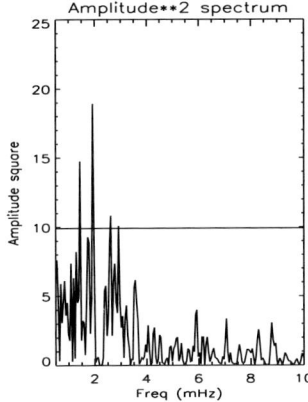

Figure 1. Power spectrum of: left panel: PG0856+121, right panel: BS16545-49. The horizontal continuous line means three times the noise level.

For the sdO subclass, we must keep in mind that the construction of stellar models involves a non-local thermodynamic equilibrium treatment due to their high temperatures. Once we have the stellar models, we use the linear adiabatic oscillation code of Christensen-Dalsgaard (Christensen-Dalsgaard and Bertomieu, 1991) to compute the theoretical frequencies. These results could be then compared with the observations and hopefully used for a seismological investigation of sds.

Elkin (1996) took circularly polarized spectra of two stars and obtained $B = -1680$ G for BD+75325, while for BD+252534 he found a variability in the range -1300 G to +1750 G. Schüssler et al. (1994) – see also Schüssler et al. (1996) –, developed a model for the treatment of magnetic flux tubes in the Sun. We are applying the same approach, based on flux-tube dynamics, to hot sds. Although the model was originally developed for the Sun, the requirements are the existence of an interface inside the star and that the stellar matter be made of ideal gas. In the solar case, the interface is the transition between the radiative envelope and the convection zone; in the case of hot sds, the interface might be the transition between the He-burning core and the Hydrogen envelope.

3. Conclusions

We have shown our first results and future plans for analysing a sample of hot sds. They are consistent with those expected for this kind of stars: multifrequency pulsations with periods of hundred of seconds. In particular, we confirmed the oscillatory nature of PG0856+121, but we need more observations to determinate accurately the periods. We also need more observations of two very promising pulsators we have found, BS16545-49 and BD+254655. With further observational and theoretical work with our sample of sds, we expect to improve our understanding of the nature of hot sds.

TABLE I

Main physical parameters of sdO stars

PG Nr.	$T_{\rm eff}$ (K)	$\log g$	Y	PG Nr.	$T_{\rm eff}$ (K)	$\log g$	Y
0208+016	45000	6.0	0.66	1544+253	55000	5.6	0.99
0240+046	37000	5.3	0.55	1708+614	55000	5.5	0.90
0838+133	55000	5.8	0.9	2129+151	49000	5.8	0.95
0902+057	43000	6.0	0.97	2201+145	45000	5.7	0.95
0921+311	41000	5.4	0.85	2244+152	43000	5.5	0.95
1011+649	57000	5.9	0.90	2326+510	43000	5.9	0.89
1020+695	53000	5.6	0.99	2352+181	51000	6.0	0.85
1230+067	43000	5.5	0.94				

Acknowledgements

This work was supported by the Spanish Ministerio de Ciencia y Tecnología under Project AYA2000-1691. R. Oreiro acknowledges CajaCanarias for its research grant. This work acknowledges computational support from the Centro de Supercomputación de Galicia.

References

Bica, E., Bonatto, C., Pastoriza, M.G. and Alloin, D.: 1996, *A&A* **315**, 405.
Christensen-Dalsgaard, J. and Berthomieu, G.: 1991: in: A.N. Cox et al. (eds.), *Solar interior and atmosphere*, Space Sci. Series, Univ. Arizona Press, p. 401.
Elkin, V.G.: 1996, *A&AL* **312**, 5.
Fontaine, G. and Chayer, P.: 1998: in: A.G. Phylip et al. (eds.), 3rd *conference on Faint Blue stars*, L. Davis press, in press.
Fontaine, G., Charpinet, S., Brassard, P., Chayer, P., Rogers, F.J., Iglesias, C.A. and Dorman, B.: 1998: in F.-L. Deubner, J. Christensen-Dalsgaard and D. Kurtz (eds.), *New Eyes to See Inside the Sun and Stars*, IAU Sym. 185, p. 367.
González Pérez, J.M. et al.: 2002: in *NATO Advance Research Workshop on White Dwarfs*, Napoles, Italy, in press.
Heber, U.: 1986, *A&A* **155**, 33.
Jeffery, C.S. and Pollacco, D.L.: 1998, *MNRAS* **298**, 179.
Kilkenny, D., Koen, C., O'Donoghue, D., van Wyk, F. and Stobie, R.S.: 1997, *MNRAS* **285**, 645.
Michaud, G., Bergeron, P., Wesemael, F. and Fontaine, G.: 1985, *A&A* **239**, 275.
Mullan, D.J. and MacDonald, J.: 2001, *ApJ* **559**, 353–371.
Piccioni, A. et al.: 2000, *A&AL* **354**, 13.
Saffer, R.A., Bergeron, P., Koester, D. and Liebert, J.: 1994, *ApJ* **432**, 351.
Saffer, R.A., Livio, M. and Yungelson, L.R.: 1998, *ApJ* **502**, 394.
Schüssler, M., Caligari, P., Ferriz-Mas, A., Solanki, S. and Stix, M.: 1996, *A&A* **314**, 503.
Schüssler, M., Caligari, P., Ferriz-Mas, A. and Moreno-Insertis, F.: 1994, *A&AL* **281**, 69.
Ulla, A., Zapatero Osorio, M.R., Pérez Hernández, F. and MacDonald, J.: 2001, *A&A* **369**, 986.

NONLINEARITY OF NONRADIAL MODES IN EVOLVED STARS

RAFAŁ M. NOWAKOWSKI[1] and WOJCIECH A. DZIEMBOWSKI[1,2]
[1] *Copernicus Astronomical Center, Bartycka 18, 00-716 Warsaw, Poland*
[2] *Warsaw University Observatory, Al. Ujazdowskie 4, 00-478 Warsaw, Poland*

Abstract. We show that in evolved stars, even at relatively low surface amplitudes, nonradial modes become strongly nonlinear in the hydrogen shell source, where the Brunt-Väisälä frequency has its absolute maximum. The measure of nonlinearity is the product of horizontal displacement times the radial wavenumber, $|\xi_H k_r|$. It becomes large already in evolved δ-Scuti stars. This nonlinearity presents a major problem for interpretations of amplitude modulation in RR Lyrae stars in terms of nonradial mode excitation.

Keywords: δ-Scuti stars, RR Lyrae stars

1. Introduction

All recent models of Blazhko-type modulation in RR Lyrae stars postulate departures from pure radial pulsation. In the oblique pulsator model (Shibahashi, 2000) the $\ell = 2$ component is induced by magnetic field. In the resonant model (Nowakowski and Dziembowski, 2001) low-ℓ modes are excited due to a resonant coupling with the radial mode. Amplitudes of postulated nonradial components are quite sizable, according to Kovács (2002) they reach up to 0.7 radial mode amplitudes.

Apart of RR Lyrae stars, nonradial oscillations with quite high amplitudes are observed in some evolved δ-Scuti stars. The best example is a post-MS star 4CVn, where several modes have been identified as those of $\ell = 1$ or 2 (Breger and Pamyatnykh, 2002).

The nonradial modes observed in RR Lyrae and evolved δ-Scuti stars are of a mixed character. While it is known that such modes at observed amplitudes are only weakly nonlinear in the p-mode cavity, the nonlinearity in the g-mode cavity, where most of the mode energy is concentrated, has never been studied carefully.

2. A Measure of Nonlinearity

We study the nonlinearity of the g-waves in the asymptotic approximation, which may be applied when the oscillation frequency is much smaller than the Brunt-Väisälä and Lamb frequencies. Then we have, approximately,

$$\vec{\xi} = A(r) \left(\cos \Phi(r) \vec{e}_r - \frac{rk_r}{\sqrt{l(l+1)}} \sin \Phi(r) \vec{\nabla}_H \right) Y_l^m(\theta, \phi) \exp(-i\omega t), \quad (1)$$

where $\vec{\xi}$ is the displacement, $A(r)$ is a slowly varying amplitude, $\Phi(r)$ is a rapidly varying phase,

$$k_r = \frac{d\Phi}{dr} = \frac{\sqrt{l(l+1)}}{r} \frac{N(r)}{\omega} \quad (2)$$

is the radial wavenumber, and $N(r)$ is the Brunt-Väisälä frequency. One can see that due to the high value of N/ω, the following strong inequalities take place:

$$k_r \gg \frac{\sqrt{l(l+1)}}{r} \equiv k_H, \quad |\vec{\xi}_H| \gg |\xi_r| \quad (\text{except near } \sin \Phi = 0). \quad (3)$$

Standard estimate of nonlinearity, i.e. comparing the $\partial \vec{v}/\partial t$ and $(\vec{v} \cdot \vec{\nabla})\vec{v}$ terms in the momentum equation, is correct providing that the curvature effect is included. Then we obtain

$$SN \equiv \max(|\vec{\xi}_H| k_r) \gtrsim 1 \quad (4)$$

as the criterion for the strong nonlinearity. This is different from that given by Kumar and Goodman (1996). The most precise way to obtain our criterion is to apply the asymptotic approximation (Eqs 1,2) to the amplitude expansion of the Hamiltonian. However, at least fourth order expansion is needed (see Van Hoolst, 1994, who considers the general case of the nonradial stellar oscillation).

3. Surface Amplitude at the Onset of the Strong Nonlinearity

As an application we considered the $\ell = 1$ and 2 modes in three models of evolved stars: a TAMS δ-Scuti star, a post-MS δ-Scuti star, and an RR Lyrae star. The two δ-Scuti models have $\log T_{\text{eff}} = 3.86$ and the RR Lyrae model has $\log T_{\text{eff}} = 3.84$.

Figure 1 shows the ratio of the Brunt-Väisälä frequency to the fundamental radial mode frequency in our models. The differences are striking. The maximum value of this ratio in the RR Lyrae model is at least by an order of magnitude larger than those in δ-Scuti models. This suggests that nonlinearity in the interior may be a greater problem in more evolved stars. To assess the problem we need an estimate of $\vec{\xi}_H$ from the observed surface amplitude. To this aim we performed linear non-adiabatic calculations for the selected three models focusing on linearly unstable modes of $\ell = 1$ and 2. In this way we derived bolometric light amplitudes corresponding to $SN = 1$ which we call the critical amplitudes.

The critical amplitudes for the TAMS star were found typically higher than 100 mmag which is well above the values observed in MS δ-Scuti stars. Thus the nonlinearity is not a problem in this case. For the post-MS model the critical

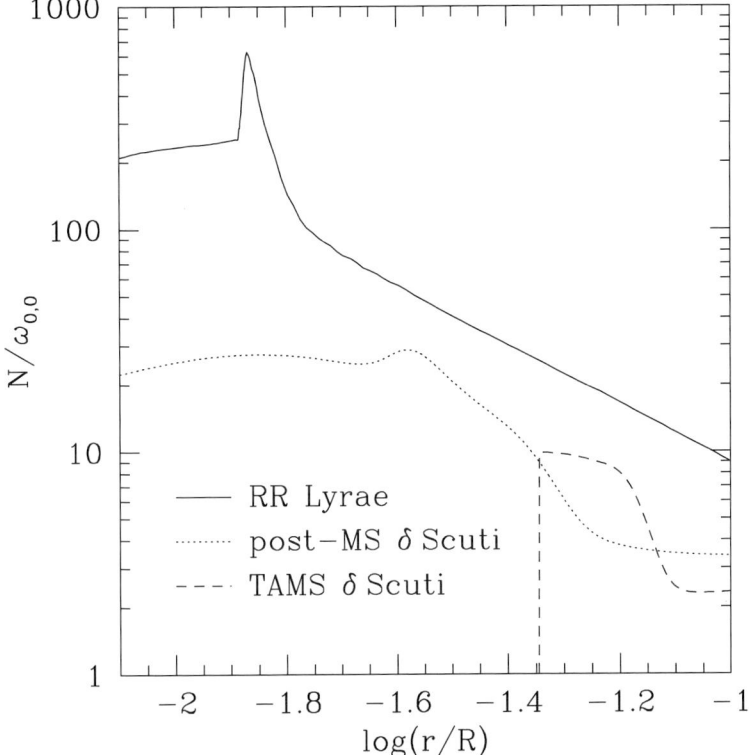

Figure 1. Ratio of the Brunt-Väisälä frequency to the fundamental radial mode frequency in three selected models.

amplitudes are much lower, often below 10 mmag. Some of the modes in 4CVn have observed amplitudes exceeding the critical values. Thus the nonlinearity in the deep interior may be a problem in this case.

In the case of RR Lyrae model, critical amplitudes, shown in Figure 2, are even lower. The highest values are found for $\ell = 1$ modes in the vicinity of the first overtone. But even in this case, the value of 0.02 mag is significantly lower than observed. We see that the critical amplitudes are much lower near the fundamental mode frequency and for the whole $\ell = 2$ sequence. It is thus clear that the observed close peaks in RR Lyrae stars cannot be interpreted in terms of linear nonradial eigenmodes.

The nonlinearity presents a problem for all models involving nonradial motion, whether it is due to a nonradial mode excitation or to a magnetically induced asphericity. A fully nonlinear treatment of the motion is needed if we want to model the observed amplitude modulation.

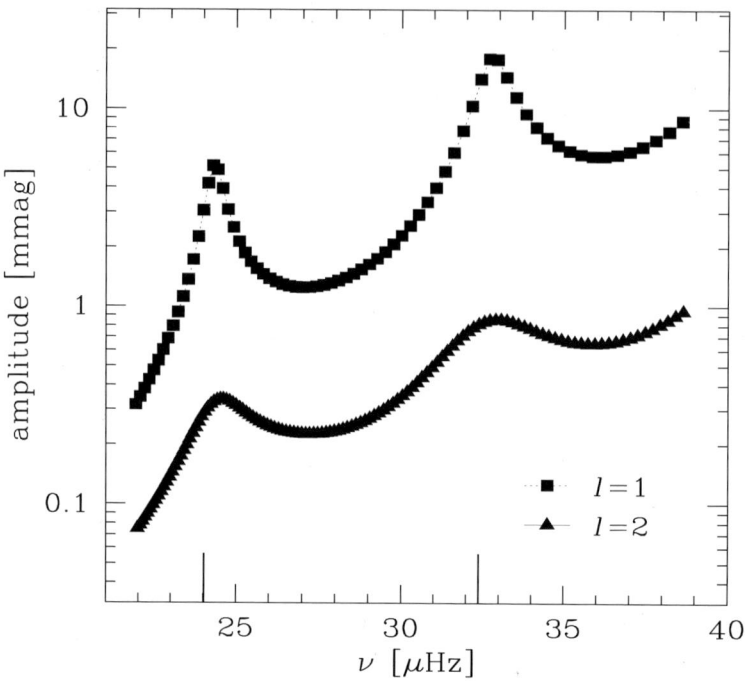

Figure 2. Critical amplitudes of the bolometric flux for nonradial modes in the RR Lyrae model. The long tickmarks denote frequencies of the lowest radial modes.

Acknowledgements

Our work is supported by the KBN grant No. 5 P03D 030 20.

References

Breger, M. and Pamyatnykh, A.A.: 2002: in: C. Aerts, T.R. Bedding and J. Christensen-Dalsgaard (eds.), *Radial and Nonradial Pulsations as Probes of Stellar Physics*, ASP Conf. Ser. **259**, 388.
Kovács, G.: 2002: in: C. Aerts, T.R. Bedding and J. Christensen-Dalsgaard (eds.), *Radial and Nonradial Pulsations as Probes of Stellar Physics*, ASP Conf. Ser. **259**, 396.
Kumar, P. and Goodman, J.: 1996, *ApJ* **466**, 946.
Nowakowski, R.M. and Dziembowski, W.A.: 2001, *Acta Astron.* **51**, 5.
Shibahashi, H.: 2000, in: L. Szabados and D. Kurtz (eds.), *The Impact of Large Scale Surveys on Pulsating Star Research*, ASP Conf. Ser. **203**, 299.
Van Hoolst, T.: 1994, *A&A* **286**, 879.

PROBLEMS, CONNECTIONS AND EXPECTATIONS OF ASTEROSEISMOLOGY: A SUMMARY OF THE WORKSHOP

JØRGEN CHRISTENSEN-DALSGAARD

Teoretisk Astrofysik Center, Danmarks Grundforskningsfond
Institut for Fysik og Astronomi, Aarhus Universitet, DK-8000 Aarhus C, Denmark

Abstract. The workshop took place at the beginning of what promises to be a golden age of asteroseismology. Ground-based instrumentation is finally reaching a level of stability which allows detailed investigations of solar-like oscillations in at least bright, slowly rotating main-sequence stars. Very extensive results are expected from the coming space missions, including data on a broad range of stars from the Eddington mission. The observational situation is therefore extremely promising. To make full use of these promises, major efforts are required towards the efficient utilization of the data, through the development of techniques for the analysis and interpretation of the data. A broad range of topics related to these issues is discussed in the present proceedings. Here I review some of the relevant problems, relate the asteroseismic investigations to broader areas of astrophysics and consider briefly the basis for our great expectations for the development of the field.

Keywords: asteroseismology – stars: structure – stars: evolution – stars: rotation

1. Introduction

Asteroseismology promises a major improvement in our understanding of the structure and evolution of stars. Observations of very accurate oscillation frequencies yield stringent constraints on stellar properties which, when combined with other types of observations, may provide information about the internal structure of the stars and the variation of rotation in stellar interiors. Also, comparison with carefully computed stellar models provides tests of the physics used in the model computation and hence information about the properties of matter under stellar conditions. The potential importance of this, in astrophysics and beyond, is obvious.

In the case of the Sun, such *helioseismic* investigations have been extremely successful in providing detailed and accurate information about the solar interior, thus testing stellar evolution theory and stellar physics to a very high level in the case of this one star (for a detailed review, see Christensen-Dalsgaard, 2002). Ever since the early days of helioseismology, the importance of extending seismic investigations to other stars has been evident. Indeed, very interesting results have been obtained for compact pulsators (planetary-nebula nuclei and white dwarfs) where extensive sets of frequencies of g modes have provided important information about the properties of these stars, including the thickness of their chemically separated outer layers (e.g. Metcalfe, these proceedings). However, a crucial but elusive goal has been to detect oscillations similar to those observed in the Sun, although necessarily restricted to low degrees. This has led to extensive efforts to reach the very high level of stability needed to observe velocity oscillations with amplitudes below 1 m s^{-1} or intensity oscillations of the order of parts per million. Despite some apparently positive results, the observational situation concerning solar-like oscillations in distant stars was until recently uncertain.

The present conference comes at a period where this situation is undergoing dramatic changes, as is evident from these proceedings. Further very substantial advances can be expected over the next few years, as ground-based observations are refined and space missions are undertaken. To make the fullest use of these possibilities the conference was intended as an opportunity to discuss how best to organize the observations, the analysis of the data and the modelling and interpretation that provides the desired physical insight. These goals of the conference were to a large extent fulfilled; in the present paper I present some of the key issues that were taken up.

Before so doing, however, a few highlights should be mentioned. As far as solar-like oscillations are concerned, the most striking development in the last couple of years has been the success in ground-based radial-velocity observations of solar-like oscillations in stars on and near the main sequence, reviewed by Bouchy (these proceedings) who, justifiably, gave prominence to the spectacular results obtained by the CORALIE instrument. These include the detection and detailed analysis of oscillations in α Cen A (Bouchy and Carrier, 2001, 2002; Ballot et al., these proceedings; Thoul et al., these proceedings; Thévenin et al., 2002), the subgiant δ Eri (Carrier et al., these proceedings), and the giant ξ Hya (Teixeira et al., these proceedings; Frandsen et al., 2002). From a personal point of view, however, perhaps the most striking result was the confirmation by Carrier et al. (these proceedings) of the oscillations in η Boo, first detected by Kjeldsen et al. (1995) through measurements of equivalent widths, but in some doubt since the failure by Brown et al. (1997) to see them in radial-velocity observations. With these results, and the observations also from the Anglo-Australian telescope (e.g., of β Hyi – Bedding et al., 2001, and α Cen A – Bedding et al., these proceedings), we now have a considerable material of solar-like oscillations over a range of stellar parameters, providing a first glimpse of the properties of these stars and a test of models of the excitation of the oscillations (see also Houdek and Gough, 2002). Indeed, the information about the variation of oscillation amplitudes with stellar parameters is invaluable for the planning of the upcoming space missions.

However, breakthroughs in asteroseismology are occurring across the HR diagram. As summarized by Kurtz (these proceedings) a long-standing mystery in the oscillation spectrum of the rapidly oscillating Ap star HR 1217 has been solved within the last year. In the observed spectrum of this star one mode did not satisfy the otherwise nearly uniform frequency spacing of approximately 34 μHz between adjacent modes, but rather differed from the previous mode by 50 μHz. Cunha (2001), based on detailed theoretical investigations by Cunha and Gough (2000), showed that this could be due the strong influence of the magnetic field and predicted the presence of an additional mode, not previously observed and possibly suppressed by magnetic effects. Further observations by Kurtz et al. (2002) have in fact triumphantly confirmed the presence of the mode, approximately at the predicted position. It is certainly satisfying that another puzzle in our understanding of pulsating stars has been solved; more importantly, however, this provides an opportunity for detailed investigations of the effects of strong magnetic fields on stellar oscillations.

In another striking development Green et al. (these proceedings) announced the detection of a new class of pulsating stars, near but somewhat cooler than the subdwarf B variables (sdBv, also known as EC14026 stars; see Fontaine et al., these proceedings), and with longer periods. Although the excitation mechanism of these longer-period stars has not yet been definitely established, it is striking that they bear a similar relationship to the sdBv as do the slowly pulsating B stars to the β Cephei stars, on the main sequence. This new class (which still needs an official designation, to replace the unofficial one widely used at the conference – see Fontaine, Brassard and Charpinet, these proceedings) will undoubtedly contribute to our understanding of late phases of stellar evolution.

At the other extreme of stellar sizes, although still in late stages of evolution, Bedding (these proceedings) presented results on semiregular variability of individual extreme red giants. These showed features similar to solar-like oscillations, although with periods of hundreds of days. This supports the analysis of a more statistical nature by Christensen-Dalsgaard et al. (2001) suggesting that semiregular variability has an origin similar to that of solar oscillations. Thus we may hope to obtain information about at least the excitation mechanism, and of some aspects of stellar structure, over a very broad range of stellar properties.

2. Problems

An important goal of the conference was to identify areas where further work is needed. This was certainly accomplished. However, rather than as problems, these areas should be regarded as challenges; and the very impressive work presented gave hope that these challenges are well on the way towards being met.

2.1. MODELLING

The interpretation of the observed frequencies in terms of the physics of stellar interiors relies on computation of stellar models. Since typically subtle effects are studied, it is important to ensure that the models faithfully represent the assumed input physics, with no significant effects of numerical errors. It is not clear that present stellar models meet this condition, even for the 'standard' calculation of spherically symmetric, non-rotating stars which form the basis of stellar-evolution theory. In the case of solar modelling, some efforts have been made to test numerical procedures through comparison of independently calculated models with the same, simplified physics; here good agreement has been achieved (Gabriel, 1991; Christensen-Dalsgaard and Reiter, 1995). However, these tests have so far not included effects of diffusion and settling, known to be highly significant at the level of helioseismic precision. Furthermore, models of other stars involve physical effects not significant in the solar case which substantially complicates the calculation. Particularly important, and troublesome, are the changes in composition associated with convective cores. In the early phases of evolution of models between 1.2 and 2 $M_\odot$ growing convective cores are typically obtained, leading to composition discontinuities, at least if diffusion is ignored. Also, the shrinking convective cores in later stages of central hydrogen burning or in more massive stars leave behind a steep gradient in the hydrogen abundance, of great importance to the oscillation properties of the star but often affected by irregularities caused by numerical inadequacies. It should also be kept in mind that numerical issues involve not only methods for solving the equations of stellar structure and evolution, but also, for example, the interpolation routines typically involved in the use of tabulated data on the equation of state or opacity. These numerical problems need to be reduced through a concerted effort, including detailed comparisons of independently calculated models.

The composition evolution associated with convective regions also leads to problems of a more physical nature: often the extent of a convective region depends on the composition profile, due to its effect on the opacity, energy-generation rate or equation of state, while the profile itself is determined by the extent of the convective mixing. Effects of this nature, often leading to what is described as 'semiconvection', have to be dealt with in the evolution of massive stars, although the physical basis for the prescriptions typically used is probably not entirely solid. Similar effects may occur at the onset of core convection during the evolution of moderate-mass stars; interestingly, they appear to affect models of α Cen A and β Hyi, at least for some choices of model parameters. Also, Bahcall et al. (2001) found evidence for effects of this nature caused by the gradient in the heavy-element abundance resulting from settling below the convective envelope of a 1 $M_\odot$ model evolved beyond solar age. Further uncertainties arise from the likely mixing beyond convective region resulting from convective overshoot, of an extent that cannot be determined theoretically *a priori*. As discussed by Demarque and Robinson (these proceedings), there is hope that sufficiently accurate frequency observations for stars where these effects are relevant might help elucidating the proper physical treatment of them.

The diagnostic use of stellar oscillation frequencies obviously requires reliable computation of frequencies for stellar models. Here I consider the multiplet frequencies ν_{nl}, corresponding to the spherically symmetric component of stellar structure, n being the radial order and l the degree. In the solar case there are substantial differences, increasing rapidly in magnitude with increasing frequency, between observed and computed frequencies. The nature of the differences strongly suggests that their causes are localized very near the solar surface. Indeed, this is a region where the physics of the model, and the physics of the oscillations, are uncertain: the structure of the star depends on the uncertain details of convective energy transport as well as on the, commonly ignored, dynamical effects of convection (often described as a turbulent pressure); also, the oscillations are nonadiabatic in this region and depend on the perturbations in the convective flux and turbulent pressure, yet the frequency calculations usually assume adiabatic oscillations and ignore all convective effects. Similar discrepancies must evidently be expected for all high-frequency modes, typical of solar-like oscillations, in stars with significant near-surface convection. Owing to the frequency dependence of the near-surface effects they also influence the so-called large frequency separation $\Delta\nu_{nl} = \nu_{n+1\,l} - \nu_{nl}$ between modes of the same degree and adjacent orders. Other surface effects on the frequencies might be induced by the large-scale distribution of active regions on the stellar surface, as is also observed in the solar case; since these depend on latitude their effect on the frequencies may depend on degree, thus also affecting the small separation $\delta\nu_{nl} = \nu_{nl} - \nu_{n-1\,l+2}$ (Gough, these proceedings; see also Dziembowski and Goode, 1997). This may have significant effects on the use of $\delta\nu_{nl}$ as an indicator of stellar evolution.

It is possible to reduce the discrepancy between the observed and computed frequencies in the solar case through modifications to the treatment of the near-surface layers or the atmosphere, although such modifications are often largely *ad hoc* and hence difficult to extrapolate to other stars. As discussed by Demarque and Robinson (these proceedings; see also Li et al., 2002) some reduction in the discrepancies can be achieved by using more sophisticated models of convection; similar results were obtained by Rosenthal et al. (1999) with a model based on an average of hydrodynamical simulations.

All stars rotate, yet effects of rotation are most often not included in stellar modelling. A relatively simple effect arises from the centrifugal force that modifies the equation of hydrostatic support. The spherically symmetric part of this can be taken into account by a simple modification of the effective gravitational acceleration. However, computation of stellar evolution with rotation must also take into account the variation of the distribution of angular momentum as the star ages. In the absence of transport of angular momentum in the stellar interior the rotation rate of the inner parts of the star generally increases during evolution, as a result of the contraction of these parts of the star, due to local conservation of angular momentum, whereas the rotation rate of the outer regions decreases as a result of the expansion. In the opposite extreme, assuming uniform rotation at any time, the increase in the overall moment of inertia leads to a decrease in rotation rate. Needless to say, the truth is likely intermediate between these two extremes. Also, the possible loss of angular momentum must be taken into account. Indeed, there is strong evidence that lower-mass stars, including the Sun, suffer a strong loss of angular momentum during main-sequence evolution, leading to the present state of slow solar rotation. In the solar case helioseismology has shown that the rotation rate in the interior is approximately uniform, indicating efficient rotational coupling between the interior and the convection zone on the time scale of solar evolution; however, we do not know the extent to which this is a general phenomenon.

The transport of angular momentum is intimately linked to the transport of elements caused by rotation. That such transport is required, at least in relatively massive stars, is immediately obvious from the fact that most stars appear to have 'normal' surface composition. As discussed by Vauclair (these proceedings) and Turcotte (these proceedings) selective settling and levitation cause dramatic changes in surface composition in stars with thin outer convection zones, unless other processes serve to counteract these effects. Mixing perhaps induced by rotational effects just below the solar convection zone is a likely explanation for the discrepancy between the helioseismically inferred sound speed in this region and that of normal, unmixed, solar models (Brun et al., 1999; Elliott and Gough, 1999). Also, mixing caused by rotation may bring products of nuclear reaction to the stellar surface (for a review, see Maeder and Meynet, 2000). The physics of rotational mixing is complex and not completely understood, although instabilities related to meridional circulation are likely to play an important part (e.g., Zahn, 1992; Maeder and Zahn, 1998). An interesting treatment of the influence of composition gradients on the instabilities, dealt with in a two-dimensional modelling, was discussed by Théado and Vauclair (these proceedings).

Rotation lifts the degeneracy of the frequencies in azimuthal order m. For slow rotation, such as in the solar case, there is a simple linear relation between the rotation rate and this rotational splitting. However, as discussed by Pamyatnykh (these proceedings) even moderate rotation causes substantial departures from this simple pattern, as a result of the centrifugal distortion of the star, as well as higher-order effects of rotation on the oscillations which include coupling between modes. Although procedures have been established to calculate these effects, they substantially complicate the frequency patterns and hence attempts to identify the modes based on their frequencies. Also, as discussed below, the perturbations to the eigenfunctions arising from rotation must be taken into account in spectroscopic or photometric mode identification.

The treatment of convection remains a serious concern in the calculation of pulsation frequencies and stability coefficients, as already alluded to. However, for stochastically excited oscillations it is at least possible to estimate the energy input to the oscillations from convection; this has been quite successful in the solar case, on the basis of hydrodynamical simulations (Stein and Nordlund, 2001), and similar calculations for other stellar parameters would be very interesting. In the case of oscillations driven by opacity mechanisms many aspects of the regions of instability are well understood (Pamyatnykh, these proceedings), although the role played by convection requires further study. On the other hand, very little is known about the mechanisms that control the final mode amplitudes, and hence the selection of modes that reach observable amplitudes, amongst the typically much larger set of modes predicted to be unstable. Nonlinear mode coupling is likely to play a major role. As discussed by Nowakowski and Dziembowski (these proceedings) strong nonlinearity in the region of the hydrogen shell source in evolved stars may cause serious problems, particularly in the case of nonradial oscillations in RR Lyrae stars.

2.2. OBSERVATIONS

As discussed in the Section 1 the observational situation of asteroseismology has improved greatly and is set to improve even further as a result of the upcoming space projects. It was noted by Kurtz (these proceedings) that in this situation some coordination is required, in order not to waste effort, for example, on ground-based studies that will be superseded in a few years by observations from space. On the other hand, there is no doubt of the continuing need for ground-based observations: space is a very valuable resource which should be used only where appropriate, and ground-based work offers, for example, more extended observations of selected objects. Even so, it is essential to make the optimal use of the ground-based efforts by coordinating the observations and data analysis; a particularly important example is the organization of observations from two or more sites to suppress the problems caused by data gaps. Also, as pointed out by Kjeldsen (these proceedings) observing from space does not solve all problems and adds new ones: these include effects of radiation on the detectors, field crowding and the need to carry out analysis on the satellite, possibly using less than optimal algorithms as a result of the limitations of onboard computing resources. Also, there may be cases where ground-based observations provide data that are better than those available from space: this could be the case for sufficiently stable radial-velocity observations, for a slowly rotating star, since the intrinsic stellar noise is generally smaller in radial velocity than in intensity (Harvey, 1988).

Despite the general need for more and better data, there is a sense in which we are already suffering from a data glut, for example from the large-scale projects to detect micro-lensing (Alcock et al., 1997, 1999; Udalski et al., 1999). This is exemplified by the exciting results presented by Alonso et al. (these proceedings) based on the STARE project designed to search for extra-solar planets. Ground-based projects like this, and even more dramatically space-based photometric exo-planet searches from Eddington and Kepler, will produce massive amounts of photometric data on all sorts of stellar variability. It is crucial that effective techniques are developed for the analysis of these data, and that scientific procedures be developed that make the most efficient use of them.

2.3. DATA ANALYSIS

The observations yield time series of the observed quantities (such as intensity, Doppler velocity, etc.) and the first step in the analysis is typically to identify the frequencies from these time series, through Fourier analysis and subsequent fitting. Appourchaux (these proceedings) discussed techniques for so doing for solar-like oscillations, where the statistical properties resulting from the stochastic excitation must be taken into account, as is done in the solar case. Based on hare-and-hounds tests on artificial data, and on the solar experience, he concluded that the techniques, based for example of Maximum Likelihood Estimation, are well in hand.

A key problem in the data analysis, much discussed during the workshop, is the identification of the observed modes, particularly the determination of their degree l and azimuthal order m which in principle is reflected in the relation between amplitudes and phases obtained with different types of observations. Given a determination of l and m, the radial order can be determined with some confidence through fits to stellar models. In simple cases, such as solar-like oscillations in a relatively unevolved star, where most of the modes in a given frequency range may be assumed to be observed, and with slow rotation so that the multiplets are well separated, the modes may be identified from the frequency pattern, as was indeed done in the solar case in the early days of helioseismology. At the other extreme, for opacity-driven pulsators only a small subset of the possible modes are typically seen and, in the common case of rapid rotation for, e.g., δ Scuti or β Cephei stars, the rotational effects may be larger than the separation between multiplets. In this case, identification of at least some of the modes is indispensable for proper asteroseismic analysis.

Reed (these proceedings) discussed mode identification using multi-colour photometry; this is based on the way that amplitude ratios and phase differences between observations in different wavebands depend on the properties of the modes. The technique goes back to an analysis of the geometrical properties of the modes by Dziembowski (1977) and was presented in considerable detail by Watson (1988), now forming the basis for most discussions of these procedures. In their simple implementation, the effects of the oscillations on the emitted radiation of the star are estimated by varying static stellar atmospheres, and furthermore assume simplified relations between the perturbation quantities in the stellar atmosphere. This is likely inadequate, particularly at high frequencies (close to the acoustical cut-off frequency) where the detailed properties of the oscillations in the atmosphere must be taken into account. This requires a full non-adiabatic calculation, including treatment of radiative effects (although a simplified procedure was presented by Dupret et al., 2002; see also Dupret et al., these proceedings). In cool stars effects of convection may be important; some progress may be made on including them through mixing-length formalisms for pulsating stars (e.g. Houdek et al., 1995), but full hydrodynamical simulations would also be very interesting. As discussed by Daszyńska-Daszkiewicz et al. (these proceedings; see also Daszyńska-Daszkiewicz et al., 2002), further complications arise in rapidly rotating stars as a result of rotationally induced perturbations to the eigenfunctions, particularly in cases of strong coupling between modes of differing degree.

Spectroscopic techniques were discussed by Telting (these proceedings) and Balona (these proceedings). Here mode identification may be achieved either from analysis of moments of the spectral line profiles or from detailed fits to the spectral lines. These techniques offer the exciting possibility of observing modes of relatively high degree through Doppler imaging in rapidly rotating stars (see also Kennelly et al., 1998). Again, a full analysis of the observations requires realistic modelling of the oscillations in the stellar atmosphere; however, since the effects on the spectral lines are probably dominated by the radial velocity, the interpretation is likely less sensitive to the uncertainties in the treatment of nonadiabatic effects than in the case of the photometric techniques (e.g., De Ridder et al., 2002). Rotational effects on the eigenfunctions, including mode coupling, are surely also important for the spectroscopic techniques.

There seems to be a tendency to treat these different techniques for mode identification separately. In reality, it seems likely that a combination of the techniques might provide a more reliable identification. This could either be achieved through the 'sanity check' that different techniques yield the same result, or by combining the data into one fit. Such a fit, ideally, should also take into account uncertainties in the modelling of the stellar atmosphere and the atmospheric oscillations, as well as the effects of stellar rotation. Furthermore, as emphasized by De Ridder et al. (these proceedings) a proper statistical analysis of the results is required.

2.4. DATA INTERPRETATION

Given observational determination of the oscillation frequencies, the goal is evidently to obtain as detailed information as possible about the star. In the case of oscillations similar to those of the Sun information about the overall properties of the star can be obtained from the large and small frequency separations $\Delta \nu_{nl}$ and $\delta \nu_{nl}$ (Gough, these proceedings), even without a detailed mode identification. However, the goal is ideally to reach a level where the properties of the physics of the stellar interior can be constrained. This typically requires detailed identification of the observed modes.

In general, the basic stellar parameters are not sufficiently well constrained by 'classical' observations, and hence determination of these parameters is part of the analysis of the frequencies. Thus a fit has to carried out in a space of several, but typically relatively few, parameters which characterize both global properties such as mass and age and selected internal properties of the star; the data should comprise both frequencies and other observed properties of the star, such as effective temperature and luminosity. A fit of this nature typically has several local optima, and hence special techniques are required to ensure that the best fit is found. Metcalfe (these proceedings) discussed genetic algorithms which have been employed with considerable success in the analysis of frequencies of white dwarfs (e.g., Metcalfe et al., 2000). An interesting possibility is to include the mode identification, in terms of radial order and, to the extent that it has not been determined independently, degree and azimuthal order as part of the fit. Indeed, it may be possible to carry out a 'global' fit that includes also those observed properties which, as discussed in Section 2.3, constrain the mode identification;

the fit could also include the global stellar parameters required in the modelling of the atmosphere, as well as the rotation properties of the star.

Fits of this nature only represent a first cut at the utilization of the oscillation data, and their success or otherwise in matching the observations obviously depends on the underlying model and parameter space. In fact, the most interesting outcome of the fit would be to find significant departures from the data, indicating the need to improve the physics of the underlying model or change some of the assumed properties of it. A very interesting example is the case of the δ Scuti star XX Pyx for which extensive sets of observed frequencies are available; also, the star is believed to be relatively unevolved and hence comparatively simple to model. However, in a very extensive series of model calculations Pamyatnykh et al. (1998) were unable to find a model and mode identification which provided a satisfactory fit. The recently found explanation of the discrepancy is dramatic: as shown by Aerts et al. (2002), XX Pyx is member of a close binary system and is strongly tidally distorted; thus it can evidently not be matched by the previously assumed spherically symmetric models. Instead, it offers a possibility for investigating the complex phenomena, related to 'quantum chaos', that are associated with oscillations of distorted stars (Perdang, 1986).

In helioseismology, linearized structure inversion has been very successful in determining the differences between solar structure and the structure of solar models. Applications to asteroseismology were discussed by Basu (these proceedings). The technique assumes that a reference model is available which is sufficiently close to the actual stellar structure that the frequency differences between the star and the model can be linearized in terms of the structure differences. The resulting linear relations can then be analyzed to determine from the frequency differences aspects of the differences in structure. A very substantial advantage of these techniques is that they provide direct information about the resolution, in terms of the so-called *averaging kernels*, and error properties of the inferences. As shown by Basu, even with the relatively limited sets of modes available for distant stars, and assuming realistic observational errors, it is possible to obtain localized information about the structure of stellar cores. However, the assumptions of linearity, amply satisfied by the small corrections typically found in the solar case, are more questionable for stars whose global parameters are less well constrained. An efficient analysis technique is likely to combine parameter fits and inverse analyses, using the former to obtain a reference model sufficiently close to the actual properties of the star and the latter to determine the corrections required, given the limited space of parameters used in the fit. As discussed by Roxburgh and Vorontsov (these proceedings) one may also use a non-linear, iterative technique where the structure of the model is recomputed based on the result of the inversion; their analysis is based on the phase of the oscillations, at the observed frequencies, in the outer parts of the model where the phase can be assumed to be a function of frequency alone.

Although several techniques are available to analyze asteroseismic data, we have barely scratched the surface of testing and optimizing them, given the wide variety of oscillation modes and stellar properties that will be involved in the investigations. As with helioseismology, different analysis techniques will undoubtedly be found to have advantages for different scientific questions. However, as a general procedure 'hare and hounds' tests should be carried out, based on realistic artificial data and with the analysis carried out without information about the underlying model beyond what would be available from the expected observations. To simulate realistically effects of observational errors these tests should ideally be based on simulated time strings, computed taking into account knowledge about the excitation mechanism for the oscillations, and analyzed with procedures corresponding to those used for real data. Tests of this nature are an important part of the preparations for the upcoming asteroseismic space missions (Berthomieu et al., these proceedings).

3. Connections

The goals of asteroseismology obviously extend beyond the observation and analysis of stellar oscillation frequencies: the true goals must be to use the information so obtained to improve our understanding of stellar structure and evolution, and of the physics of stellar interiors. In this broader sense asteroseismology must be regarded as a tool in the astrophysical toolbox, much as is (radiative) spectroscopy.

3.1. THE SOLAR-STELLAR CONNECTION

The prospects for asteroseismology must unavoidably be seen in the context of the results already obtained with helioseismology. As a seismic target the Sun stands out in several respects. There is an obvious difference in terms of access to modes of all degrees and high azimuthal order, leading to the ability to carry out very detailed inversion as a function of radius and latitude in most of the Sun. Another important advantage of the Sun is the availability of accurate determinations of the solar mass, radius and luminosity, and a good estimate of the solar age. Furthermore, the solar surface composition, relative to the hydrogen abundance, is generally taken to be known with substantial precision, although recent results (Allende Prieto et al., 2002a) cast some doubt on this assumption. The information about the global properties of the Sun provides stringent constraints on solar models, substantially enhancing the level of detail to which the observed frequencies can be used to probe the physics of the solar interior.

For the distant stars the global information is typically far less accurate than for the Sun. Indeed, one of the important uses of asteroseismology is precisely to constrain the global properties, for example stellar ages. Exceptions are provided by well-observed binaries where precise determinations of stellar masses and, for eclipsing binaries, radii, are available; here, also, the shared age and chemical composition of the system further constrain the properties of the stars. Even stronger constraints are available for stellar clusters whose members may also be assumed to have the same composition and age, the latter being reflected in the location of the turn-off from the main sequence in the Hertzsprung-Russell diagram. Also, the approximately common distance to the members of the cluster reduces the uncertainties associated with the generally poorly determined stellar distances. In such cases, at least, we can hope to learn about failures in our modelling of stellar interiors and how to improve it.

Compared with helioseismology, the main strength of asteroseismology is evidently the ability to investigate properties of stars of very different nature, across the HR diagram. In many ways, it is difficult to imagine a star simpler than the Sun: it is relatively unevolved, it has no convective core complicating the modelling of the evolution of the abundance profile, it rotates comparatively slowly, and the physical conditions, as far as the microphysics is concerned, are rather benign with, e.g., only modest non-ideal effects in the equation of state. In this sense the Sun represents an ideal point of departure for our seismic investigations of stellar interiors. However, it is evidently crucial now to address the complications encountered in other types of stars.

The properties of solar oscillations clearly also form the basis for our expectations for solar-like oscillations in other stars. This extrapolation from a single case, not atypical for astrophysics, has to a large extent been confirmed by the recent detections of solar-like oscillations, although the observed amplitudes are not always in close agreement with expectations. Thus, the solar data provide an excellent basis for testing analysis methods to be applied to other, similar stars. An example is provided by Roxburgh and Vorontsov (these proceedings) who carried out a fit for the mass and central hydrogen abundance based on solar low-degree frequencies from the BiSON group, increasing the errors to make them more representative of the expected stellar data. Also, Kosovichev (1993) made a linearized inversion of low-degree solar data to infer the depth of the solar convection zone and the envelope helium abundance. Analysis of solar data, starting from time series corresponding to what will be observed for other stars, and comparing with the results obtained from the full data available for the Sun, would evidently be very valuable. It would provide a realistic assessment of techniques for frequency determination and the effects of frequency errors on the inferences on stellar properties. Also, it may be possible to obtain solar data of a nature that would allow tests of at least some of the proposed methods of mode identification; an important example are the intensity observations in several wavelength bands from the VIRGO instrument on the *SOHO* spacecraft (e.g., Fröhlich et al., 1997).

Clearly, the insight into stellar physics obtained from the Sun will be crucial to interpreting the asteroseismic data. An important example are the near-surface effects on the oscillation frequencies, which can perhaps only be studied in detail in the solar case. By obtaining an understanding of the causes of these effects in the Sun we may have a better chance to correct for them in other stars, through a combination of modelling and observations.

3.2. GLOBAL STELLAR PARAMETERS

As discussed above, the usefulness of asteroseismic data depends crucially on the availability of other information about the stars, obtained through 'classical' observations. Photometry and spectral analysis provide information about the apparent magnitude, surface gravity and effective temperature of the star, assuming calibrations based on stellar atmospheric modelling or other observations. Given also a determination of the stellar parallax, the apparent magnitude can be converted into an absolute luminosity; interestingly, with the accurate parallax determinations from, e.g., the *HIPPARCOS* satellite, the main uncertainty in the determination of the luminosity of nearby stars is probably now the bolometric correction, relating the observed magnitude to the total energy output, rather than the parallax. High-resolution spectroscopy, combined with atmosphere modelling, gives information about the composition of the stellar atmosphere. This is required as input to the stellar modelling. Also, the abundances provide crucial information about processes, such as settling or mixing, which affect the composition profiles in the stellar interiors.

The observational basis for these investigations has been substantially improved in recent years, with the introduction of stable high-resolution spectrographs on large telescopes which allow detailed abundance investigations even on faint stars. Also, the use of interferometry, e.g., with VLTI on the European Southern Observatory Very Large Telescope, will provide information on stellar radii for a broad range of stars. Finally, much improved parallax determinations will eventually become available with the Gaia mission to be launched by ESA before 2012; this will result in accurate parallaxes also of rather distant and intrinsically bright targets for asteroseismology such as β Cephei stars, as well as for the fainter stars to be observed with the Eddington mission.

Perhaps even more important than these observational advances is the development of more realistic modelling of stellar atmospheres. Traditional stellar-atmosphere models are typically assumed to be plane parallel and static, despite the unquestionable presence of strong time-dependent flows and inhomogeneities. Some effects of these flows are included in terms of parameters such as 'micro turbulence', adjusted to obtain the correct spectral line profiles. This evidently may give rise to systematic errors in the inferences from the models, seen, for example, in the dependence for a given star of the inferred abundances on the choice of spectral lines. However, realistic hydrodynamic simulations of stellar convection, including radiative effects, now yield the observed spectral line profiles without the introduction of additional parameters (Asplund et al., 2000), and they may be expected to lead to substantially more reliable inferences of stellar abundances and other parameters (e.g. Allende Prieto et al., 2002b). Extensive calculations of this nature are required to obtain more reliable spectral diagnostics, including temperature and gravity calibrations, bolometric corrections and abundances, for a broad range of stellar parameters.

3.3. IMPROVEMENTS TO STELLAR MODELLING FROM ASTEROSEISMOLOGY

The main connection of asteroseismology is evidently to the general theory of stellar structure and evolution. Two aspects were given emphasis during the meeting: the treatment of stellar convection and the effects of stellar rotation.

Demarque and Robinson (these proceedings) reviewed the treatment of convection and the possibilities of asteroseismic diagnostics, emphasizing the seismic signatures of regions of transition between convective and radiative energy transport. This includes the possibilities for determining the depth of convective envelopes (see also Monteiro et al., 2000), and investigations of the properties of convective cores, including penetration beyond the edge of instability. Constraints on convective cores, and possibly other processes such as 'semiconvection' in massive stars, are of potentially great importance to the study of stellar evolution, including the processes involved in supernova explosions of massive stars. Demarque and Robinson also emphasized the uncertainties in the treatment of near-surface convection and the possibilities for studying these regions through direct hydrodynamical simulations. A very promising procedure is to parametrize the results of such simulations for use in general calculations of stellar evolution and oscillations. Sufficiently accurate asteroseismic determinations of properties of convective envelopes could be a powerful test of such calculations, as has already been applied in the solar case (e.g., Rosenthal et al., 1999). They also noted that, in addition to the oscillation frequencies, the rate of energy input to stochastically excited modes could be an important diagnostics of convection properties. There is obviously a close relation between such studies of convection and those involving (radiative) spectral analysis, discussed above.

The potential importance of inferences of stellar rotation was reviewed at the workshop by Pinsonneault. He emphasized that the rotation of a given star depends directly on the evolutionary history of the star, from its formation through the processes that lead to angular-momentum loss or redistribution within the star. Thus asteroseismic determinations of stellar internal rotation, particularly in the early phases of stellar evolution, may provide information about the processes that led to the formation of the star. Although we cannot expect anything like the detailed determination of internal rotation that has been possible in the Sun, even a few seismic estimates of average rotation rates, combined with the surface rate determined from intensity variations associated with star spots, would provide some idea about the possible variations of rotation through the star. Also, as noted by Pijpers (these proceedings) one may obtain a reliable estimate of the total angular momentum of a star, of substantial importance to investigating its rotational history, from measurement of rotational splittings of just low-degree modes. It is essential, however, to combine the asteroseismic inferences with other information relevant to an understanding of the evolution of stellar rotation. Detailed and accurate determinations of stellar surface abundances provide clues to the extent of mixing, likely rotationally induced, in stellar interiors. Also, rotation of evolved stars reflects their rotational history; for example, a possibly rapidly rotating core may be reflected in the surface rotation as angular momentum is redistributed by deep penetration of the outer convection zone in the red-giant phase, or may be revealed by extensive mass loss. Interestingly, Sills and Pinsonneault (2000) found evidence that the observed angular velocity of horizontal-branch stars in the globular cluster M13 indicated the presence of previously rapidly rotating cores in these stars, which may be at variance with the near-uniform rotation found in the solar interior. There is a strong variation in the surface rotation along the horizontal branch (e.g., Behr et al., 2000), cooler stars rotating relatively rapidly and hotter stars rotating very slowly. The variation in rotation appears to be correlated with a variation in the surface abundance, with the hotter stars showing strong enhancement in the heavy-element abundance (e.g., Grundahl et al., 1999); this may result from selective radiative levitation, as originally predicted by Michaud et al. (1983). A review of these results was given recently by Behr (2002). The physical relation between the abundance and rotation variations remains uncertain. Evidently, any information from asteroseismology about the interior rotation of horizontal-branch stars would be very valuable.

Stellar microphysics, i.e., the equation of state, opacity, etc., remains an area of considerable uncertainty and scientific interest. It is obvious that helioseismology is limited to testing this under the conditions found in the Sun; studies of other stars will very substantially extend this range. As an example of the power of even restricted data to provide crucial information, one may recall the prediction by Andreasen and Petersen (1988), based on periods of double-mode Cepheids and δ Scuti stars, of a substantial increase in the opacity, subsequently confirmed by the OPAL calculations (Rogers and Iglesias, 1992; see also Moskalik et al., 1992).

4. Expectations

It should be clear from this brief review, and even more from these proceedings, that asteroseismology is developing rapidly, with impressive results having already been obtained, and hopes being high and well-founded for the future development. The observational situation has improved dramatically through the recent radial-velocity observations of solar-like oscillations, and important contributions to the study of other types of stars have been made by extensive campaigns of coordinated multi-site observations. Further ground-based instrumentation will become available in the near future, including the HARPS spectrograph on the ESO 3.6 m telescope at the La Silla Observatory (Queloz et al., 2001). Also, one may expect an increasing role for automated photometric telescopes, thus alleviating the manpower problem in obtaining the long series of observations that are required. As emphasized by Kurtz (these proceedings) such extensive ground-based efforts require careful thought about the choice of targets, but there seems little doubt that they will remain an important part of asteroseismology.

A dramatic increase in the available data for asteroseismology will come with the space missions scheduled to be launched over the coming years (see Kjeldsen, these proceedings), starting with the Canadian MOST mission in the spring of 2003, and culminating with the ESA Eddington mission in 2007 or 2008 (Roxburgh and Favata, these proceedings). These will provide extensive and accurate data on a wide variety of stars including, from Eddington, solar-like oscillations for stars in open clusters, providing tight constraints on the internal properties of the stars.

Stellar modelling also shows considerable promise for improvement. Hydrodynamical calculations are addressing key uncertainties in the treatment of near-surface convection and may provide insight into other aspects of stellar hydrodynamics, including penetration beyond convective regions and effects of rotation. Evolution calculations including rotation, albeit in a somewhat approximate fashion, are providing very interesting results on the transport of chemical elements, allowing detailed comparisons with the observed surface abundances (Maeder and Meynet, 2000). At a more mundane level there remains the important task of studying the numerical accuracy of calculations of stellar models and oscillation frequencies.

The prospects for improved observations do not end with the present generation of space projects, however promising they are. In the solar case, much has been learned from very extended observations of low-degree modes, providing data with high accuracy on modes with relatively low frequency and yielding information about frequency variations associated with the solar cycle. Similar observations are certainly highly desirable for other stars; this might be achieved with a number of small satellites, perhaps similar to the Danish Rømer satellite, or a single mission with several telescopes, each observing one star for an extended period. Also, the restriction to low-degree modes is evidently a serious limitation for the presently planned observations. In a distant future (well beyond my retirement) interferometric missions will perhaps allow observations with sufficient spatial resolution to resolve oscillations in distant stars of degrees as high as 50, thus for example allowing some resolution of rotation in the region near the base of the convection zone in a star like the Sun; this would test whether the transition in rotation observed in the solar tachocline is a common phenomenon.

Whatever the details of this development, there is no doubt that the comparison between observations and models will reveal new and unexpected aspects of stellar structure and evolution. As stated optimistically by Douglas Gough during the conference, matters will be much more complicated, and hence interesting, than we currently expect.

It is evident that our expectations for the development of asteroseismology are indeed great. As shown by Dickens (1861) great expectations sometimes yield unexpected outcomes, under the best circumstances including insight into oneself and one's Estella. In the present case, unlike Dickens' Pip our expectations are surely not for riches (although possibly for fame); but we may certainly hope for a much improved insight into not one Estella, but many *estrelas*.

Acknowledgements

I am very grateful to F. Grundahl for a useful conversation on the properties of horizontal-branch stars. The preparation of this review was supported in part by the Danish National Research Foundation through the establishment of the Theoretical Astrophysics Center.

References

Aerts, C., Handler, G., Arentoft, T., Vandenbussche, B., Medupe, R. and Sterken, C.: 2002, *MNRAS* **333**, L35.
Alcock, C., Allsman, R.A., Alves, D., Axelrod, T.S., Becker, A.C., Bennett, D.P., Cook, K.H., Freeman, K.C., Griest, K., Guern, J., Lehner, M.J., Marshall, S.L., Minniti, D., Peterson, B.A., Pratt, M.R., Quinn, P.J., Rodgers, A.W., Sutherland, W. and Welch, D.L.: 1997, *ApJ* **482**, 89.
Alcock, C., Allsman, R.A., Alves, D., Axelrod, T.S., Becker, A.C., Bennett, D.P., Cook, K.H., Freeman, K.C., Griest, K., Lehner, M.J., Marshall, S.L., Minniti, D., Peterson, B.A., Pratt, M.R., Quinn, P.J., Rodgers, A.W., Rorabeck, A., Sutherland, W., Tomaney, A., Vandehei, T. and Welch, D.L.: 1999, *ApJ* **511**, 185.
Allende Prieto, C., Asplund, M., García López, R.J. and Lambert, D.L.: 2002b, *ApJ* **567**, 544.
Allende Prieto, C., Lambert, D.L. and Asplund, M.: 2002a, *ApJ* **573**, L137.
Andreasen, G.K. and Petersen, J.O.: 1988, *A&A* **192**, L4.
Asplund, M., Nordlund, Å , Trampedach, R., Allende Prieto, C. and Stein, R.F.: 2000, *A&A* **359**, 729.
Bahcall, J.N., Pinsonneault, M.H. and Basu, S.: 2001, *ApJ* **555**, 990.
Bedding, T.R., Butler, R.P., Kjeldsen, H., Baldry, I.K., O'Toole, S.J., Tinney, C.G., Marcey, G.W., Kienzle, F. and Carrier, F.: 2001, *ApJ* **549**, L105.
Behr, B.: 2002, in: Piotto, G., Meylan, G., Djorgovski, G. and Riello, M. (eds), *New Horizons in Globular Cluster Astronomy*, ASP Conf. Ser., in the press.
Behr, B.B., Djorgovski, S.G., Cohen, J.G., McCarthy, J.K., Côté, P., Piotto, G. and Zoccali, M.: 2000, *ApJ* **528**, 849.
Bouchy, F. and Carrier, F.: 2001, *A&A* **374**, L5.
Bouchy, F. and Carrier, F.: 2002, *A&A* **390**, 205.
Brown, T.M., Kennelly, E.J., Korzennik, S.G., Nisenson, P., Noyes, R.W. and Horner, S.D.: 1997, *ApJ* **475**, 322.
Brun, A.S., Turck-Chièze, S. and Zahn, J.P.: 1999, *ApJ* **525**, 1032. (Erratum: *ApJ* **536**, 1005).
Christensen-Dalsgaard, J.: 2002, *Rev. Mod. Phys.*, **74**, 1073.
Christensen-Dalsgaard, J. and Reiter, J.: 1995, in: Ulrich, R.K., Rhodes Jr, E.J. and Däppen, W. (eds), *Proc. GONG'94: Helio- and Astero-seismology from Earth and Space*, ASP Conf. Ser. **76**, 136.
Christensen-Dalsgaard, J., Kjeldsen, H. and Mattei, J.A.: 2001, *ApJ* **562**, L141.
Cunha, M.S.: 2001, *MNRAS* **325**, 373.
Cunha, M.S. and Gough, D.: 2000, *MNRAS* **319**, 1020.
Daszyńska-Daszkiewicz, J., Dziembowski, W.A., Pamyatnykh, A.A. and Goupil, M.-J.: 2002, *A&A* **392**, 151.

De Ridder, J., Dupret, M.-A., Neuforge, C. and Aerts, C.: 2002, *A&A* **385**, 572.
Dickens, C.: 1861, *Great Expectations*, Chapman and Hall, London.
Dupret, M.-A., De Ridder, J., Neuforge, C., Aerts, C. and Scuflaire, R.: 2002, *A&A* **385**, 563.
Dziembowski, W.A.: 1977, *Acta Astron.* **27**, 203.
Dziembowski, W.A. and Goode, P.R.: 1997, *A&A* **317**, 919.
Elliott, J.R. and Gough, D.O.: 1999, *ApJ* **516**, 475.
Frandsen, S., Carrier, F., Aerts, C., Stello, D., Maas, T., Burnet, M., Bruntt, H., Teixeira, T.C., de Medeiros, J.R., Bouchy, F., Kjeldsen, H., Pijpers, F. and Christensen-Dalsgaard, J.: 2002, *A&A* **394**, L5.
Fröhlich, C., Andersen, B.N., Appourchaux, T. et al.: 1997, *Solar Phys.* **170**, 1.
Gabriel, M.: 1991, in: Gough, D.O. and Toomre, J. (eds), *Challenges to theories of the structure of moderate-mass stars*, Lecture Notes in Physics **388**, p. 51, Springer, Heidelberg.
Grundahl, F., Catelan, M., Landsman, W.B., Stetson, P.B. and Andersen, M.I.: 1999, *ApJ* **524**, 242.
Harvey, J.W.: 1988, in: Christensen-Dalsgaard, J. and Frandsen, S. (eds), *Proc. IAU Symposium No 123, Advances in helio- and asteroseismology*, p. 497, Reidel, Dordrecht.
Houdek, G. and Gough, D.O.: 2002, *MNRAS* **336**, L65.
Houdek, G., Balmforth, N.J. and Christensen-Dalsgaard, J.: 1995, in: Hoeksema, J.T., Domingo, V., Fleck, B. and Battrick, B. (eds), *Proc. Fourth SOHO Workshop: Helioseismology*, ESA SP-376, vol. 2, ESTEC, Noordwijk, p. 447.
Kennelly, E.J., Brown, T.M., Kotak, R., Sigut, T.A.A., Horner, S.D., Korzennik, S.G., Nisenson, P., Noyes, R.W., Walker, A. and Yang, S.: 1998, *ApJ* **495**, 440.
Kjeldsen, H., Bedding, T.R., Viskum, M. and Frandsen, S.: 1995, *AJ* **109**, 1313.
Kosovichev, A.G.: 1993, *MNRAS* **265**, 1053.
Kurtz, D.W., Kawaler, S.D., Riddle, R.L. et al.: 2002, *MNRAS* **330**, L57.
Li, L.H., Robinson, F.J., Demarque, P. and Sofia, S.: 2002, *ApJ* **567**, 1192.
Maeder, A. and Meynet, G.: 2000, *ARA&A* **38**, 143.
Maeder, A. and Zahn, J.-P.: 1998, *A&A* **334**, 1000.
Metcalfe, T.S., Nather, R.E. and Winget, D.E.: 2000, *ApJ* **545**, 974.
Michaud, G., Vauclair, G. and Vauclair, S.: 1983, *ApJ* **267**, 256.
Monteiro, M.J.P.F.G., Christensen-Dalsgaard, J. and Thompson, M.J.: 2000, *MNRAS* **316**, 165.
Moskalik, P., Buchler, J.R. and Marom, A.: 1992, *ApJ* **385**, 685.
Pamyatnykh, A.A., Dziembowski, W.A., Handler, G. and Pikall, H.: 1998, *A&A* **333**, 141.
Perdang, J.: 1986, in: Gough, D.O. (ed.), *Seismology of the Sun and the distant Stars*, p. 141, Reidel, Dordrecht.
Queloz, D., Mayor, M., Udry, S., Burnet, M., Carrier, F., Eggenberger, A., Naef, D., Santos, N., Pepe, F., Rupprecht, G., Avila, G., Baeza, F., Benz, W., Bertaux, J.-L., Bouchy, F., Cavadore, C., Delabre, B., Eckert, W., Fischer, J., Fleury, M., Gilliotte, A., Goyak, D., Guzman, J.C., Kohler, D., Lacroix, D., Lizon, J.-L., Megevand, D., Sivan, J.-P., Sosnowska, D. and Weilenmann, U.: 2001, *ESO Messenger*, No. 105, 1.
Rogers, F.J. and Iglesias, C.A.: 1992, *ApJ* **401**, 361.
Rosenthal, C.S., Christensen-Dalsgaard, J., Nordlund, Å., Stein, R.F. and Trampedach, R.: 1999, *A&A* **351**, 689.
Sills, A. and Pinsonneault, M.H.: 2000, *ApJ* **540**, 489.
Stein, R.F. and Nordlund, Å.: 2001, *ApJ* **546**, 585.
Thévenin, F., Provost, J., Morel, P., Berthomieu, G., Bouchy, F. and Carrier, F.: 2002, *A&A* **392**, L9.
Udalski, A., Soszyński, I., Szymański, M., Kubiak, M., Pietrzyński, G., Woźniak, P. and Żebruń, K.: 1999, *Acta Astron.* **49**, 1.
Watson, R.D.: 1988, *Astrophys. Space Sci.* **140**, 255.
Zahn, J.-P.: 1992, *A&A* **265**, 115.